I0477821

HEALING HARMONIES

The Power of Frequencies in Medicine

Geoffrey Zachary

CONTENTS

Frequency

HEALING HARMONIES

The Power Of Frequencies In Medicine

PART I:
FOUNDATIONS
OF FREQUENCY
MEDICINE

CHAPTER 1: THE BASICS OF FREQUENCY AND VIBRATION

Introduction: The Symphony of Life

Imagine the human body as an intricate orchestra, where each cell, organ, and system plays its own note in perfect harmony. This symphony of vibrations and frequencies, while often imperceptible, forms the foundation of all life. Just as an orchestra depends on each instrument's resonance to create a melody, our bodies rely on cellular frequencies to maintain health, balance, and vitality. When one part of the body falls out of tune, the entire harmony is disturbed, and illness may follow. Could it be that the key to healing lies in retuning our internal orchestra?

The concept of frequency-based healing has long intrigued scientists, philosophers, and healers. It suggests that by understanding and harnessing specific frequencies, we can influence biological processes in ways that go beyond conventional medicine. To grasp the potential of frequency medicine, let's begin by exploring how frequency and vibration function in the natural world and their profound impact on living organisms.

The Science of Frequency and Vibration

At its core, frequency is the number of times a wave oscillates per second, measured in hertz (Hz). Vibration, on the other hand, is the physical manifestation of these waves. In simpler terms, every object, from a rock to a living cell, emits a unique vibrational signature based on its atomic structure. Even our organs, cells, and tissues vibrate at distinct frequencies, which contribute to our overall well-being. This phenomenon is often referred to as "cellular resonance."

Consider how tuning forks work: if you strike one fork, a nearby fork tuned to the same frequency will begin to vibrate. Similarly, each cell in our body resonates to specific frequencies, maintaining harmony with the other cells. When external frequencies—such as sound waves or electromagnetic pulses—are applied to the body, they can influence this cellular resonance, either restoring balance or exacerbating imbalance.

For example, high-energy frequencies like ultraviolet light can damage cellular DNA, while gentler frequencies, such as those used in ultrasound therapy, can promote tissue healing. By understanding the nuances of frequency, scientists have uncovered new ways to apply these principles for non-invasive healing.

A Brief History: Ancient Wisdom Meets Modern Science

Although frequency medicine is often presented as a modern breakthrough, its roots extend far into history. Ancient cultures understood the power of sound and vibration intuitively, even if they lacked the language of modern physics. Tibetan singing bowls, chanting, and rhythmic drumming were employed to heal the body and elevate the spirit. The Greeks believed in "musical medicine," and Egyptian healers used specific tones to promote physical and spiritual well-being.

In recent decades, researchers have begun to validate these ancient practices. For instance, studies on sound therapy have shown that certain frequencies can reduce stress, improve immunity, and even stimulate cellular repair. This scientific exploration has given birth to a field known as bio resonance therapy, which involves using specific frequencies to bring the body back into harmony.

The Body as a Frequency Matrix

To understand how frequency-based healing works, let's explore the human body as a complex "frequency matrix." Each organ and tissue type operates within a unique frequency range, contributing to a vibrational ecosystem that maintains health. When a specific organ becomes diseased or injured, its vibrational pattern shifts, much like an out-of-tune instrument. By applying external frequencies, practitioners aim to retune these errant cells and restore their natural rhythm.

Case Study: Vibrational Healing in Practice

Consider the story of Sarah, a 45-year-old woman with chronic back pain. Conventional treatments provided little relief, so she turned to pulsed electromagnetic field (PEMF) therapy, a technique that uses electromagnetic waves to stimulate healing. After several sessions, her pain diminished significantly. PEMF, like many frequency-based therapies, works by delivering frequencies that resonate with the body's tissues, encouraging cellular repair without invasive procedures.

Sarah's experience exemplifies how frequency medicine can offer hope for those who have exhausted traditional treatments. This approach doesn't simply mask symptoms; it targets the underlying imbalances, allowing the body to heal itself.

DIY Frequency Healing: Practical Tips

For those curious about incorporating frequency-based practices into their lives, sound therapy is an accessible starting point. Here are some practical methods:

1. Sound Baths: Lie down in a comfortable setting and listen to a guided sound bath using gongs, chimes, or Tibetan bowls. These instruments produce frequencies that help relax the nervous system and promote mental clarity.

2. Binaural Beats: This involves listening to tones at slightly different frequencies in each ear. Binaural beats are believed to synchronize brainwaves, helping to alleviate stress, improve focus, and enhance sleep.

3. Mantras and Chanting: Vocal vibrations also carry healing potential. Chanting "OM," for instance, creates a resonance that can calm the mind and balance energy within the body.

Each of these methods works by tuning your body's frequency matrix, gently encouraging your cells to resonate in harmony.

Reflecting on Health Beyond the Physical

Frequency medicine also invites us to reconsider our approach to health. In conventional medicine, treatment often targets physical symptoms, but frequency-based healing suggests that true wellness involves harmonizing body, mind, and energy. By viewing the body as a dynamic, vibrating matrix, we recognize that health is not merely the absence of disease but the presence of coherence and balance within our frequency field.

This holistic view of wellness aligns with an emerging area known as energy medicine, which explores the interactions between the body's energy field and health. Just as an orchestra sounds best when each musician is in sync, our bodies function optimally when each frequency is attuned.

The Visionary Path Ahead: Frequency Medicine as a Catalyst for Transformation

Imagine a future where frequency medicine is no longer alternative but mainstream—where a patient facing surgery could explore frequency-based treatments as a first option. This vision is already taking shape. Research in areas like low-level laser therapy, PEMF, and ultrasound surgery highlights the potential of frequencies to treat everything from chronic pain to tissue regeneration. The possibilities are both vast and inspiring, suggesting a world where the body's natural rhythm is valued as the foundation of healing.

Final Reflection: Embracing the Power of Frequencies

As we unlock the secrets of frequency and vibration, we also open a door to a new era in medicine—one that honours the wisdom of the past while embracing the innovations of the future. Frequency medicine encourages us to view the body not as a machine but as a dynamic symphony, where every cell plays a part. By learning to tune into these harmonies, we empower ourselves to take control of our health, embrace the potential of non-invasive healing, and rediscover the connection between science, nature, and self.

In embracing frequency-based healing, we are reminded that perhaps the most profound medicine of all is one that resonates with the essence of who we are. Whether through sound, light, or electromagnetic waves, frequency medicine reveals a truth that has always been present: that healing begins with harmony.

CHAPTER 2: THE PHYSICS OF SOUND IN MEDICINE

Introduction: The Healing Symphony of Sound

Imagine entering a room where a soft melody plays in the background. The rhythm soothes you, and you feel a wave of calm wash over you. Though you may not notice it, your heart rate slows, your breathing deepens, and your mind quiets. This is no coincidence; sound has the power to profoundly affect the body, mind, and spirit. Over the last century, scientists have delved deeper into the physics of sound to uncover how specific frequencies and vibrations interact with the human body in ways that go far beyond mere relaxation.

In this chapter, we will explore how sound, an ancient healer, has found a place in modern medicine. We'll discuss sound waves, resonance, and entrainment—the principles behind how sound can harmonize, repair, and heal. As we journey through these concepts, imagine sound as more than a fleeting vibration in the air; imagine it as a carrier of energy that can penetrate our cells, alter our brainwaves, and even accelerate physical healing.

The Basics of Sound Waves

Sound is essentially a wave of energy that travels through a medium, like air, water, or even human tissue. These waves are created by the vibration of molecules, and each sound wave has specific characteristics: frequency, amplitude, and wavelength. Frequency, measured in hertz (Hz), refers to the number of times a wave oscillates per second. Higher frequencies create high-pitched sounds, while lower frequencies produce deep, resonant tones.

In medicine, understanding these properties allows us to harness sound in unique ways. For example, ultrasound —an imaging technique familiar to most—relies on high-frequency sound waves to produce images of our internal organs. Ultrasound operates at frequencies above 20,000 Hz, far beyond the range of human hearing. These waves pass through our skin and interact with tissues, reflecting back images that allow doctors to see inside the body. But beyond diagnostics, researchers have discovered ways to use sound waves to heal.

Resonance: The Body's Natural Frequency

If you've ever seen a singer shatter a glass with a high note, you've witnessed resonance in action. Resonance occurs when an object vibrates at a specific frequency that matches the natural frequency of another object. When these frequencies align, the second object absorbs the sound energy and vibrates in sympathy.

In the human body, each cell, organ, and tissue has its own natural frequency. When this resonance is disrupted by injury or illness, it's akin to an instrument falling out of tune within an orchestra. By applying external frequencies, it is possible to "retune" these cells, restoring their healthy vibrations. This idea forms the foundation of vibrational medicine, which uses specific sound frequencies to bring the body back into harmony.

Case Study: A Harmonious Heartbeat

Consider the case of Tom, a 54-year-old man who struggled with irregular heart rhythms. When traditional treatments failed to provide relief, Tom tried binaural beats —a technique that plays two slightly different frequencies in each ear to create a "beat" in the brain. This rhythmic sound helped synchronize his brainwaves with a calming frequency, which, in turn, influenced his heart rhythm. After several weeks, Tom's heart rhythms stabilized, a testament to how sound can influence even the most fundamental processes in the body.

Entrainment: Syncing to the Rhythm

Entrainment is a phenomenon where two rhythmic systems align with each other. Imagine watching people dance to a drumbeat; eventually, even without instruction, they fall into sync with the rhythm. In biological terms, entrainment happens when our body systems synchronize with external frequencies, whether it's our brainwaves, heart rate, or breathing.

In frequency medicine, entrainment can be harnessed to influence mental and physical states. When we listen to rhythmic sounds like a heartbeat, the brain often entrains to that frequency, helping to calm the mind. This concept has become central to sound therapies like brainwave entrainment and sound baths.

Sound Healing in Modern Medicine: Techniques and Tools

Modern science has created tools to apply these principles in practical and effective ways. Here are some of the key technologies in frequency medicine:

1. Ultrasound Therapy: Used primarily for healing soft tissues and relieving pain, ultrasound therapy sends high-frequency sound waves into the body, creating heat that promotes tissue repair. It has been used to treat everything

from muscle injuries to arthritis.

2. Low-Frequency Sound Therapy: Using frequencies between 20 and 100 Hz, this therapy can reduce inflammation and accelerate healing. Studies have shown it can even stimulate bone growth in patients with fractures.

3. Binaural Beats and Isochronic Tones: These audio therapies create beats or tones that help the brain shift into specific states. Binaural beats have shown promise in reducing anxiety, improving focus, and even aiding sleep.

4. Music Therapy: Music's healing potential extends beyond enjoyment. Research shows that listening to or creating music can lower stress hormones, boost immunity, and improve mental health.

DIY Sound Healing: Practical Tips for Everyday Life

Sound healing is accessible to everyone, and here are simple ways to integrate it into your daily routine:

- Breathing to a Rhythm: Play soothing music and synchronize your breathing to the beat. This technique can reduce stress and increase feelings of relaxation.

- Binaural Beat Apps: Numerous apps provide binaural beats for different needs, from focus and concentration to sleep and relaxation. Use headphones for the best effect.

- Sound Baths: Attend a sound bath session where gongs, bowls, and chimes are used to create resonant sounds. This immersive experience can help calm the mind, reduce anxiety, and promote clarity.

Reflection: The Hidden Power of Sound

Sound healing encourages us to see the body not just as a physical entity but as a complex, resonant system. When our cells resonate in harmony, health flourishes; when they fall out of sync, disease can arise. Sound medicine offers a bridge between ancient wisdom and modern science, reminding us of the body's intrinsic connection to vibration and frequency.

The Future of Sound in Medicine

The journey of sound in medicine is only beginning. Imagine a world where surgeries are replaced by sound waves that penetrate the body, healing with precision and non-invasiveness. As researchers continue to unlock the secrets of sound, we may see a new era in medicine where resonance, frequency, and vibration are as integral to healing as antibiotics and surgery are today.

In embracing the potential of sound, we step into a realm where healing harmonies bring science and nature together in a profound dance of life. Sound healing is a gentle, powerful reminder of our interconnectedness to all frequencies in the universe, and its future holds the promise of a deeper, more holistic approach to health and wellness. Through sound, we may yet rediscover the true symphony of life.

CHAPTER 3: THE ROLE OF FREQUENCY IN NATURE

Introduction: The Symphony of Natural Frequencies

Picture yourself in a quiet forest, the gentle hum of insects filling the air, the rhythmic songs of birds drifting from above, and the soft rustle of leaves swaying in harmony with the breeze. There's a calmness, a sense of balance that seems to permeate every aspect of this environment. This feeling, this sense of equilibrium, arises because nature operates in frequencies. Each organism, from the smallest ant to the tallest tree, exists within its own rhythm, a unique frequency that contributes to the greater harmony of the ecosystem.

In the human body, much like in the forest, each organ and system pulses at its own rate, creating a symphony of frequencies that maintain life. From the heart's steady beat to the brain's oscillating waves, we are immersed in a natural flow of frequencies that guide our sleep, moods, and even our capacity for healing. When we understand these rhythms and how they resonate within us, we unlock the potential to tap into frequency-based medicine.

Frequency Patterns in Nature: The Universal Pulse

Frequency is the language of the universe, present in every natural system. Oceans, for instance, follow tidal rhythms influenced by the moon's gravitational pull, while the earth itself vibrates at what's known as the Schumann Resonance (7.83 Hz). This frequency, often called "the heartbeat of the Earth," is thought to support biological health, aiding in synchronizing our body's natural rhythms with those of the planet. Many people find a deep sense of peace and rejuvenation when they spend time outdoors, a likely effect of harmonizing with these natural frequencies.

Certain animals are attuned to frequencies in ways that help them survive. Dolphins, for instance, use echolocation —sound waves at specific frequencies—to navigate and communicate in the ocean. Similarly, bees perform a unique "waggle dance" to communicate the location of food sources, using vibrations to convey information. These frequencies not only aid in communication but create a resonance that integrates each creature into its ecosystem.

The Human Body's Natural Frequencies

Our bodies, too, are finely tuned to resonate with specific frequencies, each essential for health and harmony. When these natural rhythms are disrupted, we may experience stress, illness, or emotional imbalance. Let's explore some of the critical frequencies and rhythms that influence our lives.

1. Circadian Rhythms: The Body's Daily Clock

Our circadian rhythm is a roughly 24-hour cycle that governs sleep and wakefulness, driven by the brain's response to light and darkness. When we align with this natural rhythm—waking with the sunrise and resting after sunset—we experience more balanced energy, mood, and health. Disruption to this rhythm, often caused by irregular schedules or excessive exposure to artificial light, can lead to various health issues, from insomnia to mood disorders.

For example, shift workers who work at night and sleep during the day often struggle with long-term health problems. This is due to the biological discord caused by living out of sync with the natural circadian rhythm. Researchers have found that aligning with our natural sleep-wake cycle, even by as little as adjusting bedtime, can improve physical health, enhance mood, and increase mental clarity.

2. Heart Rate Variability (HRV): The Pulse of Well-being

Another vital frequency is the variability in time between heartbeats, known as heart rate variability (HRV). Rather than being strictly regular, a healthy heart adjusts its rhythm based on our needs and environment. Higher HRV is often seen as a marker of resilience and adaptability, while lower HRV is associated with stress, fatigue, and poor health.

Practices like deep breathing, meditation, and even listening to calming frequencies have been shown to improve HRV. This is because the body's rhythms are deeply interconnected. When one rhythm aligns with a calming frequency, other rhythms—such as brain waves and respiratory rate—often follow suit, creating a cascade of relaxation and healing throughout the body.

Nature's Influence on Human Rhythms: A Case Study in Resonance

Take the example of Sarah, a young woman dealing with chronic anxiety and irregular sleep. After trying various treatments, she began "earthing"—a practice where one spends time barefoot on natural ground to connect with the Earth's energy. Initially sceptical , Sarah was surprised to find her anxiety decreasing and her sleep improving. Earthing, which aligns the body with the Earth's Schumann

Resonance, helped restore her circadian rhythm and even increased her HRV.

Sarah's experience is not uncommon. Many people report a renewed sense of balance and vitality after spending time in nature, an effect that may be due to the alignment of our body's frequencies with those of the natural world.

Practical Tips for Aligning with Nature's Frequencies

If you're interested in aligning with nature's rhythms to enhance well-being, here are a few simple ways to do so:

1. Morning Sunlight Exposure: Step outside for at least 10–15 minutes of sunlight in the morning to signal to your brain that it's time to wake up. This reinforces your natural circadian rhythm.

2. Grounding or Earthing: Try walking barefoot on natural surfaces, like grass or sand, for a few minutes each day. Research suggests that grounding may reduce stress, inflammation, and improve sleep by synchronizing your body's frequencies with the Earth.

3. Breathing Exercises for HRV: Engage in slow, deep breathing exercises, such as the 4-7-8 technique (inhale for 4 seconds, hold for 7, exhale for 8). This practice can improve HRV, signalling calm and relaxation to your body.

Reflection: Harmony Beyond Health

When we step back and view the body as part of a greater ecosystem of frequencies, it becomes clear that health is more than the absence of disease; it's a state of resonance with the rhythms of life. By tuning into these rhythms, we don't merely treat symptoms—we foster balance and vitality at a deeper, more sustainable level.

The Vision of Frequency-Based Medicine: A World in Tune

As our understanding of these natural rhythms deepens, we stand on the edge of a new era in medicine. Imagine a future where hospitals incorporate nature-inspired frequencies to support healing, where "nature therapy" is prescribed as commonly as medications, and where understanding our body's rhythms is an essential part of health education. This vision brings us closer to a world where we are not only treating diseases but actively nurturing the harmony that sustains life.

Final Thoughts: Rediscovering the Symphony of Life

The role of frequency in nature reminds us of a simple truth: we are not separate from the world around us; we are a part of its rhythm, its pulse, its song. Healing frequencies offer a bridge—a way to reconnect with the life-sustaining vibrations that surround us. By embracing this approach, we move beyond seeing health as a battle against disease. Instead, we begin to understand healing as a journey to restore balance, to align with the deeper frequencies that resonate through every part of our lives.

In this symphony of life, each of us plays a part. As we learn to tune into our natural rhythms, we join a larger chorus, one where every beat and every breath reminds us that health is harmony, that healing is a resonance shared with the world.

CHAPTER 4: HISTORICAL ROOTS OF FREQUENCY HEALING

Introduction: Echoes of Ancient Wisdom

Throughout human history, the pursuit of healing has often transcended physical medicine, reaching into the realms of sound, vibration, and the mysterious resonance that lies between. In ancient civilizations, healers believed that illness often signified a disruption in the body's natural harmony, and they sought to restore balance by engaging with frequencies. The practice of frequency-based healing is not a modern invention; its roots stretch back thousands of years to societies that recognized the power of sound to touch the soul and, indeed, to heal.

Imagine stepping into an Egyptian temple chamber, where echoes of a low, powerful hum fill the space, reverberating through every bone in your body. Or envision a Tibetan monk, seated with his eyes closed, striking a singing bowl that produces a deep, resonant note that lingers, seemingly merging with the very fabric of the air. From ancient Egypt to Tibet, indigenous cultures across the world were pioneers of frequency healing, using sound as a bridge between the physical and the spiritual, the seen and the unseen.

Egyptian Sound Chambers: Healing with Resonance

The ancient Egyptians, known for their advanced knowledge in architecture, astronomy, and medicine, also explored the healing power of sound. Archaeological discoveries suggest that some Egyptian temples contained "sound chambers," specifically designed to enhance resonance. These chambers, shaped and constructed to amplify certain frequencies, were used by healers and priests in rituals aimed at restoring health and spiritual balance.

One example is the Temple of Hathor at Dendera, where sound may have been an integral part of ceremonial practices. The structure and materials of these chambers were such that certain frequencies could reverberate intensely, creating an almost hypnotic effect on those within. The priests would chant or play low-toned instruments, filling the chamber with vibrations that likely facilitated altered states of consciousness and a profound sense of tranquillity. Ancient inscriptions also mention the use of sound to "heal the spirit," indicating that they understood sound as a way to reach beyond the physical body.

Tibetan Singing Bowls: Sound as a Path to Inner Peace

In the Himalayas, monks have used Tibetan singing bowls for centuries as tools for meditation and healing. Crafted from a blend of metals, each bowl produces a unique frequency when struck or played, creating a rich and resonant sound that seems to penetrate deep into the listener's consciousness. The bowls are often tuned to specific frequencies thought to align with different chakras, or energy centres, within the body. When a bowl is played, its sound is believed to harmonize with these centres, helping to unblock stagnant energy and restore physical, emotional, and spiritual balance.

Consider the case of Mingma, a monk who spent his early years struggling with anxiety and insomnia. Through years of practicing with singing bowls, Mingma learned to calm his mind, finding peace in the resonance that filled his meditation space. He now teaches others, using bowls tuned to specific frequencies to help people manage stress, anxiety, and other ailments. For Mingma, the sound of the bowls is more than just an instrument—it's a gateway to deeper healing and self-discovery.

Aboriginal Didgeridoos: The Earth's Resonance in Sound

Indigenous Australians have long used the didgeridoo—a wind instrument made from hollowed-out tree trunks—in ceremonial healing rituals. The didgeridoo's deep, droning sound is thought to mimic the natural frequencies of the earth and sky, creating a grounding resonance that promotes physical and spiritual well-being. Played in rhythm with a healer's breathing, the didgeridoo's sound is directed at areas of the body believed to need healing.

For the Aboriginal people, healing is not limited to the physical; it encompasses the spirit, emotions, and connections to the land. By invoking these vibrations, the didgeridoo creates a resonant bridge between the body and the earth, helping the listener to reconnect with nature. Some modern studies suggest that the low-frequency vibrations of the didgeridoo may help relieve muscular tension and improve circulation, providing physical benefits in addition to the spiritual ones.

Ancient Greece: Pythagoras and the Music of the Spheres

Pythagoras, the ancient Greek mathematician and philosopher, is often credited as one of the earliest proponents of frequency-based healing in Western thought. Pythagoras believed that harmony was fundamental to the universe and that everything, from celestial bodies to the human soul, vibrated with a unique frequency. This idea became known as the "Music of the Spheres," a concept suggesting that the universe itself is a symphony of frequencies.

Pythagoras experimented with musical intervals, noting that certain harmonic relationships between notes created a sense of balance and well-being. He even established a healing school where students used music to align their minds and bodies with these universal harmonies. His followers, the Pythagoreans, used lyres and flutes to create melodies aimed at restoring harmony within the body. Pythagoras's teachings on frequency and harmony laid the groundwork for what would eventually evolve into music therapy, a field that continues to explore the healing potential of sound.

Practical Tips for Modern-Day Frequency Healing

Drawing on the ancient wisdom of these cultures, we can integrate sound healing into our own lives in simple yet profound ways:

1. Tibetan Singing Bowls: Try incorporating a singing bowl into your meditation practice. Gently strike the bowl, and let the sound resonate, focusing on the vibration as it fills the room. This can be especially helpful for clearing mental clutter and reducing stress.

2. Didgeridoo or Low-Frequency Music: For grounding and relaxation, play recordings of didgeridoo music or similar low-frequency sounds. Sit or lie down, allowing yourself to absorb the vibrations, which can help release tension and promote a sense of rootedness.

3. Personal Sound Rituals: Develop a sound ritual by using humming, chanting, or soft drumming. Let your body find a

natural rhythm, feeling how the sound interacts with your own energy and emotions.

Reflection: Rediscovering Ancient Wisdom in Modern Healing

The history of frequency healing reminds us that ancient societies understood health as a state of balance, a harmony between body, mind, and the world around us. They recognized that sound has a unique power—a power to reach into the hidden places within us, to heal in ways that conventional medicine often overlooks. While our tools have evolved, the principles remain strikingly similar. Healing through frequency is a timeless practice, one that invites us to listen, to connect, and to resonate with our deepest selves.

Looking Forward: Honouring the Past to Shape the Future

As we embrace the potential of frequency healing, we stand at a fascinating crossroads. Modern science continues to uncover the therapeutic effects of sound and vibration, yet the practices we turn to often draw from the traditions of the past. By honouring the wisdom of these ancient cultures, we can build a future where healing harmonies are woven into our everyday lives, a future where we remember that health is not simply the absence of illness but the presence of harmony.

Final Thoughts: The Timeless Symphony of Healing

In the echoes of ancient sound chambers, the quiet resonance of Tibetan bowls, and the deep hum of the didgeridoo, we find a universal truth. Healing, at its core, is a return to balance—a re-alignment with the frequencies that connect us to each other, to nature, and to the universe itself. By reconnecting with these ancient practices, we discover that the true power of healing lies not only in medicine but in music, in rhythm, and in the harmony of the frequencies that guide our lives.

CHAPTER 5: FROM ANCIENT PRACTICE TO MODERN SCIENCE

Introduction: Ancient Echoes in Modern Medicine

Imagine the hum of a Tibetan singing bowl, resonating through the air, its soundwaves touching something deep within. Thousands of years ago, healers in various cultures discovered that sound and vibration could calm the mind, soothe the body, and even restore health. Long before medical imaging machines or electromagnetic therapies existed, ancient civilizations recognized the power of frequencies to heal and balance the body.

Today, science is finally catching up with what these ancient traditions have long known. From sound baths to electromagnetic field therapies, modern medicine is drawing inspiration from centuries-old practices, transforming these ancient insights into precise, scientifically validated treatments. This chapter explores how age-old wisdom has shaped contemporary approaches to frequency-based healing, uniting ancient and modern knowledge into a new vision for health.

Ancient Practices: Foundations of Frequency Healing

Throughout history, diverse cultures used sound as a bridge between the physical and spiritual realms, aiming to restore balance and harmony within the body. These practices, though different in form and execution, share a common understanding: that everything in the universe vibrates, and by aligning with these natural vibrations, we can tap into a profound source of healing.

Tibetan Singing Bowls: The Power of Resonance

One of the oldest examples of frequency-based healing is the Tibetan singing bowl. Tibetan monks have used these bowls for centuries, striking or rubbing them to produce deep, resonant tones that vibrate at frequencies believed to align with the body's energy centres, or chakras. The sound waves emitted by these bowls are thought to calm the nervous system, clear emotional blockages, and promote a meditative state.

Consider a recent study where patients with anxiety underwent sound therapy with singing bowls. They reported a notable decrease in stress levels, and researchers found reductions in heart rate and blood pressure. Science is beginning to validate the effects of these vibrations, confirming what monks have known for centuries—that the power of sound can reach far beyond simple relaxation.

Ancient Egyptian Sound Chambers: Healing with Architecture

In ancient Egypt, sound was not only a tool but a profound healing medium. Certain temples, like the Temple of Hathor at Dendera, were designed with "sound chambers" built to enhance resonance. Priests would chant or play instruments, filling these chambers with specific frequencies intended to heal those within. The walls and materials were chosen to create intense reverberations, amplifying the sound to create a full-body experience of resonance.

Recent acoustic studies conducted within these temples suggest that the chambers were designed to resonate at frequencies between 100 and 120 Hz, aligning with the low-frequency sounds that modern science has linked to pain relief and stress reduction. Although ancient Egyptians didn't have the scientific language to describe what they were doing, they intuitively understood the healing effects of sound on the body.

The Transition to Modern Science

As humanity moved into the scientific age, the intuitive practices of frequency healing evolved into structured studies and measurable results. Scientists began to investigate the physics behind sound and vibration, uncovering how sound waves interact with the human body and brain. Concepts that were once mystical— like cellular resonance and vibrational healing—gained a scientific foundation, turning ancient practices into medical therapies.

Pioneering Research: Dr. Royal Rife and Frequency Therapy

In the early 20th century, Dr. Royal Rife, an American scientist and inventor, explored the concept of using specific frequencies to target and destroy harmful cells. Rife believed that each pathogen had a unique "mortal oscillatory rate," a frequency at which it would be destroyed. Using what he called the "Rife Machine," he exposed cells to their target frequency, successfully eliminating harmful bacteria and viruses without damaging surrounding healthy cells.

Though Rife's work was controversial at the time, modern science has revisited his theories, leading to advancements in fields like bio resonance and electromagnetic field therapy. Today, devices inspired by Rife's work are used to treat chronic pain, improve cellular health, and enhance recovery by applying targeted frequencies to the body.

Ultrasound Therapy: A Modern Application of Ancient Principles

One of the most common medical uses of frequency-based healing today is ultrasound therapy. While primarily used for imaging, ultrasound can also promote healing and relieve pain by using high-frequency sound waves to stimulate tissues deep within the body. The sound waves create tiny vibrations that encourage blood flow, reduce inflammation, and accelerate the healing process.

In a sense, ultrasound therapy is an extension of the ancient practices seen in Egypt and Tibet, but with the precision and control offered by modern technology. It's a perfect example of how science has built upon ancient insights, transforming an intuitive practice into a clinically validated treatment.

Case Studies: Frequency Healing in Action

To understand the real-life impact of frequency-based healing, consider the case of Emma, a 40-year-old woman with chronic arthritis. After years of traditional treatments with limited success, Emma turned to pulsed electromagnetic field (PEMF) therapy, a modern form of frequency medicine. By applying specific frequencies to her joints, she experienced pain relief and reduced swelling. Emma's experience demonstrates how frequency-based therapy, though rooted in ancient practices, has the potential to address modern ailments in ways traditional medicine sometimes cannot.

Similarly, consider the story of Marco, a veteran dealing with post-traumatic stress disorder (PTSD). After struggling with symptoms that conventional treatments couldn't ease, he tried sound therapy with binaural beats—a technique where two different tones are played in each ear, creating a frequency that encourages the brain to enter a relaxed state. Over time, Marco's symptoms improved significantly, proving how sound and frequency can heal not only the body but also the mind.

Practical Applications: Bringing Ancient Healing into Daily Life

For those interested in exploring the healing potential of frequencies, here are some accessible ways to integrate these practices:

- Binaural Beats: Use binaural beat audio tracks to promote relaxation, focus, or sleep. Listen with headphones to achieve the intended frequency effects.

- Sound Baths: Attend a sound bath session, where gongs, chimes, and bowls are used to create immersive soundscapes. These sessions help relieve stress and encourage mental clarity.

- PEMF Devices: For those with chronic pain, consider consulting with a health professional about PEMF therapy, a modern application of frequency healing that aligns with ancient principles.

Reflection: Honouring the Past, Embracing the Future

The journey from ancient sound chambers to modern frequency therapies highlights a deep truth about healing: that wellness is not just the absence of disease but the presence of balance and harmony. Ancient cultures understood that healing goes beyond the physical body, touching the mind, spirit, and energy that connect us to the world. By bridging the gap between ancient wisdom and modern science, frequency-based medicine invites us to see health as a symphony—one where each frequency plays a vital role.

Vision for the Future: A World in Tune with Healing Frequencies

Imagine a future where hospitals, clinics, and wellness centres integrate frequency-based healing into their offerings, allowing patients to experience therapies that are not only effective but gentle and non-invasive. As science advances, we may reach a point where sound and frequency therapies become as common as surgery and pharmaceuticals, providing options that heal in harmony with the body's natural rhythms.

By honouring the ancient wisdom that laid the foundation for these practices, we're paving the way for a future where healing is as much about resonance and balance as it is about physical repair. This vision embodies the very essence of healing harmonies—a world where science and spirit merge, where health is not merely treated but nurtured and celebrated.

Final Thoughts: The Power of Harmonizing Science and Tradition

From the reverberations of ancient temples to the hum of modern therapeutic devices, frequency-based healing reminds us that we are part of a greater symphony of life. Each frequency, each vibration carries the potential to restore us to balance, to remind us of our connection to the world, and to heal not just our bodies but our hearts and minds. By blending the best of ancient practices with modern science, we open doors to a future where the harmony of healing frequencies becomes an integral part of our journey toward wellness.

CHAPTER 6: ENERGY FIELDS AND HUMAN HEALTH

Introduction: The Body's Invisible Symphony

Imagine you're walking through a forest, feeling the warmth of the sun on your skin, the subtle hum of life vibrating all around you. You may not see it, but you're immersed in a complex web of energy. In much the same way, every human being is surrounded and permeated by an invisible field of energy—a biofield. This biofield is like an energetic blueprint that influences our physical, emotional, and mental health. While ancient healing traditions have long worked with the concept of an "aura" or "energy field," modern science is just beginning to explore the profound impact of these energy fields on our health and well-being.

The biofield concept bridges the ancient and the modern, embodying the mystery of our inner energy. As research advances, we're learning how specific frequencies can interact with the biofield, offering new avenues for healing that respect the body's natural harmony.

Understanding the Biofield: A Science of Energy

At its most basic, the biofield can be understood as an electromagnetic field that surrounds and flows through every living organism. The concept aligns with principles in quantum physics, which tell us that all matter is energy vibrating at specific frequencies. In the case of the human body, this energy exists not only within but around us, forming an invisible "bubble" of sorts. This biofield is what connects us to both our internal energy systems and the world around us.

The National Institutes of Health (NIH) even recognized the biofield in 1994 as a potential contributor to health. According to this understanding, imbalances in our biofield can manifest as physical or emotional ailments. For example, trauma, stress, or environmental toxins can create disruptions in this field, just as a rock thrown into a pond disturbs the water's calm surface. When the biofield is out of balance, we may experience fatigue, illness, or emotional disturbances.

A Case Study in Biofield Healing: Sarah's Journey

Consider Sarah, a 36-year-old woman battling chronic fatigue and depression. Traditional treatments provided little relief, so she turned to energy healing as a last resort. During her first session, the practitioner worked on her biofield, using gentle frequencies to "clear" energetic blockages around her heart and solar plexus. The experience was subtle, almost imperceptible, yet Sarah left feeling lighter and more centred.

Over a series of sessions, her physical symptoms diminished, and her mood stabilized. Sarah's story is one of many that illustrate the power of balancing the biofield through frequency-based healing. Although still new to Western science, the biofield is becoming a focal point for alternative and integrative therapies that view the body as more than just a physical organism, but as an energetic whole.

The Biofield in Science: Electromagnetic Frequencies and Cellular Health

The biofield resonates with frequencies that correspond to different aspects of our well-being. Heart rate, brainwaves, and even the electrical activity of our cells all contribute to this field, creating a unique vibrational signature. External frequencies—such as electromagnetic fields (EMFs) from electronic devices—can influence our biofield, for better or worse.

Research shows that low-frequency electromagnetic fields can enhance cellular repair and reduce inflammation, while high-frequency EMFs may interfere with cellular health. This is where frequency-based healing comes in: by applying beneficial frequencies, we can bring balance back to the biofield, promoting healing on a cellular level. Tools like PEMF (Pulsed Electromagnetic Field) devices apply specific frequencies to re-align disrupted fields, helping the body to heal itself naturally.

Techniques for Balancing the Biofield

Many techniques have emerged to harmonize the biofield. Here are a few approaches that use frequency to restore balance and vitality:

1. Reiki: Energy Healing Through Frequency Intention

Reiki is a Japanese technique for stress reduction and relaxation that also promotes healing. It operates on the principle of channelling universal life force energy through the hands, and practitioners often report sensing variations in a person's biofield. By using their hands, they aim to restore balance, clear blockages, and create a sense of harmony within the biofield.

2. Sound Therapy: Realigning the Biofield with Vibrations

Sound therapy, through tools like singing bowls, gongs, and tuning forks, offers a way to align the biofield using resonance. Each instrument produces a specific frequency, and by targeting areas of imbalance, sound therapists can clear stagnant energy. For instance, tuning forks are often used to detect and correct areas of the biofield that feel "off," acting like a tuning process for the energy body.

3. PEMF Therapy: Applying External Frequencies to Heal

Pulsed Electromagnetic Field (PEMF) therapy uses devices that emit low-frequency electromagnetic waves to stimulate cellular repair. This approach is particularly effective for injuries, pain, and inflammation. By introducing frequencies that resonate with the body's cells, PEMF therapy strengthens the biofield, aiding in faster recovery and better health.

DIY Biofield Practices: Practical Tips for Balancing Your Own Energy

You don't need professional equipment to start working with your biofield. Here are simple, accessible practices to help you stay balanced and harmonized:

- Grounding: Spending time barefoot on natural surfaces like grass or sand helps to reconnect your biofield with the earth's frequency, promoting a calming and centring effect.

- Breathing Exercises: Practice deep, rhythmic breathing. Visualize each breath as a wave of energy moving through and clearing your biofield.

- Self-Tuning with Sound: Try using a tuning fork at 528 Hz or 432 Hz (frequencies associated with harmony and healing) by striking it and holding it near different parts of your body. This can help clear stagnation and re-align your energy field.

Reflection: A New Paradigm for Health

Understanding the biofield requires a shift in perspective

—from seeing ourselves solely as physical beings to acknowledging that we are, at our core, beings of energy. Just as we care for our physical health by eating well and exercising, so too should we care for our biofield, which plays an essential role in maintaining overall well-being.

In many ways, biofield healing reminds us of the unity between the mind, body, and environment. By balancing our energy, we create space for emotional resilience, mental clarity, and physical health to flourish. Frequency-based therapies offer a way to connect with our bodies at a profound level, a level where healing can become truly transformative.

Vision for the Future: Frequency Medicine as a Holistic Path to Wellness

As research into the biofield advances, we can imagine a world where energy medicine becomes a mainstream approach to health care. Imagine a hospital where biofield therapies are as common as prescription medications, where patients receive treatments that harmonize their energy fields alongside physical care. We may one day see wearable devices that monitor our biofield health, providing real-time insights into our energetic state and alerting us to imbalances before they manifest as physical ailments.

The future of frequency medicine lies in its holistic nature—a way to treat the entire person, respecting the interconnection of mind, body, and energy. This vision moves beyond the limitations of treating symptoms, recognizing that true healing involves restoring the natural flow of energy within us.

Final Thoughts: Rediscovering the Body's Invisible Symphony

The concept of the biofield encourages us to look beyond

the physical and see the energetic fabric that sustains us. It invites us to listen to the body's subtle whispers, to the frequencies that connect each cell, each heartbeat, each thought. In the biofield, we find a bridge between ancient wisdom and modern science, a reminder that health is a dance of energies, a symphony that plays in harmony with the universe.

As we learn to understand and nurture our biofield, we unlock new possibilities for healing, both within ourselves and within the world. Frequency medicine is not just a tool for treating illness—it is a path to rediscovering our inherent connection to life, to the rhythms and vibrations that sustain us, and to the healing harmonies that lie within.

CHAPTER 7: VIBRATION AND CELLULAR COMMUNICATION

Introduction: The Body's Quiet Symphony

Think of your body as an orchestra, with each cell playing its own unique note to create a harmonious melody. This symphony of life is not always audible to us, yet each cell, each molecule, resonates with a specific frequency, constantly communicating with its neighbours. At the microscopic level, our cells engage in a silent conversation, sending and receiving messages through tiny vibrations that influence every process in our body. Understanding this cellular symphony helps us realize how our health hinges on maintaining harmony within this delicate orchestra.

In recent years, researchers have uncovered more about these invisible vibrations and how they impact our health. When the vibrational "notes" of our cells become dissonant due to stress, illness, or injury, the body's communication system is disrupted, leading to imbalance. But what if we could retune these cellular vibrations to restore harmony and health?

How Cells Communicate: The Science of Cellular Vibrations

Cells do not merely exist side by side; they are in constant communication, much like an ensemble of musicians who listen and respond to one another. This communication happens through a combination of electrical signals, chemical messengers, and, perhaps most fascinatingly, mechanical vibrations. Every cell in our body has a natural frequency, a "note" it resonates at when it is healthy. These vibrations influence the behaviour of nearby cells, promoting everything from tissue repair to immune responses.

One theory explaining this phenomenon is known as mechanotransduction. This is the process by which cells sense and respond to mechanical stimuli—essentially, vibrations and pressure. When a cell experiences stress, its vibrational frequency changes, much like a musical instrument going out of tune. In response, neighboring cells adjust their vibrations, creating a ripple effect that can either lead to healing or further dissonance if left unchecked.

Imagine, for instance, a small orchestra. If one violinist plays out of tune, the others may try to adjust their pitch, but if the dissonance continues, the whole performance suffers. Similarly, when one cell falls out of harmony, it can affect the health of surrounding cells, potentially leading to disease if the imbalance persists.

The Impact of Altered Vibrations on Health

When cells communicate harmoniously, the body operates in balance. However, various factors can disrupt this cellular communication. Factors such as stress, injury, environmental toxins, and even negative emotions can alter the vibrational frequencies of our cells, interrupting this delicate balance.

For instance, cancerous cells have been found to vibrate at different frequencies than healthy cells. These cells disrupt their environment, behaving erratically and spreading that dissonance to neighboring tissues. Scientists are now exploring ways to detect these altered frequencies early as an indicator of disease. By identifying these shifts in cellular vibrations, we can potentially diagnose illnesses long before they manifest in physical symptoms.

In another example, researchers studying people with chronic pain have found that their cells often exhibit lower vibrational frequencies in affected areas. By increasing the vibration levels in these cells, it might be possible to alleviate pain and accelerate healing. This is the basis of several emerging therapies that use specific frequencies to "re-tune" cellular vibrations and restore harmony within the body.

Case Study: Frequency Therapy in Action

Consider the story of James, a 45-year-old man with a debilitating shoulder injury. After trying various treatments, he came across a frequency-based therapy known as Pulsed Electromagnetic Field (PEMF) therapy. By applying low-frequency pulses to his shoulder, the therapy aimed to stimulate cellular communication and encourage his cells to return to their natural frequencies.

Over a period of weeks, James began to notice significant improvements. His pain subsided, and his mobility increased as the cells in his shoulder responded to the PEMF signals. By restoring the cellular vibrations to their optimal frequencies, his body's natural healing processes were able to re-establish harmony, illustrating the potential of frequency-based interventions for managing chronic conditions.

Practical Tips: DIY Frequency Practices for Cellular Health

You don't need access to a clinic to begin working with frequency practices for cellular health. Here are some ways to gently encourage healthy cellular communication on a daily basis:

1. Sound Healing: Listening to specific frequencies, such as 432 Hz or 528 Hz, is believed to promote relaxation and cellular balance. You can find these frequencies in many music streaming platforms and incorporate them into your daily routine.

2. Meditation with Tuning Forks: Tuning forks calibrated to healing frequencies can be struck and held near areas of discomfort. The gentle vibrations can help encourage cells to resonate in harmony, providing relief and promoting a sense of balance.

3. Hydration and Cellular Vibration: Proper hydration supports cellular communication, as water molecules enhance the transmission of vibrational energy. Staying hydrated allows cells to maintain their natural resonance more effectively.

4. Breathwork: Controlled, rhythmic breathing exercises can stimulate vibrational alignment within the body. Slow, deep breaths create a wave of oxygen that nourishes cells and encourages synchronized cellular communication.

The Visionary Potential of Cellular Vibration in Medicine

Imagine a future where doctors don't just rely on tests and scans, but instead measure the "notes" of your cells to assess health. If a cell's vibrational frequency starts to deviate from its natural resonance, a practitioner could intervene early, "retuning" those cells before disease manifests. This visionary approach to medicine could make frequency-based treatments a first line of defence.

In this future, we might wear sensors that constantly monitor our cellular vibrations, alerting us to potential imbalances long before we feel them physically. Early intervention could mean fewer surgeries, less reliance on medications, and a gentler approach to health care that works in harmony with the body.

Reflecting on the Symphony Within

The concept of cellular vibration invites us to reconsider what it means to be healthy. Rather than viewing health as the absence of disease, we might start to see it as a state of resonance, where every cell plays its part in a balanced symphony. Just as a symphony reaches its fullest expression when every instrument is in harmony, so does the body flourish when all cells resonate in tune with one another.

Perhaps health is not just something we treat or manage but something we cultivate and sustain through harmonious practices. By learning to listen to our bodies, by tuning into our cellular symphony, we may unlock deeper levels of healing and understanding. Frequency medicine offers us a path to rediscovering this inner harmony, where we are not just surviving but thriving in resonance with ourselves and the world around us.

Final Thoughts: Harmonizing Science and Soul

Vibration and cellular communication remind us that life is an exquisite, interconnected dance. Each cell is a note, each frequency a piece of the melody that makes up who we are. By learning to engage with these vibrations, to tune and care for them, we open doors to a future where healing is about more than treating symptoms. It becomes about restoring balance, about listening to the body's natural wisdom, and about embracing a science that recognizes the profound symphony of life within us.

As we continue exploring the power of frequency in medicine, let us remember that every cell, every vibration, is part of a much larger story—a story that celebrates the potential for harmony, health, and healing. Through frequency-based medicine, we are not only advancing science; we are honouring the music of life that has always been there, waiting to be heard.

CHAPTER 8: FREQUENCY HEALING'S PLACE IN MODERN MEDICINE

Introduction: A New Frontier in Healing

Imagine a world where healing doesn't always involve invasive surgeries or potent pharmaceuticals, but instead, taps into the natural rhythms and frequencies that already exist within and around us. This is the vision that frequency medicine offers—a field that is gaining traction in healthcare as more research supports its effectiveness and potential. While the idea of healing through sound and electromagnetic frequencies might sound futuristic, it is actually rooted in ancient practices, now refined and supported by modern technology and scientific evidence.

Frequency-based therapies are finding a place in modern medicine, offering a bridge between conventional treatments and holistic approaches. From pain management to mental health support, the applications of frequency healing are vast and continually evolving. This chapter provides an overview of some of the most widely used and scientifically supported frequency-based therapies in healthcare today.

The Science Behind Frequency Medicine

The human body operates on frequencies, from the oscillations of our cells to the rhythms of our heartbeats and brainwaves. These internal frequencies maintain the harmony required for optimal health. However, illness, injury, and stress can disrupt these natural rhythms, leading to disharmony and imbalance. Frequency medicine aims to restore the body's natural balance by using external frequencies—whether sound waves, electromagnetic fields, or light—to "tune" the body back to health.

Scientific interest in frequency medicine stems from a growing understanding of bioelectromagnetics, the study of how electromagnetic fields interact with biological organisms. Studies have shown that specific frequencies can influence cellular processes, reduce inflammation, and even promote tissue regeneration. Let's explore how these principles are applied in various frequency-based therapies.

Ultrasound Therapy: Sound Waves for Healing

One of the most familiar forms of frequency-based medicine is ultrasound therapy. While many associate ultrasound with imaging, therapeutic ultrasound has distinct applications in healing. Therapeutic ultrasound uses high-frequency sound waves that penetrate deep into tissues, creating tiny vibrations that stimulate cellular processes, increase blood flow, and reduce inflammation.

Ultrasound therapy is often used for musculoskeletal injuries, such as sprains, tendonitis, and arthritis. By promoting the repair of damaged tissues, ultrasound therapy helps reduce pain and accelerate healing without the need for invasive procedures. In studies with athletes recovering from injuries, ultrasound therapy has consistently shown faster recovery times compared to traditional therapies alone.

Pulsed Electromagnetic Field (PEMF) Therapy: Tuning Cells with Electromagnetic Fields

PEMF therapy uses low-frequency electromagnetic waves to stimulate the body at the cellular level. When cells are damaged or stressed, their electromagnetic field weakens, disrupting communication between cells and slowing down the healing process. PEMF therapy helps restore these frequencies, creating an environment that supports cellular repair and regeneration.

PEMF has been used to treat conditions such as chronic pain, osteoporosis, and even depression. In one study, patients with chronic lower back pain experienced significant pain reduction after consistent PEMF therapy. Another study found that PEMF could promote bone growth, making it a promising treatment for osteoporosis and bone fractures.

Low-Level Laser Therapy (LLLT): Harnessing Light Frequencies for Healing

Low-Level Laser Therapy (LLLT), sometimes called photobiomodulation, involves applying low-intensity lasers to injured or painful areas. These lasers emit light at specific wavelengths that penetrate tissues, stimulating cellular processes and encouraging regeneration. LLLT has shown effectiveness in treating conditions such as arthritis, muscle strains, and nerve injuries.

One of the mechanisms behind LLLT's success is its ability to increase mitochondrial activity within cells. Mitochondria, the "powerhouses" of our cells, play a crucial role in energy production. By energizing these organelles, LLLT boosts cellular activity, leading to faster recovery. This therapy is also gaining recognition in dermatology for promoting wound healing and reducing scarring.

Transcranial Magnetic Stimulation (TMS): Rewiring the Brain with Magnetic Fields

Transcranial Magnetic Stimulation (TMS) is a non-invasive therapy used primarily to treat depression and other mental health disorders. TMS works by sending magnetic pulses to specific areas of the brain, stimulating neural activity and encouraging new neural connections. By targeting the brain's natural frequencies, TMS can help restore balanced brainwave patterns.

Clinical trials have shown that TMS is particularly effective for patients with treatment-resistant depression, offering an alternative for those who do not respond to medication. Recent studies have also explored its use in treating conditions like anxiety, PTSD, and even chronic pain, with promising results.

Case Study: Healing with Frequency-Based Therapies

Consider the story of Lisa, a 52-year-old woman who had been struggling with chronic knee pain due to arthritis. Traditional treatments, including medication and physical therapy, provided only temporary relief. Seeking an alternative, Lisa turned to PEMF therapy and LLLT. Over a series of sessions, her pain levels decreased significantly, and she regained mobility she hadn't experienced in years.

For Lisa, frequency medicine offered a solution that worked with her body's natural rhythms rather than against them. Her case exemplifies how frequency-based therapies can complement conventional treatments, providing relief for those with chronic conditions who may feel they have exhausted other options.

Practical Tips: How to Incorporate Frequency Healing into Daily Life

While some frequency-based therapies require specialized equipment, there are accessible ways to incorporate the principles of frequency healing into your daily life:

1. Sound Therapy: Incorporate soothing frequencies, like binaural beats or 432 Hz music, into your routine. These sounds can help balance your mood and promote relaxation.

2. Meditation with Frequency: Guided meditations or sound baths can introduce healing frequencies into your daily practice, reducing stress and enhancing mental clarity.

3. Infrared Saunas: If accessible, an infrared sauna can help detoxify the body and promote cellular rejuvenation, using light frequencies that penetrate the skin.

4. PEMF Devices for Home Use: Portable PEMF devices are available for home use and can be helpful for managing chronic pain and enhancing recovery from exercise.

Reflecting on the Role of Frequency in Healing

The emergence of frequency medicine is a reminder that health is not solely about treating symptoms; it is about restoring harmony within the body. Frequency-based therapies offer a more holistic approach, one that respects the body's intrinsic rhythms and its capacity for self-healing. In a world where many are disillusioned with side effects and invasive treatments, frequency medicine presents an empowering alternative, allowing patients to take an active role in their own healing journey.

Vision for the Future: Expanding the Role of Frequency Medicine

Imagine a healthcare system where frequency-based therapies are integrated with traditional treatments, where patients can choose non-invasive options that align with their natural frequencies. In this vision, hospitals might have dedicated frequency healing rooms, equipped with PEMF beds, sound therapy chambers, and infrared light installations.

Such a future is not far off. As research and technology advance, we may see a shift in how we approach wellness, moving toward treatments that honour the body's inherent vibrational nature. By supporting health on a vibrational level, we're not just addressing illness; we're fostering a deeper, more sustainable state of well-being.

Final Thoughts: Rediscovering the Body's Rhythm

The power of frequency healing lies in its simplicity and its profound impact. It teaches us that healing can be as gentle as a sound wave, as subtle as a magnetic field, and as natural as light. The field of frequency medicine is young, but its potential is vast, grounded in the understanding that each of us is a symphony of frequencies. By tuning into this symphony, we have the potential to unlock new levels of health, resilience, and vitality.

As we continue exploring the power of frequency-based therapies, we are reminded that healing is not only a scientific process but a journey of rediscovery—of our body's wisdom, of our connection to nature, and of the harmonies that sustain life itself. Frequency medicine is more than a tool; it is an invitation to listen to the subtle music within us, to find balance and healing in the resonance of our own unique rhythms.

PART II: SCIENCE OF FREQUENCIES AND THE BODY

CHAPTER 9: BRAIN WAVES AND MENTAL HEALTH

Introduction: The Rhythm of the Mind

Imagine tuning a radio, adjusting the dial to find the perfect station. As you move through the frequencies, you hear different stations coming in and out of focus, each with its own unique sound and purpose. Our minds work in much the same way. Beneath our conscious thoughts, our brains constantly hum with electrical activity, producing waves that shift as we move through different states of consciousness. These brain waves—alpha, beta, theta, delta, and gamma—each have their own rhythm, affecting everything from our focus to our sleep, our creativity, and our emotional well-being.

Understanding brain waves is like having the key to our mental "tuning dial." With advances in neurofeedback and frequency-based therapies, we can now modulate these waves, helping to manage anxiety, depression, PTSD, and other mental health challenges. This chapter explores the science of brain waves and how harnessing their natural rhythms can promote mental health and emotional harmony.

The Basics of Brain Waves

Brain waves are the result of synchronized electrical pulses generated by neurons communicating with each other. Like an orchestra with multiple instruments, the brain produces different wave frequencies that correspond to various mental and physical states. Here's an overview of the primary brain waves:

1. Delta Waves (0.5–4 Hz): Delta waves are the slowest brain waves, associated with deep, restorative sleep and healing. They help the body recharge, promote immune function, and support tissue repair.

2. Theta Waves (4–8 Hz): Theta waves occur during light sleep and deep meditation. They're linked to creativity, intuition, and emotional processing, allowing us to access deeper, subconscious thoughts and memories.

3. Alpha Waves (8–12 Hz): Alpha waves are produced when we're awake but relaxed, such as during light meditation or when we're "in the zone" creatively. They foster a calm, alert mind and reduce stress and anxiety.

4. Beta Waves (12–30 Hz): Beta waves are associated with active, focused thinking and problem-solving. They're prominent when we're engaged, alert, and concentrating on a task.

5. Gamma Waves (30–100 Hz): Gamma waves are the fastest brain waves, associated with high-level cognitive functions, learning, and memory. They're less well-understood but are linked to peak mental performance and insight.

Each of these waves has its role, but issues can arise when our brain's natural rhythms become unbalanced. For example, too many beta waves can lead to anxiety, while an excess of delta waves during the day might cause fatigue and lack of focus. Frequency-based therapies aim to bring these brain waves into harmony, aligning them with the mental and emotional state we wish to cultivate.

Case Study: Reducing Anxiety with Alpha Waves

Consider Sarah, a 30-year-old graphic designer who struggled with anxiety. Her mind constantly raced, stuck in high beta-wave activity, which led to insomnia and chronic stress. Traditional treatments provided some relief, but she wanted an alternative that would allow her to address her anxiety at a deeper, neurological level. Sarah decided to try neurofeedback therapy, a technique that helps train the brain to produce desired wave frequencies.

Over several sessions, Sarah practiced producing alpha waves through guided meditation and neurofeedback exercises. By learning to increase her alpha waves, she could enter a calm, focused state more easily. After a few months, her anxiety levels significantly decreased, and she reported feeling more grounded and present in her day-to-day life.

Sarah's experience shows how adjusting brain wave patterns can address mental health challenges by working directly with the rhythms of the mind.

How Frequency-Based Therapies Modulate Brain Waves

There are several ways to modulate brain waves, from sound-based therapies to direct brain stimulation. Here are a few of the most promising approaches:

1. Binaural Beats: Sound as a Brainwave Tuner

Binaural beats use two slightly different frequencies in each ear to produce a third frequency that the brain perceives as a "beat." This beat encourages the brain to synchronize with the frequency of the desired brain wave, whether it's alpha for relaxation or theta for meditation. Studies show that listening to binaural beats can reduce anxiety, improve focus, and even enhance sleep quality.

For example, if a person wants to relax, they might listen to a binaural beat tuned to 10 Hz, an alpha frequency associated with calm and focus. After a few minutes, the brain naturally begins to match this frequency, helping the listener achieve a state of relaxed alertness.

2. Transcranial Magnetic Stimulation (TMS): Realigning Brain Rhythms

TMS is a non-invasive therapy that uses magnetic pulses to stimulate specific brain regions. By targeting areas linked to certain mental health conditions, such as the prefrontal cortex in cases of depression, TMS can help recalibrate brain wave activity, promoting healthier rhythms. This approach has proven especially effective for people with treatment-resistant depression, helping to lift mood and regulate emotions by restoring a balanced brainwave pattern.

3. Neurofeedback: Training the Brain to Self-Regulate

Neurofeedback is a form of biofeedback that involves monitoring brain wave activity in real time and providing immediate feedback to the patient. During a neurofeedback session, a person might play a simple game on a screen, with their brain waves influencing the game's progress. If they achieve the desired brain wave pattern—say, producing more alpha waves for relaxation—the game rewards them with progress. Over time, this process teaches the brain to naturally enter these beneficial states.

Practical Tips: Simple Practices to Modulate Brain Waves at Home

While some brainwave therapies require specialized equipment, there are accessible ways to support balanced brain wave activity in your daily life:

1. Meditation: Mindfulness meditation naturally increases

alpha and theta waves, helping to reduce stress and enhance creativity. Regular practice can make it easier to access these calm and focused states on demand.

2. Nature Immersion: Spending time in nature has been shown to reduce beta wave dominance, promoting more alpha wave activity. A walk in the forest or time spent near water can calm the mind and improve focus.

3. Music and Soundscapes: Listening to certain types of music or nature sounds can encourage the brain to produce different waves. For relaxation, try classical music or ambient soundscapes, which tend to stimulate alpha and theta wave production.

4. Controlled Breathing: Practices like deep breathing and diaphragmatic breathing can help reduce high beta wave activity, promoting a calmer mental state.

Reflection: Brain Waves as a Gateway to Emotional Resilience

Understanding our brain waves opens a window into our emotional and mental well-being. These rhythms are a part of us, shaping how we think, feel, and perceive the world. When we learn to influence our brain waves, we gain a new form of agency over our mental health, allowing us to move beyond reactive patterns and cultivate emotional resilience.

Imagine a future where managing anxiety or improving focus is as simple as listening to specific sound frequencies, or where treating depression doesn't solely rely on medication but includes re-aligning brain rhythms. By respecting the natural rhythms of the mind, we can foster a healthcare system that is both scientifically advanced and deeply humane.

Vision for the Future: Expanding Frequency-Based Mental Health Care

With research in frequency medicine and brain wave modulation advancing rapidly, we may soon see mental health treatments that are both highly effective and non-invasive. In this future, hospitals and clinics might offer personalized frequency therapies, allowing patients to harmonize their brain waves in ways that support mental well-being without side effects.

This vision of mental health care empowers people to become active participants in their healing journey. By giving individuals tools to balance their own brain waves, frequency-based therapies could lead to a society where mental health is approached with the same care, customization, and understanding as physical health.

Final Thoughts: Embracing the Power of the Mind's Rhythm

Our brains are constantly pulsing with electrical rhythms that influence every aspect of our lives. Frequency medicine allows us to recognize and work with these rhythms, revealing a new layer of potential for mental health. By learning to balance our brain waves, we don't just improve our focus or reduce anxiety—we tap into the full potential of the mind, experiencing clarity, creativity, and emotional stability.

The future of frequency medicine invites us to engage with the subtle, natural rhythms that govern our mental states, to harmonize our lives with the natural melodies of the mind. As we explore these possibilities, we are reminded that healing is not just about treating symptoms but about tuning in to the deep frequencies that sustain us, finding harmony within ourselves, and embracing the quiet symphony of well-being that lies within.

CHAPTER 10: NEUROPLASTICITY AND SOUND

Introduction: Sound as the Sculptor of the Mind

Imagine the brain as a vast, living network of connections —each pathway forged and strengthened by the experiences we accumulate over time. This intricate web of neural connections is not fixed; it is remarkably adaptable, continually reshaping itself in response to our actions, thoughts, and environment. This ability to adapt and reorganize is known as neuroplasticity, one of the most profound capabilities of the human brain. Neuroplasticity allows us to recover from injuries, learn new skills, and even overcome mental and emotional challenges.

In recent years, researchers have uncovered the fascinating role that sound and frequency-based therapies can play in supporting neuroplasticity. By engaging the brain in specific frequencies, sound therapy can stimulate new neural connections, enhance cognitive function, and aid in recovery from conditions such as stroke, brain injuries, and mental health disorders. This chapter explores how sound and vibration are more than mere sensory experiences; they are tools that can help us reshape and rejuvenate the mind.

The Science of Neuroplasticity

Neuroplasticity refers to the brain's ability to reorganize itself by forming new neural connections. In the past, scientists believed the adult brain was largely unchangeable, but research has since shown that it remains malleable throughout life. Neuroplasticity allows neurons to compensate for injury, adjust to new activities, and respond to environmental changes. This adaptability underlies our ability to learn, remember, and heal.

Sound has a unique way of engaging the brain because it influences our mental and emotional states. The brain processes sounds in specific areas, but those signals quickly ripple across multiple regions, engaging areas responsible for memory, emotions, and physical coordination. Sound frequency therapies tap into these pathways, encouraging the brain to rewire itself and optimize cognitive function.

How Sound Frequency Therapy Supports Neuroplasticity

Sound frequency therapy utilizes different frequencies to evoke responses in the brain that promote relaxation, focus, and healing. Through techniques like binaural beats, auditory stimulation, and rhythmic patterns, we can guide the brain into states conducive to neuroplasticity.

Binaural Beats: A Gateway to Brain Synchronization

Binaural beats involve playing two slightly different frequencies in each ear, which creates an auditory illusion of a single pulsing beat. For example, if a frequency of 300 Hz is played in one ear and 310 Hz in the other, the brain perceives a third "beat" of 10 Hz. This frequency can influence brainwaves, encouraging synchronization with the beat frequency. In this case, 10 Hz would promote alpha waves, associated with relaxation and focused attention.

Studies have shown that binaural beats can support neuroplasticity by inducing brainwave states that are favourable for learning and cognitive recovery. In one study, stroke patients who listened to binaural beats showed increased brain activity in affected areas, facilitating recovery. By "training" the brain to enter specific states, binaural beats offer a non-invasive method to support healing and cognitive enhancement.

Auditory Stimulation and Rhythm

The human brain is naturally attuned to rhythms. Auditory stimulation through rhythmic sound patterns can enhance neuroplasticity by encouraging the brain to create new neural pathways. This is particularly useful in rehabilitation, where rhythmic auditory stimulation has been used to help patients recover motor skills after strokes or injuries. By aligning the brain's internal rhythms with external sound patterns, we can "train" it to adopt new patterns, aiding in movement, balance, and coordination.

A well-known example of this is Rhythmic Auditory Stimulation (RAS), which uses rhythmic cues to improve gait and movement in patients with neurological disorders. RAS has been particularly effective in patients with Parkinson's disease, helping them establish steadier, more controlled movements by synchronizing their steps with rhythmic beats.

Case Study: Enhancing Cognitive Recovery with Sound

Consider Tom, a 45-year-old accountant who suffered a traumatic brain injury (TBI) in a car accident. The injury left him with memory issues, difficulty concentrating, and challenges in processing information quickly. Traditional rehabilitation had limitations in restoring his cognitive abilities fully, so his therapist introduced sound frequency therapy, specifically binaural beats and rhythmic auditory training.

Over several months, Tom listened to binaural beats daily, focusing on theta waves (4–8 Hz) to support relaxation and neuroplasticity. He also engaged in exercises that used rhythmic auditory stimulation to aid cognitive processing. Gradually, Tom experienced improvements in memory recall, focus, and overall mental clarity. His experience underscores how sound frequency therapy can complement traditional treatments, helping the brain rebuild and adapt after trauma.

Practical Tips: DIY Sound Practices to Enhance Neuroplasticity

You don't need a specialized clinic to start incorporating sound therapy practices into your daily life. Here are some accessible techniques to support neuroplasticity through sound:

1. Daily Binaural Beats Listening: Listen to binaural beats tuned to specific frequencies, such as theta waves for relaxation or gamma waves (30+ Hz) for cognitive performance. Many apps and online platforms offer binaural beats that can be incorporated into your daily routine.

2. Guided Sound Meditations: Sound meditations often include auditory cues that guide the brain into relaxed or focused states. Practicing sound meditation for even 10 minutes a day can support a calm, open mind, conducive to neuroplasticity.

3. Rhythmic Movement Exercises: Incorporate rhythmic movement with music, such as dancing or even simple walking to a beat. Engaging your body in rhythm enhances brain connectivity and can improve coordination and balance.

4. Listening to Uplifting Music: Music that resonates emotionally with you can evoke positive mental states and

encourage neural adaptation. Choose songs that inspire or relax you, as these emotional responses can promote mental flexibility.

Reflecting on Sound's Role in Rewiring the Brain

The idea that sound can reshape the brain is as poetic as it is scientific. Just as a sculptor moulds clay, so too can sound carve new pathways in our minds. Neuroplasticity, once thought to be limited to early development, has proven to be an enduring trait of the human brain, allowing us to learn, adapt, and heal throughout life. Sound frequency therapy serves as a gentle yet powerful tool in this process, inviting us to work with our natural rhythms rather than against them.

Imagine a world where cognitive decline is met not with resignation but with a symphony of therapeutic sounds—an approach that honours the brain's inherent ability to heal. Sound therapy invites us to think of healing not as an invasive intervention but as a nurturing, resonant experience.

Vision for the Future: The Integration of Sound Therapy in Cognitive Rehabilitation

As sound frequency therapy continues to demonstrate its potential, we may see it become a standard part of cognitive rehabilitation. Hospitals and clinics could one day offer sound chambers or individualized sound programs for patients recovering from brain injuries, strokes, or neurological disorders. By customizing sound frequencies to meet each patient's needs, practitioners could help the brain access its fullest healing potential.

Furthermore, advancements in wearable technology may allow patients to engage in sound frequency therapy from the comfort of their homes, ensuring consistent and personalized treatment. Imagine a future where every individual has access to tools that harness the power of sound to support their brain's adaptability and resilience.

Final Thoughts: Sound as a Pathway to Transformation

Neuroplasticity and sound invite us to consider healing not as a destination but as an ongoing journey. Our brains are dynamic and constantly evolving, shaped by every experience and interaction. Through sound frequency therapy, we gain a tool that resonates with the very essence of who we are, helping us become more adaptive, more resilient, and more in tune with ourselves.

The path of sound healing is more than a scientific pursuit; it's an invitation to listen deeply—to the brain, to the body, and to the subtle frequencies that shape our existence. As we unlock the secrets of neuroplasticity, we find ourselves on the threshold of a new era in medicine, one that embraces the mind's rhythm as a means to health and harmony. This journey, sculpted by sound, offers a profound reminder of our potential to change, grow, and transform with each resonant beat.

.

CHAPTER 11: FREQUENCIES AND THE NERVOUS SYSTEM

Introduction: The Rhythm of Relaxation and Stress

Picture the human body as a finely tuned orchestra, with each organ, cell, and system playing its part in a symphony of life. At the heart of this orchestration lies the nervous system, a complex network that governs our reactions to the world around us. The autonomic nervous system (ANS) is particularly influential, responsible for our fight-or-flight response when stressed and our rest-and-digest state when relaxed. These two states are constantly balanced, allowing us to adapt to changing environments. Yet, in today's fast-paced world, this delicate balance is often disrupted, leaving many of us trapped in chronic stress.

The exciting revelation is that frequencies—both sound and electromagnetic—can influence the ANS, guiding it from states of heightened stress back to calm and equilibrium. In this chapter, we'll explore how frequencies interact with the nervous system, bringing forward the science of their effects on stress and relaxation. We'll share stories of healing and provide practical tips to help readers harness the power of frequencies for well-being.

The Science of the Autonomic Nervous System

The ANS operates largely without our conscious control, divided into two branches: the sympathetic nervous system (SNS), which activates our fight-or-flight response, and the parasympathetic nervous system (PNS), which promotes rest, relaxation, and digestion. Ideally, these two branches work in harmony, adjusting based on our needs. When we encounter danger or stress, the SNS raises our heart rate, increases blood flow to muscles, and releases stress hormones like cortisol and adrenaline. When the danger passes, the PNS takes over, slowing the heart rate and allowing the body to rest and recover.

However, frequent stress can lead to an overactive SNS, resulting in chronic conditions like anxiety, insomnia, high blood pressure, and weakened immunity. This imbalance is where frequencies can play a role, helping the nervous system return to a balanced state.

Frequencies that Calm the Nervous System

Research has shown that specific frequencies—whether through sound, light, or electromagnetic fields—can encourage the body to enter a state of relaxation by stimulating the PNS and calming the SNS. Here are some of the most effective frequency-based therapies used to alleviate stress:

1. Binaural Beats and Heart Rate Variability

Binaural beats involve playing two slightly different sound frequencies, one in each ear, creating a perceived beat at the frequency difference. For example, listening to a 100 Hz tone in one ear and a 105 Hz tone in the other creates a 5 Hz beat, which encourages the brain to align with this frequency. Different frequencies have unique effects; lower frequencies like 4-8 Hz (theta waves) and 8-12 Hz (alpha waves) are particularly effective for relaxation.

One of the physiological markers of relaxation is heart rate variability (HRV), the variation in time between each heartbeat. A high HRV indicates a flexible, resilient nervous system, while a low HRV suggests chronic stress. Studies show that listening to binaural beats in the theta or alpha ranges can improve HRV, reflecting a shift toward the PNS and enhanced relaxation.

2. The Power of Solfeggio Frequencies

Solfeggio frequencies, ancient musical tones with specific healing properties, have been gaining renewed attention. Frequencies like 396 Hz (for releasing fear and guilt) and 528 Hz (for DNA repair and positive transformation) are believed to influence the body at a cellular level. Listening to these frequencies can have a soothing effect on the nervous system, encouraging relaxation and emotional healing.

Though scientific research on Solfeggio frequencies is still emerging, anecdotal evidence and case studies are compelling. Many listeners report reduced stress, a sense of peace, and even improved physical health, suggesting these frequencies may activate the PNS and foster a calming effect.

3. Pulsed Electromagnetic Field (PEMF) Therapy and Deep Relaxation

PEMF therapy uses electromagnetic fields at specific frequencies to encourage cellular repair and reduce inflammation. Frequencies within the 1-30 Hz range are most often used for relaxation, as they mimic the brain's natural alpha and theta waves associated with calm states.

PEMF therapy has shown promise for those with anxiety, chronic pain, and sleep disorders, helping the nervous system downshift from overactivity in the SNS to a more relaxed state in the PNS. Studies indicate that PEMF therapy can improve sleep quality, enhance mood, and even increase HRV, making it a promising tool for nervous system regulation.

Case Study: Healing Chronic Stress Through Frequency

Consider the case of Michael, a 50-year-old entrepreneur who struggled with chronic stress and insomnia. Years of overwork and pressure had left his nervous system in a constant state of alert, with symptoms ranging from heart palpitations to difficulty concentrating and severe insomnia. Desperate for relief, Michael turned to PEMF therapy, beginning with daily sessions that exposed him to low-frequency electromagnetic pulses.

Within a few weeks, Michael noticed a significant improvement in his sleep, with less time spent lying awake and more restful, uninterrupted sleep cycles. Over time, his HRV improved, and his anxiety levels decreased. Michael's experience is a testament to how frequency-based therapies can help restore balance to the nervous system, allowing the body and mind to recover from years of stress.

Practical Tips: Using Frequencies to Calm the Nervous System

If you're looking to incorporate frequency-based relaxation techniques into your life, consider the following approaches:

1. Binaural Beats for Relaxation: Use a pair of headphones and listen to binaural beats designed for relaxation, typically in the theta (4-8 Hz) or alpha (8-12 Hz) ranges. Many meditation apps offer binaural beat tracks; experiment to find one that resonates with you.

2. Breathwork with Solfeggio Frequencies: Practice deep breathing exercises while listening to Solfeggio frequencies. This combination can amplify relaxation, as deep breathing activates the PNS, and Solfeggio tones add a soothing

background that further calms the nervous system.

3. Daily PEMF Therapy: If you have access to PEMF devices, try daily sessions to reduce stress. Home PEMF devices are available and can be used while resting or meditating, making it easy to incorporate this therapy into your routine.

4. Sound Baths and Meditation: Attend sound baths, which use instruments like singing bowls and gongs to create healing frequencies. These sessions can help induce a meditative state, balancing the nervous system and supporting relaxation.

Reflection: Finding Harmony in Frequency-Based Healing

The relationship between frequency and the nervous system offers us a new way of thinking about health. Rather than merely treating symptoms, frequency-based therapies work with the body's natural rhythms, creating an environment in which true healing can take place. By addressing stress at its root and guiding the nervous system toward balance, these therapies allow us to live with greater peace and resilience.

In a world where stress is often unavoidable, learning how to guide our bodies back to a state of calm becomes a powerful form of self-care. Frequency-based therapies remind us that healing need not be invasive; it can be as simple as a rhythm, a sound, or an electromagnetic pulse that aligns us with our natural state of well-being.

Vision for the Future: Integrating Frequency Therapies into Holistic Health

As research continues to reveal the potential of frequency-based therapies, it is likely that they will become an integral part of holistic health practices. Imagine clinics that offer tailored sound and PEMF sessions to promote relaxation, resilience, and nervous system health. Such treatments could be used alongside conventional methods, providing a comprehensive approach that addresses both body and mind.

In the future, we may see frequency therapies prescribed as preventive care, not only to treat stress-related disorders but to foster a more balanced, peaceful society. This approach would represent a shift from symptom management to wellness promotion, embracing the power of frequencies as a tool for personal and collective well-being.

Final Thoughts: Embracing the Healing Power of Frequency

Healing through frequencies invites us to listen deeply—to ourselves, to our environment, and to the resonant rhythms that pulse through life. By tuning into these frequencies, we can find harmony, resilience, and a renewed sense of connection to our own bodies. The field of frequency medicine is young but holds profound promise for those seeking a gentle, natural path to health.

As we explore these possibilities, we are reminded that our journey to well-being is not just physical; it is a journey that touches the core of who we are. Through sound, vibration, and electromagnetic resonance, we discover the potential to heal not only our nervous system but our spirit, bringing us closer to a life of balance, peace, and enduring health.

.

CHAPTER 12: THE CARDIOVASCULAR SYSTEM AND RESONANCE

Introduction: A Heartbeat in Harmony with Frequency

Imagine the human heart as a metronome, pulsing with an innate rhythm that echoes through our bodies. Like an orchestra in perfect synchrony, every cell, organ, and tissue responds to this rhythm, forming the vital foundation for our well-being. However, much like a finely tuned instrument can fall out of harmony, the cardiovascular system can experience stress, inflammation, and dysfunction due to the pressures of modern life. What if there was a way to help recalibrate this system using the natural power of frequencies?

Emerging research suggests that certain frequencies can positively influence the cardiovascular system by enhancing circulation, reducing inflammation, and even aiding in stress relief. In this chapter, we will explore the science behind frequency therapy and its application in cardiovascular health. We'll also dive into case studies and practical techniques for incorporating frequency-based healing into daily life, empowering readers to discover their heart's natural harmony.

The Science of Resonance and the Heart

The cardiovascular system is a network of blood vessels powered by the heart's rhythmic contractions. This rhythm is controlled by electrical signals within the heart, which coordinate each heartbeat and keep our blood flowing. When we think about the role of frequency in cardiovascular health, it's essential to understand the concept of resonance —the idea that certain frequencies can naturally synchronize with the body's existing rhythms.

The concept of resonance becomes especially interesting when applied to the heart. Studies have shown that exposure to specific frequencies can synchronize with the heart's natural rhythms, stabilizing the heartbeat, improving blood flow, and reducing inflammation in arterial walls. Frequencies within the range of 7-12 Hz—also known as Schumann Resonances—mirror the earth's electromagnetic field and are believed to promote a state of relaxation and balanced cardiovascular function.

How Frequencies Influence Cardiovascular Health

1. Improving Heart Rate Variability (HRV)

Heart rate variability (HRV) measures the variation in time between each heartbeat and is a critical indicator of cardiovascular and autonomic nervous system health. High HRV reflects a resilient, adaptive system that can respond effectively to stress, while low HRV is associated with stress, anxiety, and increased risk of cardiovascular issues.

Binaural beats and low-frequency sound waves (7-12 Hz) have shown promising results in improving HRV by promoting parasympathetic activity, or the body's "rest and digest" state. For instance, studies on individuals with hypertension and anxiety showed that listening to these frequencies improved HRV and reduced blood pressure, indicating an overall calming effect on the heart.

2. Reducing Inflammation in Blood Vessels

Inflammation within blood vessels is a significant risk factor for cardiovascular diseases, contributing to plaque buildup and restricted blood flow. Low-frequency pulsed electromagnetic field (PEMF) therapy has demonstrated anti-inflammatory properties that can help improve endothelial function—the health of blood vessel walls. By delivering targeted electromagnetic pulses, PEMF can stimulate cellular repair and reduce inflammation, leading to better blood flow and potentially lowering the risk of cardiovascular events.

In clinical settings, PEMF therapy is commonly used for patients with peripheral artery disease (PAD), a condition where reduced blood flow to the extremities can cause pain and numbness. Studies have shown that patients who received PEMF therapy experienced reduced symptoms and improved circulation.

3. Enhancing Blood Flow and Circulation

One of the fascinating applications of frequency therapy is its ability to enhance blood flow. Studies have shown that exposing the cardiovascular system to specific frequencies, such as 50 Hz, can induce vasodilation, or the widening of blood vessels, allowing blood to flow more freely. Increased circulation delivers more oxygen and nutrients to tissues, supporting cell health and repair.

For patients recovering from surgery or dealing with chronic pain conditions like fibromyalgia, PEMF therapy can be an invaluable tool. It promotes circulation, accelerates healing, and offers a non-invasive alternative to traditional treatments. For those at risk of cardiovascular issues, regular use of PEMF and sound-based therapies can support a healthier heart and vascular system over time.

Case Study: Frequency Therapy for Post-Heart Attack Recovery

Consider Sarah, a 62-year-old woman who experienced a mild heart attack. While she received standard treatment, including medications and lifestyle adjustments, she sought additional support to enhance her recovery. Her doctor introduced her to PEMF therapy, using low-frequency electromagnetic pulses designed to improve circulation and reduce inflammation.

Over the course of six months, Sarah attended weekly PEMF sessions, targeting her cardiovascular system. Gradually, she noticed increased energy levels, reduced chest tightness, and an improved sense of well-being. Regular HRV monitoring showed a marked improvement, indicating her cardiovascular system was recovering its resilience. Sarah's story illustrates how integrating frequency therapy into recovery protocols can aid in faster, more holistic healing.

Practical Tips: Using Frequencies for Cardiovascular Health

Incorporating frequency-based practices into your daily routine can help promote a healthier cardiovascular system. Here are a few ways to get started:

1. Listen to Binaural Beats Daily: Spend 10-20 minutes each day listening to binaural beats in the 7-12 Hz range. Many apps and online platforms offer curated playlists designed to encourage relaxation and improve HRV.

2. Try PEMF Therapy: If you have access to a PEMF device, consider incorporating it into your routine. Some clinics offer sessions, or you may invest in a home device. Aiming for 20-30 minutes a few times a week can support blood flow and reduce inflammation.

3. Practice Breathwork with Resonant Frequencies: Combine

deep breathing exercises with a calming frequency (such as 396 Hz or 528 Hz) to create a sense of harmony in your cardiovascular system. Deep breathing activates the PNS, and the resonant frequency helps further relax the heart and blood vessels.

4. Engage in Sound Baths or Singing Bowl Sessions: Sound baths, using instruments like crystal bowls tuned to healing frequencies, can encourage profound relaxation. Sound vibrations penetrate the body, enhancing circulation and creating a calming effect that benefits cardiovascular health.

Reflection: Embracing Frequency for a Healthier Heart

When we consider the role of frequency in cardiovascular health, we are reminded of the subtle rhythms that govern our lives. The heart's beat is a timeless reminder of our connection to both body and mind, an internal metronome that responds to our thoughts, emotions, and even the sounds around us. By using frequency therapy, we have the potential to harness natural rhythms to recalibrate our heart and vascular system, promoting a state of balance and well-being.

As we look to the future, frequency therapies may well become integral to cardiovascular health practices, enabling non-invasive, accessible options for heart health management. Rather than relying solely on medications, we may find that the gentle resonance of specific frequencies offers a natural way to maintain a healthy heart and circulatory system.

Vision for the Future: A World of Resonant Healthcare

Imagine a healthcare system where frequency therapies are as common as annual check-ups. Clinics and hospitals could offer specialized frequency treatments for cardiovascular support, designed to enhance recovery, prevent disease, and encourage holistic well-being. Wearable devices might soon be able to deliver targeted frequencies directly to the heart or circulatory system, creating a seamless integration of natural and medical science.

As research into frequency-based therapies expands, so does our understanding of their potential to improve health outcomes. This growing field invites us to embrace the profound possibilities of resonance and rhythm in medicine, shifting from reactive treatment to proactive, preventative care.

Final Thoughts: Finding Harmony in Health

Frequency therapy offers a path to heart health that is both gentle and profound. By attuning to the natural rhythms within us and aligning with external frequencies, we find a powerful means of supporting our cardiovascular system. Healing, in this context, becomes less about intervention and more about resonance—a harmonious relationship between the heart, the mind, and the healing potential of sound and frequency.

In this journey toward balance, frequency therapy stands as a bridge between ancient wisdom and modern science, inviting us to listen deeply and discover the harmonies within. As we explore this approach to cardiovascular health, we may find not only a healthier heart but also a more profound connection to the symphony of life itself.

CHAPTER 13: LYMPHATIC SYSTEM ACTIVATION THROUGH VIBRATION

Introduction: The Body's Hidden River of Health

In the intricate design of the human body, there exists a silent yet essential network that plays a crucial role in maintaining our health: the lymphatic system. Often overlooked, this system operates like a hidden river, flowing quietly alongside our blood vessels, filtering out toxins, waste, and other unwanted materials. Unlike the circulatory system, the lymphatic system doesn't have a pump like the heart; instead, it relies on movement, muscle contractions, and, as modern research has revealed, even frequencies and vibrations to keep its flow steady.

Imagine the lymphatic system as a vast, delicate web in constant need of gentle stirring to prevent stagnation. Without adequate movement or stimulation, toxins accumulate, immune function declines, and our body's resilience falters. Enter vibration therapy—a modern application of an ancient principle: using rhythmic frequencies to stimulate and support the lymphatic system.

In this chapter, we explore the scientific and practical ways vibration therapies are being used to enhance lymphatic flow, support immune health, and aid in detoxification.

The Science Behind Vibration and the Lymphatic System

The lymphatic system consists of a network of vessels, nodes, and organs, including the spleen and tonsils, that work together to cleanse the body of waste. Lymph fluid, carrying white blood cells, flows through this network, clearing out toxins and supporting immunity. However, without a central pump, the system depends heavily on muscle movement and gravity, which means it's often less active during periods of inactivity or illness.

Vibration therapies introduce rhythmic movements that stimulate lymphatic flow by mimicking the natural muscle contractions the system needs. These therapies include low-frequency vibrations, targeted oscillations, and even gentle sound waves, all of which help improve lymph circulation and promote detoxification.

Frequency and Resonance in Lymphatic Stimulation

The principle of resonance—that every cell, tissue, and system has an optimal frequency at which it naturally vibrates—plays an essential role in lymphatic stimulation. Studies have shown that vibrations within the range of 10-100 Hz are effective for lymphatic drainage, helping move fluid through the vessels and clearing out toxins.

One popular device, the whole-body vibration machine, uses controlled vibrations to activate the muscles, enhancing lymph flow in people with limited mobility. By standing on a platform that vibrates at specific frequencies, users can stimulate their lymphatic system without intense physical effort, making it accessible for patients recovering from surgery or dealing with chronic illnesses.

Real-Life Application: Vibration Therapy for Immune Support

Consider the case of Tim, a middle-aged man recovering from knee surgery. The reduced mobility led to a sluggish lymphatic system, causing swelling in his legs and a weakened immune response. With the guidance of his therapist, Tim began using a low-frequency vibration device three times a week. The therapy not only reduced his swelling but also boosted his overall energy, as toxins and excess fluid were naturally cleared from his body.

Tim's experience highlights a key benefit of vibration therapy: its ability to support immune health by keeping lymph fluid in motion, helping prevent the buildup of waste that can burden the immune system. Such therapies are now being used in clinics and wellness centres worldwide, especially for patients with compromised immune systems, post-surgical swelling, and chronic conditions that limit mobility.

Practical Tips: Using Vibration to Boost Lymphatic Health

For those interested in exploring vibration therapy to support their lymphatic health, here are some practical ways to get started:

1. Whole-Body Vibration Machines: These machines are available in many wellness centres and gyms. Spend 10-15 minutes standing on a low-frequency vibration platform a few times a week to stimulate lymphatic flow.

2. Rebound Exercises: Rebounding on a mini-trampoline, even for just a few minutes daily, creates a gentle, rhythmic up-and-down movement that naturally activates the lymphatic system.

3. Targeted Vibration Therapy: Handheld vibration devices

can be used on areas prone to lymph stagnation, such as the legs and arms. Set the device to a low-frequency setting and apply it in circular motions to help move lymphatic fluid.

4. Sound Healing and Singing Bowls: Sound-based therapies using low-frequency tones, such as those produced by Tibetan singing bowls, can be placed near areas of lymph congestion. These gentle sound vibrations resonate through tissues, promoting flow and encouraging relaxation.

5. Dry Brushing with a Twist: Try dry brushing combined with sound therapy. While gently brushing the skin toward the heart, listen to low-frequency music to enhance relaxation and lymphatic activation.

Case Study: Supporting Detoxification Through Frequency-Based Lymphatic Stimulation

In one clinical study, a group of cancer patients undergoing chemotherapy participated in a frequency-based lymphatic stimulation program. They used vibration therapy and sound frequency sessions to aid in detoxification, helping the body process the drugs and eliminate toxins more efficiently. Over time, these patients reported reduced side effects, less fatigue, and faster recovery between treatments.

The success of this study showcases how targeted vibration and frequency therapies can complement conventional treatments by easing the lymphatic burden, which often becomes significant during intensive medical care. By enhancing the body's detoxification capabilities, frequency-based therapies can improve patients' quality of life and overall health.

Reflection: A New Way to Look at Detoxification

As we explore the potential of frequencies to support lymphatic health, a broader perspective emerges. Detoxification is often thought of as a separate, occasional process—a cleanse or a diet. But the truth is, our body detoxifies continuously, relying on systems like the lymphatic network to keep us balanced and healthy.

Vibration and frequency therapies invite us to shift our perspective from sporadic detox routines to an integrated approach that supports natural detoxification every day. By keeping the lymphatic system in motion, we're not only removing waste but actively contributing to our immune resilience, energy levels, and overall sense of well-being.

Vision for the Future: Integrating Frequency Therapies in Lymphatic Health

In the coming years, we may see frequency and vibration therapies become a standard aspect of lymphatic health management. Imagine wearable devices programmed to deliver gentle vibrations throughout the day, especially useful for individuals with sedentary lifestyles or those recovering from illness. Hospitals and clinics could offer frequency-based lymphatic support as a routine part of post-surgical recovery, helping patients heal faster and reducing complications from swelling.

The future of frequency-based lymphatic therapy holds the promise of personalized, accessible options for everyone. By combining ancient principles of sound healing with modern technology, we can create a harmonious approach to health that encourages the body's natural processes to thrive.

Final Thoughts: Embracing Vibrational Health

The lymphatic system, though often forgotten, plays a powerful role in our body's ability to heal, rejuvenate, and maintain balance. By embracing vibration therapies, we tap into a method of support that is as gentle as it is profound, aligning with the body's natural rhythms and promoting holistic health.

Frequency-based therapies offer more than just physical relief; they encourage us to reflect on the interconnectedness of our body systems and remind us of the beauty in our own complexity. Through intentional, rhythmic support, we find that healing becomes less about intervention and more about harmony—working with the rhythms of life to nurture our deepest well-being.

As we move forward in understanding how vibration and frequency influence the lymphatic system, we take another step toward a future where medicine honours both the science and the spirit of healing, bringing the best of both worlds together for a healthier, more resilient human experience.

CHAPTER 14:
SOUND'S ROLE IN
PAIN MANAGEMENT

Introduction: Tuning Out Pain Through Frequency

Imagine the human body as a finely tuned orchestra, each cell, nerve, and organ playing its own note to create harmony. Pain, in this metaphor, is the discordant note—a harsh, jarring sound that disrupts the balance and demands our attention. Conventional medicine has long tried to silence pain through medications, but what if there was another way? What if, instead of muting the discord, we could re-tune the body back to harmony?

This is the promise of frequency-based pain management. By using specific frequencies, sound waves can disrupt pain signals and induce analgesic effects, offering a non-invasive alternative to traditional painkillers. This chapter dives into the science, history, and real-life applications of sound as a tool for pain relief, revealing how ancient wisdom and modern science converge to reimagine our relationship with pain.

The Science of Sound and Pain

Pain signals travel through the nervous system as electrical impulses, sending alerts from the affected area to the brain. Certain types of pain, like chronic pain, can become ingrained, as if the body were replaying the same discordant note repeatedly. Frequency therapies, particularly sound-based approaches, work by interfering with these signals, effectively disrupting the "pain loop."

Research has shown that low-frequency sounds, around 40 Hz, can alter brainwave activity and reduce pain perception. When these frequencies are introduced to the body, they resonate with cells and nerves, disrupting the signals sent to the brain. Unlike painkillers that temporarily block pain receptors, sound therapy aims to address the root of the problem, guiding the body back to its natural equilibrium.

The Role of Brainwaves in Pain Perception

Brainwaves play a crucial role in how we experience pain. Pain often elevates brainwaves to higher frequencies, typically in the beta range (14-30 Hz), which is associated with alertness and tension. Lower-frequency brainwaves, like alpha (8-13 Hz) and theta (4-7 Hz), are associated with relaxation and deep healing states. By using sound frequencies to encourage the brain to shift from beta to alpha or theta states, patients can experience reduced pain perception and a greater sense of calm.

Studies in sound therapy have shown promising results in lowering beta activity and boosting alpha waves in chronic pain sufferers, offering them a respite from the relentless high-frequency state that pain triggers. This technique is non-invasive, free from side effects, and can be tailored to individual needs—making it an ideal solution for long-term pain management.

Case Study: John's Journey with Sound Therapy for Chronic Back Pain

Consider the case of John, a construction worker who suffered from debilitating back pain after an injury.

Painkillers provided only temporary relief, and physical therapy offered little improvement. Desperate for an alternative, John turned to sound therapy. His therapist introduced him to low-frequency sessions with binaural beats, specifically designed to stimulate alpha and theta brainwaves.

Within weeks, John began to notice changes. His pain episodes became less intense, and his tolerance for physical activity improved. Over time, he found he could manage his pain without relying on medication. John's story reflects a broader trend in frequency-based medicine: using sound not just as a passive treatment but as an active tool for reclaiming quality of life.

Practical Applications: Techniques for Pain Relief Using Sound

For readers interested in exploring sound therapy for pain management, here are practical techniques to consider:

1. Binaural Beats: Listening to binaural beats, where two slightly different frequencies are played in each ear, can encourage the brain to enter a relaxed state. This approach is particularly effective for managing stress-related pain, as it helps shift the brain to alpha and theta states.

2. Frequency-Specific Music: Music created with specific frequencies, like 174 Hz and 396 Hz, is known to have analgesic properties. Listening to these frequencies for 15-20 minutes a day may help to alleviate mild to moderate pain.

3. Sound Baths: Attending a sound bath session, where instruments like crystal bowls and gongs produce low-frequency sounds, can be a soothing experience. These sounds envelop the body, creating a resonant field that can disrupt pain signals and promote relaxation.

4. Personalized Frequency Therapy: Many therapists now

offer frequency-specific pain management sessions. This involves tuning devices to individual needs, adjusting frequencies to target pain directly.

5. DIY Sound Therapy: For those looking for an at-home solution, try creating a playlist with healing frequencies or nature sounds. While it may not replace medical treatment, it can serve as a supportive practice to reduce stress-related pain.

Real-Life Impact: Sound Therapy and Migraine Relief

In a clinical study on migraine sufferers, patients participated in a sound therapy program that utilized frequencies between 40-100 Hz. After four weeks, over 60% of participants reported a reduction in migraine intensity and frequency. This outcome is significant, as migraines are notoriously resistant to traditional pain treatments and often require heavy medication.

Sound therapy's success in addressing migraines hints at its potential for managing other types of pain. By modulating brainwave activity, these frequencies help reset the nervous system, reducing hyperactivity that contributes to pain cycles. For many, sound therapy offers a safe and sustainable alternative to pharmaceuticals, paving the way for more holistic pain management solutions.

Reflection: Rethinking Pain with Frequency-Based Therapies

The success of sound therapy in pain management invites us to reexamine our understanding of pain. Pain, rather than being an enemy to eliminate, becomes a signal that can be re-tuned, redirected, and ultimately harmonized. Frequency-based therapies shift the focus from suppressing pain to transforming it—returning the body to a state of natural rhythm where discomfort does not dominate.

Pain is a complex phenomenon, shaped by emotional,

physical, and neurological factors. Sound therapy respects this complexity, offering a multidimensional approach that aligns with the body's rhythms instead of working against them. As we continue to uncover the ways frequencies impact pain, we're reminded of the importance of addressing the root rather than the symptoms.

Vision for the Future: Integrating Sound Therapy in Pain Management Protocols

Imagine a future where hospitals, clinics, and wellness centres integrate sound therapy as a core aspect of pain management. Pain-relief suites could feature soundscapes tailored to patients' needs, allowing them to experience comfort and tranquillity without pharmaceuticals. Devices could be developed to deliver customized frequencies directly to affected areas, offering localized pain relief.

As the field of frequency medicine grows, the possibilities expand. Advances in neuroscience, biofeedback, and digital health could make personalized sound therapies accessible to everyone. This vision of frequency-based pain management is not just about relieving pain but transforming our relationship with it, creating an approach that is as gentle as it is powerful.

Conclusion: Embracing Sound as a Path to Relief

Sound-based pain management is not simply about quieting pain but restoring harmony within the body. By embracing the therapeutic power of sound, we step into a realm where healing is as natural as music, as profound as a melody that speaks directly to our cells. This approach invites us to redefine pain relief as an art—one that resonates deeply with the body's innate wisdom.

The future of pain management lies in these harmonious frequencies that align with our physical and emotional well-being. Through sound, we discover that healing need not be invasive or forceful; it can be gentle, resonant, and, above all, deeply human. In a world where silence often feels elusive, sound therapy offers a sanctuary—a place where the discord of pain fades, leaving us with the quiet strength to embrace life fully once again.

CHAPTER 15: ENHANCING RESPIRATORY HEALTH THROUGH FREQUENCY

Introduction: The Breath of Life, Tuned to Harmony

Breath is our most vital rhythm, marking each moment of life. From our first inhale to our final exhale, the act of breathing sustains us, feeds our cells, and connects our mind and body. Yet, in the modern world, where stress, pollution, and respiratory issues are on the rise, our breath often becomes shallow and strained. In recent years, science has explored how sound frequencies, by resonating with our respiratory system, can help improve breathing patterns and support respiratory health. Imagine being able to enhance your breath not by simply breathing deeper but by harmonizing your entire respiratory system with the natural frequencies of wellness.

This chapter delves into the surprising connection between sound frequencies and respiratory health. Through real-life stories, practical applications, and cutting-edge research, we'll explore how frequency medicine can aid respiratory health and give us tools to restore our most fundamental rhythm—the breath.

The Science of Sound and Breathing

The relationship between sound frequencies and the respiratory system begins with the autonomic nervous system (ANS), which controls involuntary actions, including breathing. When we are calm, our ANS encourages slower, deeper breaths; under stress, it prompts rapid, shallow breathing. Sound frequencies, particularly low and calming ones, have a unique ability to stimulate the parasympathetic branch of the ANS, promoting relaxation and, in turn, supporting healthier breathing patterns.

Scientific studies have shown that specific sound frequencies, such as those found in the range of 5-15 Hz, can entrain the body to a state of relaxation. These frequencies help to reduce tension in the respiratory muscles, allowing the lungs to expand more fully and promoting deeper, slower breaths. By lowering stress and tension, sound frequencies create an optimal environment for the respiratory system to function at its best, potentially aiding in the management of conditions like asthma, chronic obstructive pulmonary disease (COPD), and other respiratory issues.

Case Study: Anna's Journey with Asthma and Sound Therapy

Anna, a 32-year-old artist, had struggled with asthma since childhood. Her breathing issues limited her activities, and the stress of managing her condition only made her symptoms worse. After reading about sound therapy, Anna was intrigued but sceptical . Still, she decided to try a frequency-based breathing program that incorporated binaural beats and calming soundscapes at low frequencies.

Over several weeks, Anna noticed something remarkable: during her sound therapy sessions, her breathing became easier and deeper. The tension she often felt in her chest lessened, and her asthma attacks decreased. Sound therapy didn't cure her asthma, but it gave her a way to manage it, helping her breathe with less effort and stress.

Anna's experience highlights the potential of frequency-based approaches for respiratory health, particularly for those with chronic conditions. By creating an environment of calm and relaxation, sound therapy provided a pathway for Anna to reconnect with her breath, easing her symptoms and empowering her with a new way to care for her body.

Practical Applications: Using Sound Frequencies for Better Breathing

For those interested in exploring frequency-based techniques to support respiratory health, here are some effective practices:

1. Binaural Beats for Relaxed Breathing: Binaural beats can help shift the brain from high-frequency beta waves (associated with stress) to lower alpha or theta waves, which promote relaxation. Try listening to binaural beats in the 5-10 Hz range while focusing on your breath, allowing the calming frequencies to encourage deeper, slower breathing.

2. Frequency-Based Breathing Exercises: Certain sound frequencies, such as 528 Hz, have been associated with relaxation and respiratory health. Incorporate these frequencies into breathing exercises. For example, play a 528 Hz track and inhale deeply to the rhythm, holding the breath momentarily before exhaling. Repeat for a few minutes to promote relaxation.

3. Guided Sound Meditation for Breath Awareness: Guided sound meditations often incorporate soothing frequencies and rhythms that can guide you to pay closer attention to your breathing. These meditations may also encourage diaphragmatic breathing, which expands lung capacity and

improves oxygen intake.

4. Personal Soundscapes: Create your own soundscapes using calming sounds like ocean waves, wind, or even a gentle humming at specific frequencies. These natural sounds resonate with the body's rhythm and can serve as a backdrop for practicing mindful, slow breathing.

5. Tuning Fork Therapy: Tuning forks at specific frequencies, like 174 Hz and 285 Hz, are known to have calming effects on the body. Placing them near the chest can help resonate with the respiratory muscles, promoting relaxation in this area and encouraging fuller breaths.

The Role of Sound in Respiratory Health: Scientific Insights

A study conducted by researchers on the effects of low-frequency sound therapy on asthma patients demonstrated significant improvements in lung function. Patients exposed to low-frequency sound waves reported decreased shortness of breath and a greater sense of relaxation. The frequencies used in these studies worked by enhancing the body's natural rhythms, supporting the muscles and tissues involved in breathing, and helping patients experience fewer asthma symptoms.

Sound therapy's effects on respiratory health extend beyond chronic respiratory conditions. In hospital settings, some patients undergoing surgery or facing high stress levels have benefitted from sound therapy sessions. These sessions, focused on low-frequency sounds, are shown to help lower the heart rate and encourage deep breathing, which not only calms the patient but also improves oxygenation—a critical aspect of recovery and immune function.

Reflection: Breath, Sound, and the Art of Harmonizing Life

Breath and sound are two elemental forces of life, each with the power to transform the other. When we explore frequency-based therapies, we begin to see breathing not just as a biological necessity but as a rhythm that can be harmonized, tuned, and supported. Through sound, we learn that breathing can be a healing act—a means of connecting with the deepest parts of ourselves.

Consider how, in moments of stress, simply listening to a calming frequency or a familiar piece of music can guide you back to your breath. This experience reminds us that healing is often a matter of tuning into our body's natural rhythms rather than fighting against them. By blending breath and sound, we unlock the potential to elevate our health in subtle, profound ways.

Vision for the Future: Frequency Medicine and Respiratory Health

The future of respiratory health could see frequency-based therapies integrated into mainstream care. Imagine a wellness centre where individuals with respiratory conditions are guided through personalized sound therapies designed to improve lung function. Portable sound therapy devices tailored to different frequencies could one day support individuals with asthma or COPD, providing relief and relaxation wherever they go.

As advancements in technology bring us closer to this vision, the possibilities for frequency medicine continue to grow. We may soon witness a world where respiratory health is not solely managed through medication but supported by the healing power of sound, reconnecting patients with the ancient wisdom of rhythm and breath.

Conclusion: Rediscovering Breath through Sound

Enhancing respiratory health through frequency is a practice rooted in simplicity but backed by science. Sound therapy offers a gentle, supportive approach to respiratory wellness, allowing us to tap into our body's rhythms and restore balance. With each breath we take to the beat of healing frequencies, we rediscover an age-old truth: that health is not merely the absence of illness but the presence of harmony.

In our fast-paced world, where the simple act of breathing often feels rushed or neglected, sound therapy invites us to slow down, to listen, and to breathe with intention. As we embrace the possibilities of frequency medicine, we reconnect with the breath as a source of strength, healing, and life itself. Through sound, we remember that every breath can be a step toward a more harmonious, healthier life.

PART III:
TECHNIQUES
IN FREQUENCY
HEALING

CHAPTER 16: BINAURAL BEATS AND BRAIN SYNCHRONIZATION

Introduction: The Hidden Power of Sound in the Brain

Imagine your brain as a symphony orchestra, each section representing a different mental state: relaxation, focus, creativity, and rest. Like an orchestra, each part of your brain produces its own unique rhythm, or frequency, at any given time. The science of binaural beats taps into this symphony, subtly influencing our brainwaves and creating harmony within the mind. Binaural beats are a remarkable example of how sound frequencies can interact with our biology, unlocking states of deep relaxation, enhanced focus, and even bursts of creativity. But how do they work, and what is their true potential in healing and mental wellness?

This chapter takes you through the science of binaural beats, the history of their use, and real-life applications. We'll also explore the concept of brainwave synchronization and how this phenomenon can offer a non-invasive way to support emotional and cognitive health.

What Are Binaural Beats?

Binaural beats occur when two slightly different frequencies are played separately in each ear. For example, if you hear a frequency of 200 Hz in one ear and 210 Hz in the other, your brain will perceive a beat of 10 Hz, the difference between the two frequencies. This perceived beat is not an actual sound but rather a byproduct of how the brain processes these two frequencies, merging them into one "beat" in the mind.

This phenomenon, known as brainwave entrainment, allows the brain to synchronize with the frequency of the binaural beat. Since different brainwave frequencies are associated with different mental states, binaural beats can help guide the brain toward a specific state. For instance, a beat in the 8–14 Hz range (alpha waves) promotes relaxation and calm, while a beat in the 14–30 Hz range (beta waves) can enhance alertness and concentration.

The Science of Brainwave Entrainment: Tuning the Mind

The human brain operates across a spectrum of frequencies, each associated with a particular state of consciousness:

1. Delta Waves (0.5–4 Hz): Linked to deep sleep and restorative rest.
2. Theta Waves (4–8 Hz): Associated with meditation, creativity, and deep relaxation.
3. Alpha Waves (8–14 Hz): A relaxed state of wakefulness, ideal for mindfulness and focus.
4. Beta Waves (14–30 Hz): Linked to active thinking, focus, and decision-making.
5. Gamma Waves (30 Hz and above): Often observed in moments of high-level cognition and problem-solving.

Binaural beats aim to influence brainwaves by gently guiding them into a desired frequency range, facilitating a shift in consciousness. The ability to entrain brainwaves through binaural beats offers exciting possibilities for mental wellness and cognitive performance.

Case Study: Sarah's Journey with Anxiety

Sarah, a 45-year-old teacher, had struggled with anxiety for years. Traditional meditation practices had not been effective for her, as her mind would often race even when trying to sit still. After learning about binaural beats, Sarah decided to try a 20-minute session each morning using binaural beats in the alpha range (10 Hz) to encourage relaxation.

Within a few weeks, she noticed a change. Her mornings felt calmer, and her racing thoughts began to slow. The binaural beats helped her access a state of calm that had once felt out of reach. For Sarah, binaural beats provided a gateway into relaxation, allowing her mind to follow the frequency into a more peaceful state.

How to Use Binaural Beats for Brain Synchronization

For those interested in exploring the benefits of binaural beats, here are some guidelines for effective use:

1. Choose the Right Frequency: Select a frequency range based on your desired mental state. For relaxation, look for binaural beats in the alpha range (8–14 Hz). For focus, try beta range beats (14–30 Hz), and for deep meditation, theta (4–8 Hz) can be helpful.

2. Use Headphones: Since binaural beats require different frequencies in each ear, headphones are essential. Over-ear headphones are often recommended for a more immersive experience.

3. Set a Consistent Routine: Binaural beats tend to be most effective when used consistently. Set aside 15–30 minutes daily to incorporate binaural beats into your routine, whether as a morning ritual or part of an evening wind-down.

4. Experiment with Session Length: Some people respond well to short sessions, while others benefit from longer exposure. Start with 15 minutes and adjust as needed based on how you feel.

5. Combine with Meditation or Visualization: Binaural beats pair well with meditation, deep breathing, or visualization techniques. Using them alongside other practices can deepen the sense of relaxation and focus.

Real-World Applications and Benefits

Binaural beats are now widely used in various fields, from mental health to productivity. Here's how they're making an impact:

- Mental Health: Studies have shown that binaural beats can reduce symptoms of anxiety and depression by promoting relaxation and encouraging positive mental states. For individuals with PTSD or chronic stress, theta and alpha beats can offer a safe way to access calm and relief.

- Sleep: Delta beats, in the 0.5–4 Hz range, can promote deep, restful sleep. By entraining the brain to slow down, binaural beats can help those with insomnia or poor sleep quality achieve more restful nights.

- Focus and Creativity: Many professionals and creatives use binaural beats to enhance focus and spur creative thinking. Beta and gamma frequencies are particularly useful for concentration and innovation, supporting tasks that require high mental engagement.

- Pain Management: Emerging research suggests that binaural beats may have analgesic effects by reducing the brain's perception of pain. Patients with chronic pain conditions have reported reduced discomfort after using binaural beats in their routines.

Reflection: Synchronizing the Mind's Symphony

The science of binaural beats reveals a profound truth: our brains are wired to respond to rhythm and frequency. By understanding and harnessing this rhythm, we can create harmony within the mind, accessing states of calm, focus, or creativity as needed. The brain, in many ways, is like a musical instrument, capable of being tuned to resonate with our desires and needs.

Imagine a world where binaural beats are as common as coffee for those who need focus, as beloved as a bedtime story for those seeking sleep, and as grounding as meditation for those in search of calm. In this world, sound becomes not just a background experience but an intentional practice, a tool for tuning our inner states.

The Future of Binaural Beats and Brain Synchronization

As research into binaural beats advances, we may see applications beyond personal wellness, with potential uses in education, therapy, and even rehabilitation. Personalized binaural beat programs, tailored to individual brainwave patterns, could offer targeted treatments for mental health disorders, cognitive enhancement, and neurological conditions.

Imagine an app that detects your mental state in real-time, using binaural beats to elevate or calm your mind as needed. Or a classroom where students listen to focus-enhancing binaural beats before exams, creating an environment conducive to learning and memory retention.

Conclusion: Embracing Sound as a Pathway to Wellbeing

Binaural beats are a powerful reminder of how sound, when used intentionally, can create profound shifts in our mental and emotional landscapes. By listening, we can engage with our mind in a new way, bridging science with intuition, logic with creativity.

As we learn to harness the gentle power of binaural beats, we open ourselves to the idea that the mind, like any instrument, can be guided toward harmony. With each beat, we attune ourselves to the natural rhythms of calm, focus, and creativity, rediscovering the simplicity and elegance of sound as a force for healing and transformation.

CHAPTER 17: SOLFEGGIO FREQUENCIES AND HEALING POTENTIAL

Introduction: The Resonance of Ancient Sound

Imagine walking into an ancient temple, where chants reverberate against stone walls, creating a rich tapestry of sound that feels almost tangible. These sounds aren't random—they're based on precise frequencies, ancient tones that followers believed had the power to connect the human spirit to divine energy, to heal, and to bring balance to the soul. These are the Solfeggio frequencies, a set of tones that have transcended time, carrying with them a legacy of healing and harmony.

In this chapter, we explore the fascinating world of Solfeggio frequencies, tones said to resonate deeply with both the body and spirit, bridging science with spirituality. Let's delve into how these frequencies work, their potential for emotional and spiritual wellness, and how they're being rediscovered today as a powerful tool for healing.

The Science of Frequency and Resonance

Sound is vibration, and vibrations impact the body at a cellular level. Every cell in the human body, every organ, and even our brainwaves operate at specific frequencies. Just as we tune an instrument to achieve harmony, our bodies and minds can be "tuned" by resonating with certain frequencies. This phenomenon, known as entrainment, allows sound frequencies to synchronize our brainwaves and cellular energy, leading to profound physiological effects.

Solfeggio frequencies operate on this principle. These frequencies, set in a specific range, are thought to align with the body's natural frequencies, offering a gateway to restore balance and harmony. Though modern science has only recently begun to explore their impact, ancient cultures from the Gregorian chants of monks to the sounds used in Sanskrit chanting have long utilized similar frequencies for healing and spiritual practices.

What Are the Solfeggio Frequencies?

The Solfeggio scale consists of six main frequencies, each said to have unique healing properties:

1. 396 Hz – Liberating Guilt and Fear: This frequency is believed to target the root cause of negative emotions, releasing feelings of guilt and fear that may be stored within the subconscious mind. It's seen as a way to ground oneself, letting go of past trauma and finding inner stability.

2. 417 Hz – Facilitating Change: Often associated with transformation and renewal, the 417 Hz frequency is said to help clear old patterns, bringing clarity and a readiness to embrace positive change. It's particularly helpful for those feeling stuck or resistant to moving forward.

3. 528 Hz – Miracles and DNA Repair: Known as the "Miracle Frequency," 528 Hz is one of the most widely recognized and studied Solfeggio tones. Researchers suggest that this frequency can influence DNA repair and cellular regeneration. It's considered to be a frequency of profound

transformation and spiritual awakening.

4. 639 Hz – Healing Relationships: This frequency resonates with love and relationships. It's believed to foster communication, understanding, and harmony within oneself and with others, creating a bridge for healing interpersonal connections.

5. 741 Hz – Awakening Intuition: Known as the frequency for clarity and intuition, 741 Hz is said to cleanse the body from toxins and electromagnetic pollution, making it especially useful for those exposed to a high-stress environment. It's also linked to creativity and intuition.

6. 852 Hz – Returning to Spiritual Order: Finally, the 852 Hz frequency is associated with returning to a state of spiritual alignment. It's believed to help balance the mind, foster spiritual clarity, and deepen one's connection to the universe. Story of Healing: Emma's Transformation through Solfeggio Frequencies

Emma, a 32-year-old artist, had struggled with chronic anxiety and a sense of disconnection from her creative self. Traditional therapy had provided some relief, but something still felt missing. She began to explore sound healing and was particularly drawn to Solfeggio frequencies.

Starting with 528 Hz—the frequency known for DNA repair and transformation—she would listen to it every morning during her meditation. Gradually, she introduced other frequencies, particularly 417 Hz to help release old patterns and 852 Hz for spiritual connection. Over months, Emma noticed profound changes. She felt more grounded, her creative ideas flowed more freely, and her anxiety became much more manageable. For her, the Solfeggio frequencies provided a sense of harmony she hadn't felt in years.

How to Use Solfeggio Frequencies in Your Daily Life

Interested in exploring Solfeggio frequencies for yourself? Here are some practical tips to get started:

1. Set an Intention: Before listening, take a moment to set a clear intention. Are you seeking peace, release from past hurts, or a deeper connection with yourself? Let this intention guide your experience.

2. Find a Quiet Space: For best results, listen to Solfeggio frequencies in a quiet, distraction-free space. Headphones are optional but can help enhance the experience by isolating the sound.

3. Focus on Your Breath: Breathing deeply while listening to the frequencies can help deepen their impact. With each inhale, visualize healing energy filling your body; with each exhale, imagine releasing negativity.

4. Experiment with Different Frequencies: Try out various frequencies to see which resonates best with you. Each frequency has unique effects, so find the ones that feel most aligned with your current needs.

5. Incorporate into Meditation or Sleep: Solfeggio frequencies can be a powerful addition to meditation or can be played softly during sleep to promote a calming atmosphere.

The Mystery and Power of Sound Healing

The resurgence of Solfeggio frequencies in modern wellness circles reminds us of the intrinsic power of sound. When we listen to Solfeggio frequencies, we are not merely hearing tones; we are inviting our bodies to enter a harmonious state, one where healing becomes possible. Many practitioners believe that these frequencies tap into the "blueprint" of our DNA, helping us return to a state of natural well-being and alignment.

Imagine each frequency as a distinct key that unlocks

a different door within the human experience—one that allows us to explore realms of healing, self-discovery, and spiritual growth. While scientific evidence around Solfeggio frequencies is still evolving, countless personal experiences suggest that these tones resonate deeply with those seeking holistic wellness.

Reflecting on the Future: Embracing Ancient Wisdom in Modern Healing

Solfeggio frequencies bridge the ancient and the modern, blending time-honoured sound practices with current wellness approaches. As research in sound healing grows, we may come to understand more fully why these particular frequencies have such profound effects on our mind, body, and spirit. Perhaps, in the near future, we'll witness a deeper integration of Solfeggio frequencies into therapeutic settings, as they offer a non-invasive, cost-effective, and accessible means of supporting mental and emotional health.

Consider the possibilities: Imagine therapists integrating Solfeggio frequencies into sessions for trauma recovery, hospitals playing 528 Hz in recovery rooms, or meditation apps providing Solfeggio playlists to enhance daily practice. This fusion of sound and healing holds immense potential.

Conclusion: Returning to Harmony

At their core, Solfeggio frequencies invite us back to harmony—not only within our minds and bodies but also within our relationships, creativity, and spirit. They remind us of the profound truth that sound is more than just vibration; it is a language that communicates directly with our very essence.

By exploring Solfeggio frequencies, we open ourselves to a pathway that our ancestors understood and cherished. In embracing these sounds, we honour a legacy of healing, allowing us to tap into the ancient wisdom that resonates in each note, harmonizing the symphony within us and offering a gateway to profound wellness and inner peace.

CHAPTER 18: ULTRASOUND THERAPY IN MEDICAL PRACTICE

Introduction: The Hidden Power of Sound Waves

When most people think of ultrasound, they envision a medical tool that reveals images of a growing baby in the womb. But ultrasound waves are far more than just a way to visualize life's first moments. In recent years, ultrasound has become a powerful, versatile therapeutic tool, offering non-invasive solutions for diagnostics, pain relief, and even tissue regeneration. This chapter explores how ultrasound waves, a type of high-frequency sound, are transforming medical practice by healing tissues, enhancing recovery, and providing doctors with a precise method to treat patients in ways once thought impossible.

Ultrasound Technology: Sound Waves that Heal

Ultrasound waves are sound waves with frequencies above the range of human hearing—typically above 20,000 Hz. These high-frequency waves penetrate tissues, vibrating at a level that can trigger healing processes. Imagine a sound so subtle that it remains inaudible but powerful enough to affect cells and tissues deep within the body. Much like an invisible hand, ultrasound waves can stimulate blood flow, reduce inflammation, and help cells repair themselves without the need for incisions.

At the heart of ultrasound therapy is a principle called "mechanical oscillation." When ultrasound waves penetrate tissues, they cause microscopic vibrations. These vibrations create gentle heat in deep tissue layers, enhancing circulation and promoting healing by delivering oxygen and nutrients directly to damaged areas.

The Science Behind Ultrasound Healing

The science of ultrasound therapy is rooted in physics. When ultrasound waves enter the body, they encounter various tissues—skin, muscles, bones—each with its density and structure. As waves pass through, some energy is absorbed, and this absorption generates heat and vibration in the targeted tissue.

Therapeutic Mechanisms of Ultrasound:
1. Thermal Effects: The gentle heating caused by ultrasound waves helps increase blood flow, relieve muscle spasms, and reduce pain.
2. Non-Thermal Effects: Even without heat, ultrasound can improve tissue healing. Through a process known as "cavitation," it creates small gas bubbles that expand and contract within tissues, encouraging cellular activity and facilitating the body's natural healing processes.
3. Mechanical Effects: Ultrasound waves can reduce inflammation by stimulating the lymphatic system, which helps clear fluid build-up, reducing swelling and encouraging drainage.

Real-World Applications: Ultrasound Therapy in Action

The therapeutic applications of ultrasound extend far beyond imaging. Here are a few of the remarkable ways ultrasound is used in modern medicine:

1. Pain Management and Inflammation Reduction

One of the most common uses of ultrasound therapy is in managing pain, especially for musculoskeletal conditions such as arthritis, tendinitis, and back pain. For example, a patient with chronic knee pain from arthritis might find relief through ultrasound therapy. By targeting deep tissues with ultrasound waves, physical therapists can stimulate blood flow, reduce inflammation, and relax muscle spasms— all without any medication.

A real-life case involves a young athlete, Anna, who suffered from a persistent shoulder injury. She underwent ultrasound therapy as part of her rehabilitation, where her physical therapist applied ultrasound waves to the affected area for several sessions. Over time, Anna experienced a marked reduction in pain and an increased range of motion in her shoulder, allowing her to return to sports with greater strength and flexibility.

2. Tissue Healing and Regeneration

Ultrasound therapy is also a promising tool in tissue regeneration. For patients recovering from surgery or injury, ultrasound waves can accelerate cellular repair by promoting collagen production, a vital component of tissue regeneration. In treating bone fractures, for instance, ultrasound has shown promising results in enhancing healing. Low-intensity pulsed ultrasound (LIPUS) is used to stimulate bone healing by increasing blood flow and encouraging cell activity in the fracture site, leading to faster recovery times.

Consider the case of James, who suffered a complex fracture in his leg. His doctor incorporated LIPUS into his treatment plan, applying low-frequency ultrasound waves to the fracture site. The therapy stimulated healing and reduced recovery time significantly, allowing James to return to his regular activities sooner than expected.

3. Tendon and Ligament Repair

Ultrasound has also proven effective in treating tendons and ligaments, which can be slow to heal due to limited blood supply. By enhancing circulation in these dense tissues, ultrasound therapy accelerates healing, providing relief for conditions like Achilles tendinitis and rotator cuff injuries.

In another example, Sarah, an avid runner, developed Achilles tendinitis, which threatened to sideline her from her passion. After weeks of ultrasound therapy focused on her Achilles tendon, she experienced a significant reduction in pain and inflammation, allowing her to return to running with improved stability and resilience.

Practical Guide: How Ultrasound Therapy is Performed

For those considering ultrasound therapy, here's an overview of what a typical session looks like:

1. Preparation: The therapist will apply a special gel to the treatment area, which helps conduct the ultrasound waves efficiently.
2. Treatment: Using a handheld device known as a transducer, the therapist moves it in circular motions over the affected area. The device emits high-frequency sound waves that penetrate into the targeted tissue.
3. Duration: Sessions usually last between 5 to 10 minutes, depending on the severity and location of the injury.
4. Frequency: Most treatments require multiple sessions, with results often noticeable after several weeks of regular therapy.

Ultrasound therapy is generally painless, though some patients may feel a slight warmth during treatment. The procedure is safe and well-tolerated, making it a popular choice for patients seeking non-invasive pain relief.

The Future of Ultrasound in Regenerative Medicine

As technology advances, ultrasound therapy holds even greater promise. Imagine a future where ultrasound is used to deliver medication directly into tissues, enhancing absorption and effectiveness. Researchers are currently exploring the potential of "focused ultrasound," a technology that can target specific areas with even greater precision. In this vision, ultrasound could serve as a "scalpel of sound," performing non-invasive procedures that replace traditional surgery.

Ultrasound therapy's applications extend into oncology as well. Scientists are investigating the potential of high-intensity focused ultrasound (HIFU) to destroy cancer cells without damaging surrounding tissues. In this method, ultrasound waves are concentrated to produce intense heat, targeting and destroying tumours without invasive procedures.

Reflections on Ultrasound as a Tool for Healing

Ultrasound therapy exemplifies the harmonious relationship between sound and healing. It's a reminder that sound waves, though invisible, carry the power to heal and restore. By working with the body's natural processes, ultrasound therapy offers a gentler approach to pain management, recovery, and regeneration. In a world increasingly seeking alternatives to surgery and pharmaceuticals, ultrasound therapy shines as a non-invasive solution rooted in science yet evoking the ancient power of sound.

Closing Thoughts: Embracing Sound-Based Medicine

Ultrasound therapy reminds us that healing doesn't always

require force. Sometimes, gentle vibrations—sound waves in their purest form—can catalyse remarkable transformations within the body. As we look to the future, we find ourselves standing on the edge of a new era, one where sound and frequency redefine the boundaries of medicine. In embracing ultrasound therapy, we honour the wisdom of frequencies and open a door to endless possibilities in healing.

This chapter is just one note in the symphony of frequency medicine. But with each discovery, we come closer to understanding the true potential of sound waves—a power that can transform not only medicine but the very essence of health and well-being.

CHAPTER 19: LOW-FREQUENCY STIMULATION FOR CHRONIC PAIN

Introduction: Understanding Chronic Pain Through a New Lens

Chronic pain affects millions worldwide, often becoming a constant shadow that limits life's joys and daily activities. For many, traditional pain management options, like medications, physical therapy, or even surgery, offer limited relief or come with unwanted side effects. But what if relief could come from something as subtle as low-frequency electrical stimulation? This chapter explores the transformative potential of low-frequency treatments, such as Transcutaneous Electrical Nerve Stimulation (TENS), and how these approaches can alleviate pain and improve mobility by harnessing the natural language of our nervous system.

The Science Behind Low-Frequency Stimulation

Imagine your body as a network of finely tuned instruments, each nerve and cell vibrating with its unique frequency. When pain disrupts this harmony, low-frequency stimulation can help restore balance by speaking directly to the nerves in their own language. Low-frequency treatments, such as TENS, work by sending gentle, rhythmic pulses of electrical energy to affected areas, effectively "distracting" the nerves from the pain signals they typically transmit to the brain.

How Low-Frequency Stimulation Alleviates Pain

1. Gate Control Theory of Pain: The TENS unit operates on the principle that pain signals travel along nerve fibres to the brain. By introducing electrical pulses, the TENS unit creates a new set of signals that intercept pain pathways, effectively closing the "gates" to pain. Think of it as a traffic jam on the pain highway, preventing the pain signals from reaching the brain.

2. Endorphin Release: Another benefit of low-frequency stimulation is its ability to stimulate the body's natural painkillers, known as endorphins. These endogenous opioids help alleviate pain naturally, promoting relaxation and easing muscle tension.

3. Increased Blood Flow and Reduced Inflammation: Low-frequency pulses also enhance blood circulation to the affected area, which speeds up the removal of inflammatory toxins and increases the delivery of oxygen and nutrients. This process can aid in tissue repair and relieve pain over time.

Real-Life Success Stories in Low-Frequency Therapy

Low-frequency stimulation has transformed the lives of many individuals who once felt trapped by chronic pain. Let's dive into some inspiring examples that highlight the potential of these therapies:

Case Study: Sarah's Journey with TENS for Fibromyalgia

Sarah, a 38-year-old teacher, struggled with fibromyalgia, a condition that caused widespread pain and fatigue. She tried various treatments, from medications to physical therapy, but her pain remained relentless. A friend recommended TENS therapy, and although sceptical at first, Sarah decided to try it. Her physical therapist applied TENS electrodes to her back, shoulders, and neck, and within minutes, Sarah felt a tingling sensation that offered temporary relief. Over the following weeks, with regular TENS sessions, she noticed her pain levels decreasing, and her sleep improving. For Sarah, TENS was not a cure, but it became a crucial tool in managing her pain, giving her the strength to return to her classroom with renewed energy.

Case Study: Mark's Return to Mobility After Knee Surgery

Mark, a retired athlete, faced severe knee pain after surgery. Though the procedure was successful, scar tissue and inflammation left him with chronic stiffness and discomfort. Mark's physical therapist introduced him to TENS therapy as part of his recovery. By using TENS before his stretching exercises, Mark found his muscles relaxed more easily, reducing his discomfort and enhancing his range of motion. Gradually, he regained much of his mobility and felt more confident in his physical activities.

Practical Guide: Using TENS at Home

TENS devices are compact, affordable, and easy to use, making them accessible for home treatment. If you're new to TENS, here are some simple guidelines to start your journey:

1. Choose Your Device: Several TENS units are available, ranging from simple to advanced models. A basic device with adjustable settings is often enough to begin.
2. Placement of Electrodes: Place the TENS electrodes near, but not directly on, the source of pain. For instance, if you have lower back pain, position the electrodes on either side of the pain source. Many TENS units come with guides for proper electrode placement.
3. Adjusting Frequency and Intensity: Start with a low

frequency and gentle intensity. Gradually increase the frequency as you become accustomed to the sensation.

4. Session Duration: Typical sessions last around 15–30 minutes. Use the device as often as needed, but limit it to a few times per day.

5. Consultation and Safety: Although TENS is generally safe, it's essential to consult your doctor before starting if you have heart problems, epilepsy, or are pregnant.

Beyond TENS: Other Low-Frequency Techniques

While TENS is among the most popular low-frequency therapies, it's not the only option. Here are a few other low-frequency treatments that can complement TENS or serve as standalone therapies:

1. Electrical Muscle Stimulation (EMS)
   - EMS works similarly to TENS but is designed to stimulate muscles rather than nerves. It's commonly used in sports medicine and rehabilitation to help strengthen muscles, prevent atrophy, and improve circulation.

2. Pulsed Electromagnetic Field Therapy (PEMF)
   - PEMF therapy uses magnetic fields at low frequencies to stimulate cellular repair. It has shown promise in treating conditions like osteoarthritis, fractures, and chronic pain. PEMF devices are often available for at-home use and have become popular among individuals seeking alternative pain relief methods.

3. Low-Frequency Laser Therapy (LLLT)
   - LLLT, or "cold laser therapy," uses low-frequency light to penetrate tissues and stimulate cellular regeneration. Commonly applied to joints, muscles, and tendons, LLLT has been shown to reduce pain and inflammation, making it a valuable tool for chronic pain management.

Future of Low-Frequency Therapy: A World of Potential

Imagine a world where managing chronic pain is no longer reliant on medications or invasive procedures. With ongoing research, scientists are exploring new frontiers in low-frequency therapy, combining advanced biofeedback with precision-targeted stimulation. In the future, patients might have access to custom-tailored devices that adjust in real-time to their body's needs, offering personalized pain relief solutions that adapt as the body heals.

Philosophical Reflections: Redefining Pain and Healing

Low-frequency therapy invites us to rethink our approach to pain and healing. Traditional medicine often seeks to "block" pain, while frequency-based treatments aim to "transform" it by working in harmony with the body's natural rhythms. Perhaps the future of pain management lies not in silencing pain but in reshaping our perception of it, using tools like TENS and PEMF as pathways toward understanding our bodies on a deeper level.

Embracing Frequency Medicine in Everyday Life

Frequency-based therapies encourage a mindset of empowerment, where patients are no longer passive recipients of pain management but active participants in their healing journey. By incorporating low-frequency devices like TENS into daily routines, individuals can experience a greater sense of control, aligning their bodies with the healing power of subtle vibrations.

Final Thoughts: Unlocking the Healing Power Within

The journey of low-frequency therapy is a testament to the untapped power within the body and mind. With tools like TENS, PEMF, and EMS, we're discovering that healing doesn't have to be harsh or intrusive; sometimes, it's as gentle as a pulse, as natural as a rhythm. As you explore the possibilities of frequency medicine, remember that everybody is unique, resonating at its own frequency. By embracing this journey, we're not only alleviating pain but reconnecting with the body's intrinsic wisdom.

Let this be a reminder: healing, at its core, is a harmonious process. Through low-frequency stimulation, we are not only addressing pain but also rediscovering the body's profound potential to heal itself. In the subtle vibrations of these frequencies, we find the rhythm of life, the pulse of recovery, and the promise of relief.

CHAPTER 20: ELECTROMAGNETIC FIELD (EMF) THERAPY FOR BONE HEALING

Introduction: A New Era of Healing Broken Bones

Bone fractures have long been a symbol of pain, disruption, and slow recovery. For thousands of years, treatments for fractures relied solely on immobilization and time. But imagine if, alongside the cast or splint, we could use invisible waves of energy to speed up healing, reducing recovery time and pain. This is the promise of Electromagnetic Field (EMF) Therapy—a non-invasive, frequency-based approach that uses electromagnetic waves to stimulate the body's natural bone regeneration processes.

In this chapter, we explore the science behind EMF therapy, its application in orthopaedic medicine, and the profound potential it holds for revolutionizing bone healing. Through accessible language and inspiring case studies, we'll delve into the world of healing frequencies and discover how EMF therapy could soon become a mainstream complement to traditional bone treatments.

The Science of Bone Healing and Electromagnetic Fields

Bones are dynamic tissues, constantly undergoing a cycle of breakdown and renewal. When a bone breaks, the body activates an intricate healing process that involves inflammation, cell proliferation, and, ultimately, the formation of new bone tissue. EMF therapy works by enhancing this process at a cellular level, using low-frequency electromagnetic waves to stimulate the activity of osteoblasts—cells responsible for bone formation.

How EMF Therapy Works
1. Cellular Resonance: EMF therapy delivers low-frequency waves that resonate with the natural frequencies of bone cells, boosting their activity and encouraging cell growth. Think of it as "tuning" the cells, amplifying their ability to rebuild and repair.
2. Increased Calcium Uptake: Calcium is a crucial component of bone. EMF therapy enhances calcium ion movement, promoting faster and stronger bone mineralization at the fracture site.
3. Improved Blood Flow: EMF waves stimulate blood flow, delivering oxygen and nutrients essential for healing and removing metabolic waste that can slow down recovery.

Evidence of Success
Studies have shown that EMF therapy can accelerate bone healing by up to 30%. In clinical trials, patients treated with EMF showed reduced healing times and reported less pain compared to those who received conventional treatments alone. While EMF therapy is not a replacement for surgery or casting in severe fractures, it is emerging as a powerful tool to supplement traditional approaches.

Real-Life Success Stories in EMF Therapy

The impact of EMF therapy on bone healing becomes even

more compelling when we consider real-life cases. Here are some stories of individuals who have experienced the benefits of EMF therapy firsthand:

Case Study: John's Journey to Recovery from a Tibia Fracture

John, an avid runner, suffered a tibia fracture after a fall during a race. His doctors recommended a combination of immobilization and EMF therapy. Within weeks, he noticed a reduction in pain and increased strength in his leg. EMF sessions became a daily ritual, with low-frequency electromagnetic waves applied directly to the fracture site. Just two months into his recovery, John was amazed at the speed of his healing, which allowed him to resume training earlier than expected.

Case Study: Maria's Resilience with Osteoporosis-Related Fractures

Maria, a 65-year-old woman with osteoporosis, faced frequent fractures due to weakened bones. Traditional treatments alone weren't enough to prevent prolonged recovery. Her orthopaedic specialist suggested adding EMF therapy to her regimen. Over time, Maria noticed a significant improvement in bone density, as confirmed by X-rays, and her fractures began to heal faster. EMF therapy became a cornerstone in her management of osteoporosis, improving her quality of life and restoring her confidence.

Practical Guide: EMF Therapy for Bone Healing

If you're considering EMF therapy to aid in bone healing, here are some practical insights to help you make an informed decision:

1. Consultation with a Specialist: Always start by consulting your healthcare provider. EMF therapy is most effective when integrated into a broader treatment plan tailored to your specific fracture type and health condition.

2. Home vs. Clinical Devices: EMF devices come in various forms, from portable home devices to advanced clinical machines. Home devices may be suitable for minor fractures, while more severe injuries benefit from professional-grade units.

3. Treatment Duration: A typical EMF session lasts around 30–60 minutes and is repeated daily or as advised. Consistency is key to achieving optimal results.

4. Pain and Sensation: EMF therapy is painless and non-invasive, with most users experiencing a gentle warming sensation, if any, during treatment.

5. Monitoring Progress: Regular follow-ups with your doctor, including imaging like X-rays, will help track the progress of bone healing and adjust the treatment plan as needed.

Beyond Fractures: EMF Therapy in Broader Orthopaedic Care

While bone fractures are a primary focus, EMF therapy's applications extend beyond fractures into broader orthopaedic care. Conditions such as delayed union (when a bone heals more slowly than expected), spinal fusions, and even chronic arthritis have shown positive responses to EMF therapy.

1. Delayed Union and Non-Union Fractures
   - In cases where bones struggle to heal on their own, EMF therapy provides a non-surgical intervention. By enhancing cellular communication and boosting metabolic activity, EMF can help bridge the gap in stubborn fractures that traditional methods fail to address.

2. Spinal Fusions
   - Spinal fusions, a common surgery for back pain, involve fusing two or more vertebrae. Studies show that EMF therapy can increase the success rate of spinal fusions by promoting bone growth around the implanted devices, reducing the risk of surgical failure.

3. Osteoarthritis Relief

- Though not directly related to bone fractures, EMF therapy's anti-inflammatory effects make it beneficial for managing joint pain and slowing cartilage degeneration in conditions like osteoarthritis.

The Future of EMF Therapy in Orthopaedics

As technology advances, we are seeing new possibilities for EMF therapy, with ongoing research exploring how to customize EMF frequencies and intensities for individualized care. Imagine a world where fractures that once required months to heal could be treated in weeks, allowing patients to return to their lives more quickly and with less pain. In the near future, EMF therapy could even become a staple in every orthopaedic treatment room, part of a comprehensive approach that combines modern technology with the body's inherent ability to heal.

Reflecting on the Healing Power of Frequency Medicine

EMF therapy invites us to rethink how we approach the recovery process. Traditional medicine often seeks to "fix" the body, while frequency medicine encourages the body to heal itself. This difference is subtle yet profound, allowing us to reconnect with our bodies' innate wisdom. With EMF therapy, we're not just treating a bone fracture; we're honouring the body's capacity to recover, using the gentle pulse of electromagnetic fields to amplify its regenerative potential.

Empowering the Healing Journey

For those suffering from fractures, the journey of recovery can feel isolating and slow. But EMF therapy empowers patients, turning healing into an active, participatory process. By incorporating EMF sessions, patients can feel a greater sense of agency in their recovery, aligning with the rhythms of their body's natural healing frequencies.

Final Thoughts: A New Chapter in Medicine

EMF therapy represents a groundbreaking shift in how we think about bone healing. By integrating it with traditional treatments, we're opening a new chapter in orthopaedic care, one that blends science and nature. As we continue to unlock the secrets of frequencies, we may soon find that fractures, once a source of fear and frustration, become opportunities to witness the body's incredible ability to heal, guided by the subtle, invisible harmonies of electromagnetic fields.

In this, we find hope and promise—not only for faster bone healing but for a future where medicine moves in concert with the body's natural rhythms. The age of healing harmonies has begun, and it's reshaping what we once thought possible.

PART IV:
APPLICATIONS
IN MENTAL AND
EMOTIONAL HEALTH

CHAPTER 21: FREQUENCY AND PTSD RECOVERY

Introduction: A New Path to Healing Trauma

Post-Traumatic Stress Disorder (PTSD) is often described as an invisible wound, one that cuts deeply into the psyche and lingers long after the physical scars have faded. Traditionally, treatments for PTSD have focused on therapy, medication, and, in recent years, mindfulness and holistic practices. But what if sound, something so fundamental yet often overlooked, could be a powerful tool in the recovery from trauma? This chapter explores the use of specific frequencies in addressing PTSD, with an emphasis on calming the nervous system, promoting resilience, and offering relief from trauma-related symptoms.

Sound has always played a role in human history as a source of comfort, expression, and healing. From ancient chants to the soothing notes of a lullaby, certain sounds have the ability to create a profound sense of calm. Recent advances in sound therapy reveal that certain frequencies may help the brain process and release trauma, offering hope to those who have struggled with PTSD.

The Science of Sound and Trauma

PTSD impacts the brain on multiple levels, particularly

affecting the amygdala (the fear centre), the hippocampus (responsible for memory), and the prefrontal cortex (the rational centre). These parts of the brain become hyper-alert, often "stuck" in a loop of fear and stress, making it difficult for individuals to feel safe or grounded. Sound therapy works by gently influencing the brain's frequency patterns, gradually guiding it back to a state of equilibrium.

Understanding Frequency's Effect on the Brain

1. Brainwave Entrainment: The brain naturally synchronizes its electrical activity with external rhythms, a phenomenon known as entrainment. Calming frequencies, particularly in the theta (4-8 Hz) and alpha (8-13 Hz) ranges, encourage the brain to move away from beta waves, which are associated with stress, and toward states associated with relaxation and healing.

2. Activation of the Parasympathetic Nervous System: Certain frequencies stimulate the parasympathetic nervous system, which promotes rest and relaxation. This is crucial for PTSD sufferers, as they often experience an overactive sympathetic (fight-or-flight) response.

3. Reduced Cortisol Levels: Studies show that calming frequencies can lower cortisol, the hormone associated with stress. Elevated cortisol levels over time can exacerbate PTSD symptoms, so any tool that naturally reduces this hormone offers promise.

Case Studies: Real Stories of Healing through Sound

To understand the impact of frequency therapy on PTSD, let's look at some real-life examples of individuals who have experienced transformative effects through sound therapy:

Case Study: Mark's Journey with Theta Healing Frequencies

Mark, a veteran struggling with PTSD, experienced

frequent nightmares and heightened anxiety in daily life. His therapist suggested incorporating sound therapy, specifically theta frequencies, to aid in calming his mind before sleep. After several weeks of listening to 6 Hz theta tones each evening, Mark noticed a decrease in his nightmares and a greater sense of peace before bed. Over time, he felt more grounded during the day, as the practice helped train his brain to shift away from a constant state of vigilance.

Case Study: Sarah's Path to Peace through Binaural Beats

Sarah, a survivor of a traumatic accident, dealt with intrusive flashbacks and a general sense of unease. She began using binaural beats in the alpha frequency range to help alleviate her symptoms. By listening to 10 Hz binaural beats daily, Sarah's flashbacks gradually lessened, and she reported feeling a deeper sense of relaxation, which made her more receptive to other therapeutic techniques.

Practical Guide: Using Sound Frequencies for PTSD Relief

If you're interested in trying sound therapy for PTSD relief, here are some practical tips and techniques that may help:

1. Choose Calming Frequencies: Start with frequencies in the theta (4-8 Hz) or alpha (8-13 Hz) range, which are known for their relaxing properties. These can often be found as binaural beats, isochronic tones, or even certain types of soothing music.

2. Create a Relaxing Environment: Find a quiet space where you feel safe and comfortable. Reducing external distractions can enhance the effects of the sound therapy.

3. Consistency is Key: Set aside a few minutes each day to listen to these frequencies. Consistent practice helps reinforce neural pathways that promote relaxation, helping you train your brain over time.

4. Combine with Other Therapies: Sound therapy can be especially powerful when used alongside other forms of PTSD treatment, such as talk therapy, meditation, or guided imagery.

5. Mindfulness with Sound: Focus on each note, allowing yourself to feel the vibrations and gently bringing your awareness back when your mind wanders. This can enhance the therapeutic effects and deepen the relaxation response.

A Holistic Approach to Trauma Healing

For many, PTSD isn't just about a single traumatic event but a prolonged battle against recurring memories, fears, and physical responses. Sound therapy offers a unique approach by addressing the physiological response to trauma. When calming frequencies are introduced, the mind can finally begin to disentangle itself from the grip of past events.

Creating New Neural Pathways

Trauma can create deeply entrenched neural pathways that keep the brain in a loop of fear and hypervigilance. Sound therapy, however, has the potential to help rewire these pathways. As the brain is gradually exposed to calming frequencies, it begins to form new connections associated with safety and relaxation rather than threat.

Fostering Resilience

Sound therapy doesn't erase trauma, but it can enhance resilience, helping individuals develop a sense of calm and control over their responses. By using sound as a grounding tool, those with PTSD can build confidence in their ability to manage stress, creating a foundation for long-term recovery.

Reflection: Imagining a World of Healing Through Sound

Imagine a world where hospitals, therapy offices, and homes

integrate frequency-based healing as part of their approach to PTSD. Sound therapy doesn't carry the side effects of medication, nor does it require extensive resources. It's simple, accessible, and safe. By harnessing the body's natural responses to certain frequencies, we can envision a society where trauma is addressed with compassion, innovation, and a profound respect for the mind-body connection.

The Future of Frequency Medicine for PTSD

The future of frequency medicine in PTSD treatment holds enormous potential. With ongoing research, we are discovering the unique effects of specific frequencies and refining our understanding of how they impact the brain. As this field evolves, sound therapy could become a mainstream intervention, with individualized soundscapes designed for each person's unique trauma response.

Final Thoughts: Healing with Harmony

Sound is one of the most primal forces on Earth, shaping not only our lives but also our healing. PTSD is complex, but the simplicity of sound therapy reminds us that sometimes the most profound treatments are those rooted in nature's rhythms. When used mindfully, frequencies become more than just sounds—they become pathways to peace, connection, and self-healing.

As we close this chapter, let us remember that healing isn't just about silencing the echoes of trauma. It's about creating a new harmony, one that resonates within and empowers us to face the future with strength and hope. In the powerful, gentle embrace of sound, the journey of healing harmonies truly begins.

CHAPTER 22:
DEPRESSION AND
SOUND THERAPY

Introduction: The Subtle Power of Sound in Healing the Mind

Depression, often described as a "dark cloud" or a "silent struggle," affects millions of people worldwide. For many, it's not simply a fleeting sadness but a pervasive emptiness that colours their entire experience of life. Modern treatments, including medication and therapy, provide relief for some, yet others continue to search for more holistic and accessible ways to find peace. Enter sound therapy, a subtle yet profound approach to mental well-being that taps into the ancient and fundamental power of vibration and frequency.

Imagine the mind as a complex symphony, where each thought, feeling, and memory contributes to a larger harmony. Depression, in this analogy, is like a discordant note, disrupting the symphony and creating internal dissonance. Sound therapy aims to restore balance, aligning the mind's "frequencies" to create a state of calm and resilience. By using sound to influence brainwave patterns, modulate emotional responses, and stimulate neural pathways, sound therapy opens a pathway to mental healing that is accessible, non-invasive, and remarkably effective for many.

How Frequency Impacts Mental Health

Sound therapy operates on the principle of entrainment—where the brain naturally aligns its electrical activity to external rhythmic patterns. This entrainment affects our brainwaves, which are linked to different mental states:

1. Delta Waves (0.5–4 Hz): Associated with deep sleep, restoration, and healing. Essential for rest and repair, they can benefit those experiencing exhaustion due to depression.
2. Theta Waves (4–8 Hz): Connected with deep relaxation and creativity, these frequencies encourage introspection, self-reflection, and processing of emotions.
3. Alpha Waves (8–13 Hz): The frequency of calm and relaxation, alpha waves are beneficial for reducing stress and anxiety, common symptoms accompanying depression.
4. Beta Waves (13–30 Hz): Linked with active thinking, focus, and problem-solving, beta waves are essential but can become overactive in people with depression, contributing to rumination.

By using specific frequencies, sound therapy encourages the brain to shift from high-alert beta states to more relaxed alpha and theta states, offering a break from depressive thought patterns and creating a "reset" for the mind.

Case Studies: Stories of Healing Through Sound

Case Study 1: Anna's Journey with Binaural Beats

Anna, a young woman struggling with chronic depression, found it difficult to focus, sleep, and experience joy. After traditional therapies only partially alleviated her symptoms, she began incorporating binaural beats into her daily routine. Listening to theta (6 Hz) and alpha (10 Hz) frequencies helped her experience moments of calm and clarity. Over time, Anna noticed an improvement in her sleep, and her depressive thoughts softened, making them more manageable.

Case Study 2: Jake's Experience with Music Therapy

Jake, a retired teacher, had long battled with depression, especially during the winter months. His therapist recommended music therapy sessions focused on rhythmic drumming, where Jake was encouraged to play along with specific beats. The active engagement allowed Jake to channel his emotions, and the repetitive beats helped his mind to relax. Within a few weeks, Jake felt a lightness he hadn't experienced in years, finding joy in music again and noticing an improvement in his mood.

Types of Frequency-Based Therapies for Depression

1. Binaural Beats: By playing slightly different frequencies in each ear, binaural beats create an auditory illusion of a single, pulsing tone. For depression, theta (4–8 Hz) and alpha (8–13 Hz) frequencies are most effective. Listening to these beats for 15-30 minutes a day has been shown to reduce stress, improve sleep, and enhance emotional regulation.

2. Music Therapy: In music therapy, practitioners use structured musical activities to influence emotional responses. Instruments like drums, chimes, and singing bowls produce rhythmic vibrations that resonate with the body, encouraging relaxation and emotional release.

3. Isochronic Tones: Unlike binaural beats, which require headphones, isochronic tones use distinct beats at regular intervals to create specific brainwave patterns. For people dealing with depression, sessions with alpha and theta tones can enhance calmness and reduce anxiety.

Practical Tips for Using Sound Therapy for Depression

If you're considering sound therapy as part of your approach to managing depression, here are some practical steps to get started:

1. Find Your Frequency: Experiment with different tones and frequencies. Apps and online platforms provide access to a variety of binaural beats and isochronic tones. Try listening to a few and pay attention to which frequencies resonate with you.

2. Establish a Routine: Consistency is key to reaping the benefits of sound therapy. Set aside 10–15 minutes each morning or evening to listen to your chosen frequency. Many people find it especially helpful before bed to promote relaxation and better sleep.

3. Combine with Mindfulness: As you listen to the beats or tones, practice mindful breathing, or visualize a place that brings you peace. This pairing enhances the effect of sound therapy by focusing your mind on calming imagery.

4. Document Your Experience: Keep a journal to track how you feel before and after each session. This practice allows you to observe any patterns or improvements over time, helping you refine your approach.

5. Seek Guidance: If possible, work with a therapist who specializes in sound or music therapy. A trained practitioner can help tailor your sound experience to your specific needs, ensuring you get the most out of each session.

Reflection: The Profound Simplicity of Sound

Sound therapy's beauty lies in its simplicity and accessibility. Unlike medication or talk therapy, sound doesn't require active participation; it meets the listener where they are, offering subtle shifts without demanding immediate change. For those burdened by depression, this gentleness can be a welcome contrast to more invasive treatments.

There is something ancient, almost primal, about the effects of sound on our psyche. It speaks to a part of us that existed long before the complexities of modern life—a part that understands healing not through words but through resonance, rhythm, and harmony.

Vision: Sound Therapy as a Mainstream Treatment for Depression

As research into sound therapy grows, we may soon witness its integration into mainstream mental health treatments. Imagine clinics where binaural beats and music therapy sessions are offered alongside cognitive behavioural therapy, or mental health retreats where soundscapes are crafted to guide participants through their journey of healing.

Sound therapy has the potential to democratize mental health support, offering a low-cost, non-invasive, and easily accessible option for those who may not have access to traditional therapies. In a world where mental health resources are stretched thin, sound therapy represents a hopeful and inclusive alternative.

Final Thoughts: Harmony Within

Depression can feel like a silence, a world without music or colour. Sound therapy, however, offers a way to reintroduce harmony into a seemingly discordant life. By carefully tuning into the frequencies that encourage calm and resilience, those suffering from depression can find moments of peace amid their struggle.

In this journey through sound, healing harmonies do more than alleviate symptoms; they offer a reminder that even in the darkest moments, there is still rhythm, still melody, still life waiting to be embraced.

CHAPTER 23: ANXIETY RELIEF WITH FREQUENCY THERAPY

Introduction: The Sound of Calm in an Age of Anxiety

In a world filled with constant noise and demands, anxiety has become an almost ubiquitous part of modern life. Many of us have felt the racing heart, the tightening chest, and the overwhelming need to escape when anxiety takes hold. For some, these experiences are fleeting, but for others, they become a persistent backdrop to daily life. In recent years, scientists and therapists alike have turned to frequency therapy as a powerful, non-invasive tool to alleviate anxiety and restore a sense of calm.

Imagine the brain as an intricate orchestra, with each area harmonizing to create our thoughts, emotions, and reactions. When anxiety strikes, it's as though one section of the orchestra begins playing out of tune, causing tension and disharmony. Frequency therapy seeks to retune the brain's natural rhythms, helping it return to a state of equilibrium. Through specific sound frequencies, we can guide the brain toward states associated with relaxation, presence, and peace.

How Sound Frequencies Influence Anxiety

Our brains operate on various frequency patterns, known as brainwaves, which correspond to different states of consciousness. For example, when we're fully alert, beta waves (13–30 Hz) dominate, facilitating active thought and concentration. However, anxiety often involves excessive beta activity, pushing the brain into overdrive. Frequency therapy can help by gently guiding the brain toward slower, calming wave patterns, such as alpha and theta.

1. Alpha Waves (8–13 Hz): These are associated with relaxation and reduced stress. Alpha waves are the bridge between alertness and calm, creating a state ideal for rest and recovery.
2. Theta Waves (4–8 Hz): Linked to deep relaxation, meditation, and creativity, theta waves are especially effective for easing the mental noise that often accompanies anxiety.
3. Delta Waves (0.5–4 Hz): Found during deep sleep, delta waves promote healing and restoration, beneficial for those whose anxiety disrupts restful sleep.

By using sound frequencies to influence these brainwave states, we can counteract the heightened beta activity often found in individuals with anxiety, fostering a more balanced and peaceful mind.

Case Studies: Transforming Lives with Frequency Therapy

Case Study 1: Lisa's Path to Peace

Lisa, a 35-year-old teacher, had struggled with chronic anxiety for years. Traditional treatments, including medication and therapy, offered temporary relief but failed to address her underlying sense of unease. At her therapist's suggestion, she began incorporating alpha-wave binaural beats into her routine, listening to an 8 Hz frequency each morning. Over time, Lisa noticed a significant reduction in her anxiety levels, especially in situations that once triggered panic. Frequency therapy became a grounding practice, enabling her to reclaim her life with a newfound sense of calm.

Case Study 2: James and Theta Waves for Deep Relaxation

James, a software engineer, found himself constantly anxious about work deadlines and expectations. When he discovered frequency therapy, he began using theta-wave beats (5 Hz) to calm his mind before bed. Within a few weeks, his sleep improved, and his anxiety became more manageable. His morning routine now includes a few minutes of theta waves, setting a peaceful tone for the day ahead.

Types of Frequency Therapy for Anxiety Relief

1. Binaural Beats: Binaural beats work by delivering slightly different frequencies to each ear, which the brain interprets as a single, pulsing tone. This method is especially effective for generating alpha and theta brainwave states. Binaural beats in the range of 4–12 Hz are particularly beneficial for anxiety relief, as they help shift the brain into relaxed and meditative states.

2. Isochronic Tones: Unlike binaural beats, isochronic tones do not require headphones. These distinct pulses are spaced at regular intervals and can induce similar brainwave states. Isochronic tones are a versatile option for anxiety, allowing individuals to listen at home or in an office environment without specialized equipment.

3. Sound Baths and Singing Bowls: Sound baths involve immersing oneself in sound waves created by instruments like Tibetan singing bowls, chimes, and gongs. The resonance from these instruments produces frequencies

that help synchronize brainwave activity, encouraging a deep sense of relaxation. Many people find the immersive nature of sound baths particularly soothing for anxiety relief.

4. Breath-Focused Sound Therapy: By combining sound frequencies with breathing exercises, practitioners can deepen the effects of frequency therapy. Slower frequencies, paired with deep breathing, can quickly reduce tension and increase oxygen flow, calming both mind and body.

Practical Tips for Using Frequency Therapy to Alleviate Anxiety

If you're new to frequency therapy, here are some practical tips to incorporate these techniques into your daily routine:

1. Choose the Right Frequency: Start with alpha or theta frequencies (4–12 Hz) to promote relaxation and reduce mental chatter. Online platforms and apps offer a variety of soundscapes and binaural beats that can help you explore and find what resonates with you.

2. Set a Routine: Frequency therapy works best when practiced consistently. Try listening to calming frequencies for 10–15 minutes each morning or evening to establish a grounding ritual.

3. Combine with Meditation: Deepen the experience by practicing mindful breathing or meditation while listening. Focus on your breath as the sound guides your mind to a calmer state.

4. Create a Sanctuary: For best results, find a quiet, comfortable space free from distractions. Dim the lights, close your eyes, and allow the sound to envelop you, guiding you toward relaxation.

5. Document Your Experience: Keep a journal to track your

progress. Note how you feel before and after each session, and observe any patterns over time. This practice can help you fine-tune your approach and identify which frequencies have the most significant impact on your anxiety.

Reflection: Rediscovering Inner Harmony

In today's fast-paced world, anxiety often feels like a natural response. Yet, our minds long for harmony, for moments of quiet that allow us to reconnect with ourselves. Frequency therapy offers a pathway back to this inner peace. By tuning into calming vibrations, we can soothe our overactive minds, creating a refuge from the noise of daily life.

There's something profoundly moving about the idea that sound, one of the oldest forms of communication, can guide us back to balance. Frequency therapy doesn't require complex technology or costly equipment; it relies on our innate responsiveness to rhythm and resonance, showing us that healing can be as simple as listening.

Vision: The Future of Frequency Therapy for Anxiety

As we look toward the future, it's easy to imagine a world where frequency therapy is an integral part of mental health care. In schools, workplaces, and hospitals, dedicated spaces could allow individuals to use sound as a tool for emotional regulation. This low-cost, accessible form of therapy could offer relief to millions, providing a sustainable and non-invasive way to manage anxiety.

Final Thoughts: Embracing the Power of Sound

Anxiety may be an inevitable part of life, but it doesn't have to dominate our experiences. Through frequency therapy, we can reclaim our sense of calm and resilience. Sound frequencies remind us of a deeper truth—that our bodies and minds are designed to resonate with harmony. By embracing the healing power of sound, we are not just reducing anxiety;

we are reconnecting with our natural state of balance and well-being.

CHAPTER 24: IMPROVING FOCUS AND COGNITIVE HEALTH

In today's digital age, focus and clarity often feel elusive, swept away by a tide of constant notifications, multitasking, and mounting demands. Yet, amidst this whirlwind, a surprising answer emerges from the world of frequencies. Scientists are discovering that sound frequencies—specific rhythms that interact with our brainwaves—can be harnessed to sharpen focus and enhance cognitive function, offering new hope for conditions like ADHD and cognitive decline. It's a reminder of the deep, intrinsic connection between the rhythms in the world around us and the rhythms within us.

The Science of Frequencies and Focus

To understand how frequencies influence focus, imagine the brain as an orchestra, where each musician represents a different function—attention, memory, problem-solving, and so forth. When all musicians are in sync, we experience peak mental performance. But disruptions, like ADHD or even a stressful day, cause certain parts to play out of tune, creating mental "noise" that distorts our focus and clarity.

The human brain operates on various brainwave frequencies that correspond to different mental states:
- Beta Waves (13–30 Hz): Found in states of alertness and concentration, beta waves help us focus on tasks and make decisions.
- Alpha Waves (8–13 Hz): These waves are associated with a relaxed but alert state, ideal for creativity and open-minded problem-solving.
- Theta Waves (4–8 Hz): Linked to deep relaxation and meditation, theta waves help with memory consolidation and emotional processing.

Frequency therapies, like binaural beats, stimulate these brainwave patterns, enabling the brain to synchronize with the desired frequency. For focus, a beta frequency is typically used to "entrain" the brain into alertness and attention, much like tuning an instrument to a specific pitch. By aligning the brain's rhythms with these frequencies, individuals can experience sharper mental clarity, improved attention, and a greater sense of calm.

Case Study: Frequency Therapy for ADHD

Take, for example, 12-year-old Leo. Diagnosed with ADHD, he found it challenging to complete homework or focus on classroom tasks without becoming overwhelmed or distracted. Traditional medications left him feeling drowsy and disconnected, so his family explored alternative treatments, eventually discovering frequency therapy. By listening to beta-frequency binaural beats for 15 minutes each morning, Leo gradually experienced improved concentration and reduced impulsivity. After several weeks, his teachers and parents noticed he could engage in tasks longer and with greater ease—a transformation sparked not by medication but by sound.

Leo's story reflects a broader trend in ADHD management,

where parents and professionals increasingly turn to non-invasive solutions. Frequency therapy, especially at beta frequencies, encourages the brain to stabilize its attention networks, reducing hyperactivity and impulsivity. For many children and adults alike, it's a revelation that the simple act of listening can gently guide the brain toward a more focused, balanced state.

How to Use Frequencies to Enhance Focus

If you're looking to improve focus, whether to boost productivity or support cognitive health, consider the following methods and tips:

1. Choose the Right Frequency: For focus, beta frequencies (around 14–20 Hz) are often most effective. These frequencies enhance alertness and help the brain engage with tasks.
2. Practice Consistency: Frequency therapy isn't a quick fix; it requires consistency. Set aside 10–15 minutes daily to listen to your chosen frequency, preferably at the same time each day.
3. Create a Focus-Friendly Environment: Use headphones in a quiet space to maximize the effect of binaural beats or isochronic tones. Dimming lights or practicing deep breathing can enhance the experience.
4. Combine with Physical Activity: Studies show that short bursts of exercise can improve brain function. Try listening to focus-inducing frequencies after physical activity to enhance the effects.
5. Pair with Mindfulness: Pairing frequency therapy with mindfulness techniques can improve its impact. Close your eyes, focus on your breathing, and let the sound gently guide your attention.

Frequency Therapy for Cognitive Health in Aging

Frequency therapy is not only a tool for focus and ADHD

management but is also showing promise in cognitive health for aging populations. For example, Alzheimer's disease and other forms of dementia are characterized by disruptions in brainwave activity. Research suggests that certain frequencies, particularly gamma waves (30–40 Hz), might promote memory and cognitive function, making them a potential aid for early-stage Alzheimer's.

One groundbreaking study from the Massachusetts Institute of Technology (MIT) found that exposure to 40 Hz sound could stimulate gamma activity in the brain, reducing toxic proteins associated with Alzheimer's in animal models. While these findings are preliminary, they represent a hopeful frontier in non-invasive cognitive therapies, where sound frequencies might one day slow or prevent cognitive decline.

Real-Life Application: A Case of Cognitive Clarity

Consider Jane, a 68-year-old retired teacher who began experiencing mild memory lapses. Concerned about the early signs of cognitive decline, she explored various brain-boosting exercises and eventually discovered gamma frequency therapy. After incorporating 40 Hz frequencies into her daily routine, she found not only improvement in memory recall but also a boost in her mental clarity. Her sense of mental agility returned, allowing her to enjoy her hobbies and stay socially engaged without fear of losing herself to age-related decline.

Reflecting on the Power of Sound

The idea that sound can reshape cognition is a profound revelation. It suggests that we can actively participate in our mental well-being, not solely relying on external interventions but instead tuning into frequencies that exist naturally in the world around us. It evokes a timeless connection between humanity and sound, one that ancient civilizations may have understood intuitively. Today, science is catching up, revealing that specific rhythms and frequencies—once confined to meditation and spiritual practices—hold genuine potential for cognitive enhancement.

Vision: A Future of Frequency Medicine

Imagine a future where frequency therapy becomes a mainstream tool in educational and clinical settings. Schools might have dedicated sound therapy rooms, where students with ADHD or focus challenges can spend a few minutes listening to carefully curated frequencies before class. In elder care, residents might listen to gamma waves to stimulate brain activity, reducing the risk of cognitive decline. It's a vision of health that doesn't rely solely on medication but instead harnesses the therapeutic power of sound.

Closing Thoughts: Rediscovering the Symphony Within

In this exploration of frequency and focus, we see that our minds are as flexible and adaptive as the frequencies we absorb. Just as a musician tunes an instrument, we, too, can fine-tune our minds. For individuals with ADHD, cognitive decline, or even everyday challenges with focus, frequency therapy opens a door to balance, clarity, and cognitive vitality.

The beauty of frequency therapy lies in its accessibility; it doesn't require complex technology or invasive procedures. It relies only on sound and our natural resonance with it. Through dedicated listening, we can awaken the symphony within, fostering a more focused, present, and resilient mind. Frequency therapy invites us all to explore a new frontier where science and sound converge, illuminating pathways to mental clarity, emotional stability, and cognitive health that we are only beginning to understand.

As we continue to learn and refine these techniques, frequency therapy holds the promise of reshaping our approach to mental well-being—reminding us that within the frequencies around us lie the harmonies of health and healing.

CHAPTER 25: SOUND THERAPY FOR SLEEP DISORDERS

Sleep is an essential rhythm of life, an intricate process that restores our mind and body each night. Yet, for millions of people, restful sleep remains an elusive goal. From insomnia to interrupted sleep patterns, sleep disorders have profound effects on overall health, mood, and cognitive function. But what if sound frequencies could help lull the brain into a state of rest and deep relaxation, supporting the natural flow of sleep? In recent years, scientists and practitioners alike have turned to sound therapy as a promising solution, particularly the use of delta and theta frequencies to support sleep.

The Science of Sound Frequencies in Sleep

Imagine the brain as a symphony, its various stages of consciousness represented by different frequencies. Just as each movement in a symphony has a tempo, the brain's electrical activity fluctuates in frequency, producing waves that correspond to our level of alertness or relaxation:
- Delta Waves (0.5–4 Hz): These slow waves are associated with deep, restorative sleep, a state in which the body heals and rejuvenates.
- Theta Waves (4–8 Hz): Present during light sleep and deep relaxation, theta waves facilitate creativity, dreaming, and

emotional processing.

These waves create the neural backdrop for sleep, shifting in dominance as we move through the stages of the sleep cycle. When people struggle with insomnia or disrupted sleep, it is often because their brainwave patterns are not aligning with these natural rhythms. Sound therapy, through techniques like binaural beats and isochronic tones, can help "entrain" the brain—meaning it nudges the brain's frequencies toward delta and theta waves, inviting a state of relaxation conducive to sleep.

A Journey Through Frequency-Based Sleep Therapy

Consider Sarah, a busy professional who struggled with insomnia for years. Traditional sleep aids left her groggy, and her stress only exacerbated her restlessness at night. Eventually, she encountered sound therapy and began experimenting with delta wave frequencies. Each night, Sarah would set aside 30 minutes to listen to low, rhythmic delta frequencies as she prepared for bed. Within weeks, she noticed an improvement: falling asleep became easier, her sleep felt deeper, and she awoke refreshed. For Sarah, the soft, resonant frequencies became a nightly sanctuary.

This real-world example highlights how sleep therapy with sound is more than a passive experience. Sound frequencies actively work to reset the brain's rhythms, encouraging a harmonious flow between body and mind that many who suffer from sleep disorders find transformative.

How to Use Delta and Theta Frequencies for Sleep

If you're seeking to improve your sleep quality with sound therapy, here are some actionable steps you can follow:

1. Set the Environment: A quiet, comfortable space is essential. Use headphones to listen to binaural beats or speakers for isochronic tones, depending on what suits you

best.

2. Choose the Right Frequency: Start with delta frequencies (around 2–4 Hz) for deep sleep or theta frequencies (4–8 Hz) if you need help easing into sleep.

3. Create a Routine: Consistency matters. Try listening to your chosen frequencies at the same time each night as you wind down. This helps signal to your brain that it's time to transition into sleep.

4. Limit Distractions: Turn off electronic screens and dim the lights while listening to the frequencies. Light and blue light can interfere with melatonin production, making it harder to relax.

5. Combine with Mindfulness: Close your eyes, focus on your breath, and allow the sound frequencies to guide you into relaxation. Visualizing peaceful images or focusing on positive thoughts can amplify the effect.

The Role of Delta Waves in Deep, Restorative Sleep

During the deepest stages of sleep, the brain produces delta waves, which support essential functions like cell regeneration, immune system strengthening, and energy restoration. This stage of sleep, also known as "slow-wave sleep," is critical for physical and mental health. When people fail to reach this stage, they miss out on much of sleep's rejuvenating benefits.

Delta wave sound therapy helps induce this state by promoting a brainwave pattern associated with deep sleep. Studies have shown that listening to delta frequencies before and during sleep can not only improve sleep duration but also enhance sleep quality, allowing for more time spent in the deep, restorative stages.

One study conducted at the University of Toronto involved participants who listened to delta wave frequencies over several weeks. Results showed that participants spent a higher proportion of their sleep in deep stages and experienced fewer interruptions. Participants also reported feeling more refreshed upon waking—a testament to the power of delta waves in supporting restorative sleep.

Theta Waves: Easing into Sleep and Dreaming

While delta waves are essential for deep sleep, theta waves play a vital role in the transition to sleep and the dreaming state. Theta waves allow the mind to enter a state of relaxed awareness, where thoughts drift without attachment. For those who experience difficulty falling asleep, theta frequencies can be especially helpful in calming the mind, relieving mental tension, and reducing anxiety.

Incorporating theta frequencies into a pre-sleep routine can be a gentle way to bridge the wakeful and sleep states. By listening to theta frequencies in the hour before bed, the mind naturally relaxes, making it easier to drift into sleep. These waves also enhance REM sleep, where dreaming occurs, supporting emotional processing and psychological well-being.

Reflecting on the Transformative Power of Sound

The idea that sound can restore sleep is both ancient and groundbreaking. For centuries, people have used natural rhythms—the rustling of leaves, the sound of waves— to lull themselves to sleep. Modern frequency therapy is a continuation of this practice, refined through science to target the specific frequencies that support each stage of sleep. It reminds us that healing isn't always about intervention but about creating harmony between our bodies and the natural rhythms around us.

Envisioning the Future of Frequency-Based Sleep Therapy

Imagine a world where sound therapy is a standard treatment for sleep disorders, where patients in hospitals can listen to carefully calibrated frequencies to aid recovery and restorative rest. Picture students learning to harness sound therapy techniques to manage stress and sleep well before exams. As research advances, the potential for frequency therapy to become a mainstream tool for sleep support grows increasingly likely.

Final Thoughts: Sound as Medicine for the Mind

Sound therapy for sleep represents a powerful, non-invasive tool for those seeking relief from insomnia, disrupted sleep, or stress-induced restlessness. By entraining the brain to align with delta and theta frequencies, individuals can experience the deep, restorative sleep that allows for healing, clarity, and rejuvenation.

The journey to sleep is a deeply personal experience, shaped by lifestyle, stress, and internal rhythms. Sound therapy offers an invitation to reconnect with those rhythms, to tune into the natural flow of rest and renewal. For anyone struggling with sleep, these frequencies hold the potential to transform nights of restlessness into restful slumber, supporting not only physical health but also mental and emotional well-being.

In a world where restful sleep is becoming increasingly rare, the rediscovery of sound as a sleep aid is both timely and transformative. It's a reminder that some of the most profound healing tools are the simplest, awaiting only our openness to experience them. Through the power of sound frequencies, we can reconnect with the essential, nurturing rhythms that define a life lived in harmony.

PART V: FREQUENCY HEALING IN PHYSICAL REHABILITATION

CHAPTER 26: BONE REGENERATION WITH FREQUENCIES

Healing and regeneration have long fascinated medical professionals and researchers alike, especially when it comes to the human body's ability to repair complex structures like bones. But what if frequencies could be used to accelerate and enhance this regenerative process? Imagine a scenario where a bone fracture, instead of requiring weeks of healing, could recover more swiftly, painlessly, and even stronger than before. Welcome to the emerging field of frequency-based therapies for bone regeneration, where science meets the resonance of the body's own natural processes.

The Science Behind Bone Healing and Frequency Stimulation

Our bones are not static structures; they are living tissues that constantly renew and adapt. The process of bone healing after a fracture typically involves three main stages: inflammation, repair, and remodelling. During these phases, cells known as osteoblasts (bone-building cells) and osteoclasts (bone-resorbing cells) play a critical role in repairing the damaged tissue and restoring bone strength.

In traditional treatment, bones heal over time, relying on rest, proper alignment, and the body's intrinsic repair mechanisms. But researchers have discovered that specific electromagnetic and vibrational frequencies can stimulate cellular activity, particularly in the osteoblasts and osteoclasts, accelerating bone repair. The frequency-based therapy involves using pulses or specific vibrational patterns to encourage the bone cells to regenerate faster, optimizing the natural healing process.

Case Study: Accelerated Healing through Pulsed Electromagnetic Field (PEMF) Therapy

Consider the story of James, a professional cyclist who fractured his tibia in an accident. Frustrated by the prospect of being sidelined for months, James sought out alternative treatments to speed up his recovery. He was introduced to Pulsed Electromagnetic Field (PEMF) therapy, a technique that uses low-frequency electromagnetic fields to stimulate bone healing.

After incorporating PEMF sessions into his treatment plan, James's healing process accelerated. His doctors noted a surprising improvement in his bone density and cellular activity around the fracture site, significantly shortening his recovery time. For James, the therapy offered not only a physical recovery but a restored sense of hope and resilience.

PEMF therapy is now recognized in various clinical settings for its ability to aid bone regeneration. By delivering electromagnetic pulses, this therapy stimulates the growth of new bone cells and enhances blood flow to the affected area, which is vital for delivering the nutrients necessary for healing.

How Frequency-Based Bone Regeneration Works

1. Pulsed Electromagnetic Field (PEMF) Therapy: PEMF operates by emitting low-frequency electromagnetic pulses that penetrate deep into the bone tissue. These pulses stimulate osteoblast activity, encourage cellular repair, and boost local circulation, which helps bring essential nutrients

to the damaged area.

2. Low-Intensity Pulsed Ultrasound (LIPUS): Another technique, LIPUS, uses sound waves at low intensities to stimulate bone formation. This method works on a similar principle as PEMF, but instead of electromagnetic pulses, it uses ultrasound waves. Clinical studies have shown that LIPUS can reduce healing time in fresh fractures and is especially effective in cases where traditional healing is slow or challenging.

3. Electric Field Stimulation (EFS): EFS is less commonly used but has shown promise in stimulating bone growth. By applying a low electrical field to the injured area, EFS encourages cellular activity and helps osteoblasts deposit new bone tissue. It's a promising area for further research and development in frequency-based bone healing.

Practical Tips for Exploring Frequency Therapy for Bone Healing

If you're interested in frequency-based bone regeneration therapies, here are some steps to consider:

- Consult a Medical Professional: Before starting any new therapy, especially with bone injuries, consult an orthopaedic specialist or physical therapist familiar with frequency-based treatments.
- Research Device Availability: PEMF and LIPUS devices are becoming more accessible, but make sure to use devices approved by health authorities.
- Create a Treatment Routine: Consistency is key. Follow the therapy protocol as recommended, typically involving daily or regular sessions.
- Track Your Progress: Maintain a record of pain levels, mobility, and any sensations around the injury site. Frequency therapies work subtly, so tracking progress can help measure effectiveness over time.

The Promise of Frequency-Based Therapies in Bone Regeneration

With frequency-based therapies, we are tapping into the body's intrinsic capacity to heal. By stimulating cellular processes that already occur during bone healing, these therapies amplify and support the body's natural rhythm, minimizing the need for invasive procedures. In the future, we may see these therapies not only aiding in fracture repair but also preventing osteoporosis and other degenerative bone conditions.

Reflection: Rethinking the Body's Healing Potential

The body, like a finely-tuned orchestra, resonates with frequencies that influence every cell, tissue, and organ. When we align external frequencies with this natural resonance, we create an environment where healing can happen at an accelerated pace. Frequency-based bone regeneration challenges us to rethink the boundaries of what is possible in medicine, blending ancient concepts of vibrational healing with cutting-edge science.

Looking Forward: The Future of Bone Healing with Frequency Medicine

Imagine a future where a fractured bone doesn't sideline someone for months but instead heals in weeks. Or where patients with osteoporosis can rely on non-invasive treatments to regenerate bone density. As frequency medicine advances, this vision of healing could become a reality.

Bone regeneration with frequency-based therapy is just one aspect of a broader movement toward using sound, vibration, and electromagnetic fields in medicine. As research deepens, the possibilities will continue to expand, unlocking new potential for frequency-based healing across various aspects of health.

Final Thoughts: Embracing a Harmonious Approach to Healing

Frequency therapy for bone regeneration is a testament to the body's innate wisdom and the incredible potential that lies within aligning with natural forces. As you explore this therapy, remember that healing is a holistic journey. The harmonies created by these frequencies do not just mend bones—they reconnect us with a profound truth: that within us, there is a capacity for renewal that can be nurtured, accelerated, and celebrated.

In the words of one pioneering researcher in the field, "Frequency-based healing isn't just about fixing a fracture; it's about reawakening the symphony of life that flows through us." With each pulse, each wave, and each resonance, we move closer to a world where the body's natural rhythms are not only respected but enhanced, enabling us to live stronger, healthier, and more harmonized lives.

CHAPTER 27: USING FREQUENCIES IN PHYSICAL THERAPY

Imagine your body as a symphony orchestra. Each muscle, each cell, and every tissue resonates at a unique frequency, contributing to a larger harmony. But what happens when injury strikes, disrupting this orchestration? Physical therapy, often involving traditional exercises and routines, can be enhanced by an innovative approach: vibrational therapy using specific frequencies to accelerate recovery, improve mobility, and promote muscle regeneration. This chapter dives into the science and secrets of using frequencies in physical therapy, revealing how vibrational therapy could transform recovery.

The Science Behind Frequencies and Muscle Recovery

Every part of the human body, from muscles to bones to organs, vibrates at a unique frequency. When muscles are strained or injured, their normal frequency patterns are disturbed. Vibrational therapy, a method of applying specific frequencies to affected areas, can encourage tissues to return to their natural state, boosting recovery and relieving pain. This is not just theoretical: studies show that applying targeted frequencies can increase blood flow, reduce inflammation, and enhance cellular repair in damaged tissues.

To understand how this works, think of muscle fibres as tightly woven threads. Injury can create small tears, inflammation, and stiffness, making it hard for the muscle to relax. By applying specific frequencies, we can stimulate these muscle fibres to relax, realign, and heal more effectively, reducing recovery time and restoring mobility.

Case Study: A Runner's Return to the Track

Consider the story of Sarah, a professional runner who suffered a hamstring injury during training. Traditional therapies seemed slow and frustrating, as the injury kept her off the track for weeks. Determined to recover faster, Sarah turned to vibrational therapy under the guidance of a physical therapist trained in frequency-based healing.

Through regular sessions using low-frequency vibrations, Sarah noticed reduced pain, improved flexibility, and a quicker return to her normal range of motion. What's more, the vibrational therapy appeared to help prevent scar tissue buildup, a common complication of hamstring injuries. Her recovery was swift, allowing her to return to training weeks sooner than anticipated.

Sarah's story exemplifies the growing body of evidence supporting vibrational therapy as a powerful addition to conventional physical therapy techniques. In many cases, it accelerates recovery, alleviates pain, and even prevents future injuries by promoting a more complete healing process.

How Frequencies Facilitate Muscle Recovery

There are several types of vibrational therapies used in physical therapy today, each with unique applications and benefits:

1. Whole Body Vibration (WBV): This therapy involves

standing, sitting, or lying on a machine with a vibrating platform. WBV stimulates muscle contractions and can improve strength, flexibility, and blood flow. Studies suggest that WBV can help reduce pain and improve balance, making it particularly beneficial for athletes and elderly individuals with mobility challenges.

2. Localized Vibrational Therapy: Focused directly on an injured area, this therapy targets specific muscle groups. By applying vibrational frequencies in the range of 20-50 Hz, therapists can stimulate blood flow, enhance lymphatic drainage, and reduce swelling. Localized therapy is often used for muscle strains, joint pain, and even chronic conditions like arthritis.

3. Electrical Muscle Stimulation (EMS): EMS uses electrical pulses to elicit muscle contractions. While technically not vibrational therapy, EMS operates on a similar principle by using frequencies to stimulate muscles. EMS is often used to improve muscle strength, prevent atrophy, and aid in post-surgical recovery.

4. Ultrasound Therapy: While commonly associated with imaging, therapeutic ultrasound uses sound waves at high frequencies to promote muscle healing. The vibrations penetrate deep into the muscle tissue, increasing blood flow and reducing inflammation.

Practical Tips: Incorporating Vibrational Therapy into Daily Recovery

For those interested in exploring vibrational therapy as part of their recovery or physical fitness routine, here are some tips:

- Start Slowly: Begin with short sessions at a low frequency, especially if you're new to vibrational therapy. Over time, as your body adapts, you can increase both the frequency and duration.

- Combine with Stretching: Using vibrational therapy before or after stretching can enhance flexibility by relaxing tight muscles, allowing for a deeper stretch.
- Stay Consistent: Like any physical therapy routine, vibrational therapy is most effective when performed regularly. Create a schedule and stick to it, even if it's just a few minutes each day.
- Listen to Your Body: Pay attention to how your body responds to different frequencies. Some people may experience slight discomfort initially, but the therapy should not cause pain. Adjust as needed.

Future Horizons: Frequency Therapy in Physical Rehabilitation

The potential of frequency therapy is far from being fully realized. In the future, it could be tailored to target not just muscles but also the brain and nervous system, facilitating recovery from neurological conditions. Imagine a world where stroke patients regain muscle control or where individuals with degenerative conditions like Parkinson's experience relief through targeted vibrational therapy.

Emerging technologies, including wearable vibrational devices, promise to make frequency therapy even more accessible. These devices, designed to deliver specific frequencies to muscle groups, could allow patients to continue their therapy at home, making recovery more convenient and personalized.

Reflecting on the Potential of Frequency Healing

When we think of healing, it's easy to focus on medications and surgeries. However, frequency therapy reminds us of a gentler, non-invasive path, tapping into the body's own rhythms and resonance. Just as a musician tunes their instrument, frequency-based therapy aims to "retune" the body, aligning it with its natural harmony.

The use of frequencies in physical therapy invites us to rethink healing as an art and a science. It challenges traditional perspectives, offering a method that is both scientifically grounded and philosophically profound. For those open to exploring the possibilities, frequency-based therapy provides not only physical benefits but also a sense of empowerment over one's healing journey.

Final Thoughts: Embracing Harmony in Movement

Frequency-based therapies represent an exciting frontier in physical therapy, where the body's natural vibrations are harnessed to enhance healing and restore movement. For individuals recovering from injury, athletes seeking optimal performance, and anyone striving to maintain mobility, vibrational therapy offers a transformative approach. As this field continues to grow, we may find ourselves on the brink of a new era, one where the healing power of frequencies becomes as essential as any conventional therapy.

In the words of a physical therapist deeply involved in vibrational therapy research: "Frequency therapy doesn't just treat the muscle; it resonates with the whole person, inviting every part of them to heal." This philosophy—treating the body as an orchestra of frequencies—captures the promise of vibrational therapy and the hope it offers for a healthier, more harmonious future.

CHAPTER 28: FREQUENCY THERAPY FOR JOINT HEALTH

Imagine each joint in the body as a hinge. Over time, with wear and tear, the smooth movement of this hinge can become stiff, uncomfortable, or even painful. For people suffering from arthritis and other joint issues, this discomfort can be debilitating, impacting mobility and quality of life. Now, consider the possibility of easing this pain and restoring fluidity using frequencies. Just as an instrument can be retuned to its original harmony, the body's joints may also respond to therapeutic frequencies, resonating in ways that promote healing, reduce inflammation, and increase flexibility.

The Science Behind Frequency Therapy and Joint Health

The human body is a vast network of frequencies. Each organ, muscle, and joint has its own natural resonance. When joints become inflamed or damaged, their natural frequency may become dissonant, disrupting the overall harmony of the body. Frequency therapy works by reintroducing the body to specific vibrational patterns, allowing the damaged tissue to resonate with frequencies that encourage repair and regeneration.

One of the primary benefits of frequency therapy for joint health lies in its ability to stimulate cellular repair processes. By applying low-frequency electromagnetic fields (EMFs) or sound waves, we can enhance blood flow, reduce inflammation, and promote the production of cartilage and other vital joint tissues. Additionally, frequency therapy can stimulate the body's endorphin release, which provides natural pain relief without the need for medications that may carry side effects.

Case Study: Reclaiming Mobility with EMF Therapy

Consider David, a 65-year-old retired teacher with osteoarthritis in both knees. After years of struggling with limited mobility and chronic pain, he was introduced to frequency therapy by his physical therapist. David's regimen included pulsed electromagnetic field (PEMF) therapy sessions, which he underwent three times a week. Over several months, David noticed a significant improvement in his mobility and a reduction in his pain levels. His once-stiff joints moved more freely, and he was able to enjoy long walks for the first time in years.

David's story highlights the transformative potential of frequency therapy for those suffering from joint issues. By introducing targeted EMF frequencies, his body was able to activate its own natural healing processes, addressing the inflammation and tissue degeneration at the core of his arthritis.

Types of Frequency Therapy for Joint Health

There are several forms of frequency therapy commonly used in joint health management:

1. Pulsed Electromagnetic Field (PEMF) Therapy: PEMF therapy is one of the most researched and widely used forms of frequency therapy for arthritis and joint pain. By

delivering pulsed EMFs at specific frequencies, this therapy can stimulate the regeneration of joint tissue, reduce inflammation, and enhance circulation. Studies show that PEMF therapy can reduce symptoms of arthritis, providing relief for patients with both osteoarthritis and rheumatoid arthritis.

2. Low-Level Laser Therapy (LLLT): Also known as cold laser therapy, LLLT uses light at specific frequencies to penetrate the skin and stimulate cellular repair. The light's energy promotes circulation and reduces inflammation in the affected joints, making it a popular choice for arthritis management.

3. Ultrasound Therapy: Though commonly associated with imaging, ultrasound therapy also has therapeutic applications. By using sound waves at frequencies that penetrate deep into the tissue, ultrasound can reduce pain and inflammation in joints. The vibrations generated by ultrasound improve blood flow, delivering oxygen and nutrients essential for tissue repair.

4. Vibroacoustic Therapy: Vibroacoustic therapy uses sound waves to create vibrations in the body, which can relieve joint pain and improve flexibility. By targeting joints with specific frequencies, this therapy can alleviate pain and stiffness in people with arthritis.

Practical Tips: Using Frequency Therapy for Joint Health

If you're interested in exploring frequency therapy for joint issues, consider these practical tips:

- Consult a Professional: It's essential to work with a trained physical therapist or medical professional experienced in frequency-based therapies to determine the best treatment plan for your specific needs.
- Consistency Is Key: Frequency therapy often requires

repeated sessions to yield the best results. Consistency will help reinforce the benefits of each treatment, allowing the joints to adapt to the therapy.

- Pair with Gentle Exercise: Low-impact exercises like swimming or cycling can enhance the effects of frequency therapy by maintaining joint mobility without adding strain.

- Follow Up with Anti-Inflammatory Foods: Consider supplementing your treatment with an anti-inflammatory diet. Foods like turmeric, ginger, and leafy greens can support the body's response to frequency therapy, helping to reduce inflammation further.

Looking Forward: The Future of Frequency Therapy in Joint Care

The applications of frequency therapy in joint health represent a growing field with exciting potential. Advances in technology are enabling the development of portable frequency-based devices that patients can use at home, making therapy more accessible and consistent. As research continues, there is hope for even more tailored frequencies and methods designed to target specific forms of arthritis and joint issues.

Imagine a future where, instead of facing surgery or long-term reliance on painkillers, patients can tune their bodies back to health using frequencies that restore harmony at a cellular level. We may soon see a world where surgery becomes a last resort, replaced by non-invasive therapies that respect and harness the body's own healing abilities.

Reflecting on the Synergy Between Body and Frequency

At its heart, frequency therapy is about resonance. Just as a pianist tunes each key to create harmony, frequency therapy aims to retune the body's tissues, restoring them to their natural, healthy state. This approach acknowledges that healing isn't always about drastic measures; sometimes, it's about subtle interventions that gently guide the body back to balance.

For joint health, this means addressing the root of the pain—whether it's inflammation, degeneration, or stiffness —by nurturing the tissues involved and encouraging their regeneration. Frequency therapy invites us to view healing not as a forceful intervention but as a partnership between the body's own rhythms and those of the frequencies applied.

Final Thoughts: Embracing Harmony for Healthier Joints

In the journey of exploring frequencies, we've found a profound tool for managing joint health. Frequency-based therapies offer a gentle, non-invasive way to alleviate pain, reduce inflammation, and promote joint health. For those with arthritis or other joint conditions, this approach provides not only physical relief but also a sense of hope and empowerment. Frequency therapy honours the body's resilience, working with its natural rhythms to restore health and mobility.

In embracing the potential of frequencies, we're reminded of the harmony inherent in all things—even within our joints. As we tune into the possibilities of frequency-based healing, we open ourselves to a new understanding of health, one that is deeply connected, reverent, and resonant with the rhythms of life itself.

PART VI: CASE STUDIES IN HEALING

CHAPTER 29:
CASE STUDY:
CANCER RECOVERY THROUGH FREQUENCY HEALING

Imagine walking into a room where soothing tones hum gently through the air, creating an atmosphere of calm and rejuvenation. This isn't a spa or a relaxation centre—it's a therapeutic setting for a new approach to cancer recovery. Here, sound waves and electromagnetic frequencies become powerful allies in the healing journey, complementing traditional cancer treatments to enhance resilience, ease side effects, and nurture the body's natural healing processes.

In this chapter, we'll explore a remarkable case study of frequency-based therapies in cancer recovery. With a blend of scientific insight, personal narrative, and practical takeaways, we'll uncover how specific frequencies may work alongside chemotherapy and radiation, offering patients a fuller, more holistic healing experience.

Introduction: A New Frequency in Cancer Recovery

Cancer treatments often come with side effects that challenge not only the body but also the spirit. For many patients, conventional therapies can feel harsh, leaving them physically and emotionally exhausted. This is where frequency therapy—using sound waves, electromagnetic fields, or light frequencies—enters the picture as a complementary treatment. Rather than directly targeting cancer cells, frequency-based therapies aim to fortify the body's immune system, reduce pain, manage nausea, and relieve the mental strain of long-term treatments.

Much like tuning an orchestra, frequency therapy seeks to bring harmony back to the body, restoring it to its natural rhythm. In cancer recovery, this harmony doesn't mean eliminating the disease overnight; rather, it supports the body's resilience, making the journey to health more tolerable and even empowering.

Case Study: Emma's Journey to Recovery with Frequency Healing

Emma was a 45-year-old mother of two when she was diagnosed with stage III breast cancer. After her diagnosis, Emma's treatment regimen included aggressive chemotherapy and radiation therapy. However, within the first few months, she began experiencing significant side effects, including fatigue, severe nausea, and bouts of depression. Emma's oncologist recommended looking into complementary therapies that might ease these symptoms.

Emma decided to incorporate frequency therapy into her treatment plan. She started with regular sessions of pulsed electromagnetic field (PEMF) therapy, which involved exposing her body to low-level electromagnetic fields that are thought to stimulate cellular repair. She also tried sound therapy using binaural beats and Tibetan singing bowls to alleviate her stress and anxiety.

Over several months, Emma found that these therapies did not replace her conventional treatment but complemented it effectively. Her PEMF sessions helped reduce pain and inflammation in her body, particularly in areas affected by radiation. Sound therapy sessions provided her with mental clarity and an emotional release, helping her navigate the psychological challenges of cancer recovery.

The Science Behind Frequency-Based Cancer Support

So, how exactly did these frequencies aid Emma's recovery? Research into frequency therapy in cancer recovery suggests that certain treatments can support cellular repair and immune function, although they are not cures for cancer itself. Here's how a few of these therapies operate:

- PEMF Therapy: This therapy delivers electromagnetic pulses to the body, which can penetrate deep into tissues and promote cell repair. For cancer patients, PEMF may assist in reducing inflammation and pain, making post-treatment recovery more comfortable. Additionally, PEMF is thought to enhance blood circulation, which can support the body in managing the physical strain of chemotherapy.

- Sound Therapy: Sound therapy uses specific frequencies to influence the brain's electrical activity. Binaural beats, for example, create an effect by playing two slightly different frequencies in each ear, encouraging brain waves to align to a frequency that promotes relaxation or focus. This effect can ease anxiety and elevate mood, addressing the mental strain that accompanies cancer treatment.

- Low-Level Laser Therapy (LLLT): In some cases, low-level lasers emitting particular frequencies are used to stimulate mitochondrial function in cells, which may boost energy levels and speed up tissue repair. Although more research is needed, early findings indicate that LLLT could be beneficial for managing chemotherapy-induced fatigue.

Practical Tips: Integrating Frequency Therapy into a Cancer Recovery Plan

If you or a loved one are considering frequency therapy as part of a cancer recovery strategy, here are some practical steps:

1. Consult with Your Healthcare Team: Always talk to your oncologist or healthcare provider before adding new therapies to your treatment plan. Complementary therapies should align with your existing medical plan.

2. Start with Short Sessions: Many patients find benefit in starting with brief, focused sessions and gradually increasing the duration. Listen to your body's responses to gauge whether a particular therapy is right for you.

3. Create a Calming Environment: Sound therapy can be enhanced by a calm, quiet setting. Using headphones with binaural beats or meditative music at home can turn any space into a supportive healing environment.

4. Track Your Progress: Keep a journal to document your physical and emotional responses. This feedback can help tailor the therapy sessions to suit your needs and monitor improvements in symptoms like pain, fatigue, and mood.

Future Implications: What Frequency Therapy Could Mean for Cancer Care

While frequency therapy is still emerging as a complementary treatment in cancer care, its potential is vast. Imagine a future where every cancer centre offers patients access to non-invasive, frequency-based therapies designed to harmonize the body, mind, and spirit. As research continues, we may find more targeted frequencies capable of not only alleviating side effects but also promoting long-term health and resilience after recovery.

Reflecting on the Healing Power of Harmony

Frequency therapy doesn't claim to be a cure for cancer, but it opens doors to a new way of supporting the body. By using natural, resonant frequencies, we respect the body's inherent rhythms and give it the resources to withstand aggressive treatments. For Emma, this approach meant finding relief, clarity, and even moments of peace amid her journey to recovery.

At its core, frequency therapy invites us to imagine health as a state of resonance. Just as each note in a song holds power to evoke emotion, each frequency used in healing holds the potential to restore balance. Emma's experience reflects a profound truth: healing isn't solely about eradicating illness but about nurturing harmony in every layer of our being.

Closing Thoughts: Embracing Frequency Medicine in Cancer Recovery

Frequency therapy offers hope and comfort to cancer patients, not as a replacement but as a partner to conventional medicine. It acknowledges that healing is a multifaceted journey that encompasses the body, mind, and spirit. As we embrace these therapies, we step closer to a world where healing is gentle, harmonious, and holistic.

For those on their own healing paths, remember: the body, like a finely tuned instrument, responds to frequencies that restore its innate resilience. In moments of difficulty, these therapies remind us of the beauty and strength within us all, making the road to recovery just a little smoother, a little brighter, and much more resonant.

CHAPTER 30: CASE STUDY: CHRONIC PAIN MANAGEMENT THROUGH FREQUENCY-BASED TREATMENTS

Chronic pain can feel like an endless loop—an invisible burden that disrupts the rhythm of life. Traditional treatments, from medications to physical therapy, often focus on masking symptoms rather than addressing the root causes of pain. Frequency-based therapies, however, approach pain from a different perspective, aiming to restore balance at the cellular level by aligning the body's frequencies with its natural rhythm. In this chapter, we'll explore a remarkable case of chronic pain relief through frequency-based treatments, examining the science behind this approach, sharing practical insights, and envisioning a future where frequency medicine offers a new pathway to relief.

A Journey from Pain to Harmony: Sarah's Story

Sarah, a 52-year-old teacher and avid gardener, developed

chronic pain after a car accident that injured her lower back and left her with lasting nerve pain in her legs. After years of physical therapy, medications, and limited mobility, Sarah was exhausted, discouraged, and seeking alternatives. Her doctors recommended surgery, but Sarah hesitated, concerned about the risks and long recovery period. That's when she discovered frequency-based therapy.

Through an integrative pain clinic, Sarah began a combination of pulsed electromagnetic field (PEMF) therapy and low-frequency sound therapy. PEMF therapy uses electromagnetic waves to stimulate cells, while low-frequency sound therapy involves sound waves designed to resonate with specific areas of the body. For Sarah, these therapies provided gentle yet transformative effects over several months, helping her reduce her reliance on medication and increase her mobility.

The Science Behind Frequency Therapy for Pain Relief

Frequency-based treatments for chronic pain operate on a unique principle: they stimulate cellular repair and improve the flow of energy in the body. Here's how PEMF and sound therapy work to alleviate pain:

- PEMF Therapy: When cells are exposed to specific electromagnetic fields, it can enhance cell function and encourage repair, particularly in damaged or inflamed areas. PEMF therapy is believed to increase blood circulation, reduce inflammation, and improve cellular oxygenation, all of which are essential for pain management. Studies suggest that regular PEMF therapy can support long-term pain relief by helping cells regenerate more effectively, promoting healthier tissue over time.

- Sound Therapy: Sound waves at low frequencies resonate with the body's tissues, penetrating deep into muscles and nerves. For chronic pain patients, sound therapy may act as a soothing balm, encouraging relaxation and helping reduce muscle tension. Certain frequencies can even stimulate the release of endorphins, the body's natural painkillers, which makes sound therapy particularly effective for managing pain without medication.

Together, these therapies form a holistic approach to pain management that aligns with the body's natural healing processes rather than overpowering them with external interventions.

Real-World Impact: Sarah's Progress and Reflections

After only a month of frequency therapy sessions, Sarah noticed a reduction in pain intensity and a newfound ability to sleep through the night—something that had been challenging due to her chronic discomfort. Her case demonstrates the potential of frequency-based treatments to support the body in achieving a state of equilibrium, particularly in patients who have not found relief through conventional methods.

By the third month, Sarah's daily pain levels had decreased by nearly 40 percent, and she had begun walking short distances without assistance. Her sleep improved, her mood stabilized, and her overall energy levels rose. Sarah's story reflects how frequency therapies don't just alleviate pain but also encourage a profound shift in quality of life by addressing the body's pain response holistically.

Practical Insights for Incorporating Frequency Therapy into Pain Management

If you or someone you know is considering frequency therapy as an option for chronic pain, here are some practical tips:

1. Consult a Specialist: Work with a healthcare provider or specialist who understands frequency-based therapies and

can tailor treatments to your specific needs. Chronic pain varies widely in cause and intensity, so individualized care is essential.

2. Start Slowly: Begin with shorter sessions and gradually increase the duration and intensity. Many patients report that their bodies need time to adjust to frequency-based treatments, especially if they've been dealing with pain for years.

3. Combine Therapies: Frequency therapies can be effective on their own, but they often work best as part of a comprehensive treatment plan. For instance, PEMF therapy may be combined with sound therapy or low-level laser therapy for synergistic effects.

4. Track Your Symptoms: Keeping a journal of pain levels, mood, sleep quality, and other factors can help you understand how frequency therapy is impacting your condition and identify any necessary adjustments.

5. Stay Consistent: As with many holistic treatments, the effects of frequency therapy often accumulate over time. Regular sessions can lead to long-term improvement, so it's essential to commit to a consistent schedule.

Future Implications: A New Horizon in Pain Management

As research in frequency medicine progresses, we may see frequency-based pain management becoming more accessible and widely accepted. Imagine a future where frequency therapy clinics are as common as physical therapy centres, offering chronic pain sufferers new pathways to relief without the need for invasive procedures or dependency on medication. As more people experience the benefits, we're likely to see even more innovative uses of frequencies in managing pain and other health conditions.

Philosophical Reflections: Pain, Healing, and Resonance

Chronic pain is a deeply personal experience, often leading individuals on journeys of self-discovery and resilience. Frequency-based therapies offer a unique opportunity to rethink how we view pain and healing. Instead of perceiving pain as an external force to be eradicated, frequency medicine encourages us to view it as a disharmony in the body's natural resonance—a disruption that can be gently realigned.

In Sarah's case, this realignment brought not only physical relief but also an emotional release, reminding her that healing isn't about overpowering the body but about nurturing it back to its natural state of balance. This approach resonates with an ancient understanding of health as a state of harmony, echoing philosophies that view the body as a finely tuned instrument. Just as a skilled musician can retune a violin, frequency therapies offer us the tools to retune our own bodies.

Conclusion: Embracing Frequency-Based Therapies for a Pain-Free Future

Sarah's journey is just one of many examples demonstrating the transformative potential of frequency-based treatments in chronic pain management. By blending science with the art of resonance, these therapies open doors to new possibilities, offering a gentle, non-invasive, and holistic approach to pain relief. For chronic pain sufferers seeking alternatives, frequency medicine may provide hope where conventional methods have fallen short.

As we continue exploring the healing power of frequencies, we stand at the threshold of a new era in medicine— one where sound, light, and electromagnetic fields become as vital to healing as traditional medicine. For those who are willing to embrace these new approaches, the path to recovery may no longer be a battle against pain but a journey back to harmony, revealing the profound secrets held within the body's natural resonance.

PART VII: EXPLORING SOUND'S IMPACT ON MENTAL WELLNESS

CHAPTER 31: SOUND BATHS AND THEIR CALMING EFFECTS

Imagine yourself enveloped in a wave of harmonious sounds, each note gently washing over you and releasing tension you may not even realize you're holding. This is the experience of a sound bath, a practice that combines ancient wisdom and modern science to induce a profound state of relaxation. Often seen as a doorway to tranquillity, sound baths use carefully crafted tones and frequencies to promote mental clarity, emotional release, and physical healing. In this chapter, we'll explore the science, tradition, and power of sound baths, showing how this seemingly simple practice resonates deeply with the body and mind, unlocking a sense of peace that is often elusive in today's fast-paced world.

The Concept of a Sound Bath: Immersing in Harmony

Unlike a traditional water bath that cleanses the body, a sound bath is a journey into frequencies that "bathe" the participant in healing vibrations. Originating from ancient cultures that recognized the therapeutic effects of sound, sound baths typically use instruments like Tibetan singing bowls, crystal bowls, gongs, and chimes. Each instrument is tuned to specific frequencies that target different areas of the body and mind. During a session, participants lie comfortably as the sound waves flow through the room, creating a "sonic cocoon" that envelops and heals.

To help visualize the effect, consider the body as an orchestra. Each cell and organ vibrates at its unique frequency, yet in a healthy person, these frequencies harmonize, like instruments in an orchestra playing in tune. However, stress, illness, or emotional imbalance can disrupt this harmony, creating dissonance in the body. Sound baths seek to retune these frequencies, helping restore the natural resonance of each part, so they play together in harmony again.

The Science Behind Sound and Relaxation

The physiological impact of sound baths lies in their effect on brainwaves. Typically, our minds operate in beta waves (13-30 Hz), associated with active thinking and problem-solving. However, deep relaxation, creativity, and even healing states are achieved at alpha (8-12 Hz), theta (4-8 Hz), and delta waves (0.5-4 Hz). Sound baths guide the brain into these lower frequencies, allowing the body to access states of rest and renewal that are difficult to achieve in daily life.

Research into sound therapy has shown that certain frequencies can stimulate the parasympathetic nervous system, responsible for our body's "rest and digest" functions. When activated, this system counteracts the stress-driven sympathetic nervous system ("fight or flight"), reducing heart rate, blood pressure, and levels of cortisol, a stress hormone. Sound baths provide a gentle way to invite the body into this relaxed state, making it easier to experience mental clarity, emotional balance, and even improved immunity over time.

Case Study: Sarah's Healing Sound Bath Journey

Consider Sarah, a corporate executive struggling with chronic stress and insomnia. After years of trying various methods to manage her stress, she attended her first sound bath, albeit with some scepticism. Within moments, the resonant hum of the crystal bowls started to calm her racing thoughts. She felt as if each note was dissolving layers of tension she had carried unknowingly for years.

After a few sessions, Sarah noticed significant changes. Not only was she sleeping better, but she also felt more focused and less reactive to daily stressors. The deep relaxation she experienced during her sound baths carried into her work life, transforming her approach to challenges. Sarah's story exemplifies the power of sound baths to bring the body and mind into harmony, allowing participants to manage stress and restore balance naturally.

Practical Tips: How to Incorporate Sound Baths into Your Life

For those interested in exploring the benefits of sound baths, there are several ways to begin:

1. Find a Local Practitioner: Many wellness centres offer sound bath sessions led by experienced practitioners. Attending a session in person allows you to experience the full resonance of the instruments in a communal, immersive setting.

2. Create a DIY Sound Bath at Home: While live sound baths offer the best experience, you can recreate some of the effects at home. Many online platforms provide sound bath recordings, featuring crystal bowls, gongs, and chimes. Find a quiet space, use high-quality headphones, and create a comfortable environment to relax fully.

3. Experiment with Different Frequencies: Each frequency affects the body differently. For deep relaxation, look for sound baths with lower frequencies that promote delta and theta waves. For focus and creativity, try alpha frequencies.

4. Set an Intention: Before a sound bath, set a personal

intention, such as "I release stress" or "I welcome clarity." Intentions can guide the experience, allowing the sound bath to resonate on a more personal level.

5. Embrace Regular Practice: Like most healing practices, the benefits of sound baths accumulate over time. Try integrating sound baths into your weekly routine to help reinforce relaxation, emotional balance, and mental clarity.

The Future of Sound Baths: A Vision for Healing

As more research emerges, the role of sound baths and similar vibrational therapies in medical settings may expand. Already, some hospitals use sound and music therapy to support cancer patients, helping alleviate pain and anxiety. In the future, sound baths could become part of integrative health programs, supporting treatments for conditions like anxiety, PTSD, and chronic pain.

Imagine a world where sound baths are as common as meditation classes, where every individual has access to the harmonizing effects of sound to heal both mind and body. It's a vision where sound becomes as essential to healthcare as traditional therapies, offering a gentle, non-invasive method to reset and recharge.

Reflection: The Symphony Within

Reflecting on sound baths reveals a profound truth: we are inherently rhythmic beings. From the beat of our hearts to the breath flowing in and out, our bodies are symphonies of sound and vibration. Sound baths tap into this inherent harmony, reminding us that we are connected to a greater universal rhythm.

In our fast-paced lives, we often become disconnected from this inner harmony, pulled in countless directions that disrupt our natural flow. Sound baths offer a chance to realign with our true rhythm, not by doing more, but by simply being and listening. They invite us to remember that healing isn't about fixing broken parts; it's about tuning in and allowing ourselves to resonate with the beauty within.

Conclusion: Embracing the Power of Sound Baths

Sound baths may be an ancient practice, but their relevance today is more significant than ever. They provide a bridge between the modern search for stress relief and ancient wisdom that understands healing as a journey back to balance. By immersing ourselves in sound, we can transcend the mental clutter and physical tension that obscure our natural harmony.

As you embark on your journey with sound baths, may you experience not only relaxation but a deeper sense of self, a resonance that aligns your body, mind, and soul. Embrace the power of sound as a healing tool, and let it guide you to a place of profound peace and rejuvenation. In the end, the secret of sound baths lies not in their mystical tones but in their ability to bring us back to our own, original song.

CHAPTER 32: THE ROLE OF NATURE SOUNDS IN HEALING

Imagine stepping into a quiet forest. The soft rustling of leaves in the breeze, the gentle trickle of a nearby stream, and the occasional call of a distant bird. These sounds, which might seem ordinary, carry profound effects on our well-being. Our connection to nature runs deep, and as science reveals, natural sounds can have tangible healing effects on our bodies and minds. From reducing stress to enhancing focus and promoting restful sleep, the sounds of nature harmonize with our internal rhythms, offering a powerful form of frequency-based therapy.

The Science Behind Nature's Sound Frequencies

In our fast-paced lives, we often take sounds for granted, but everything we hear impacts us at a biological level. Natural sounds, particularly, seem to engage our brain in unique ways. Studies on the brain's reaction to natural sounds reveal that they promote alpha waves, associated with calmness and relaxation, and theta waves, linked to restorative states like sleep and meditation. Just as music at specific frequencies can influence our mood and cognition, the frequencies found in nature carry an innate capacity to align our mental and physical states.

One reason natural sounds are so powerful is their rhythm

and unpredictability. Unlike the steady hum of urban noise, natural sounds flow in irregular patterns, such as the random patter of raindrops or the diverse calls of birds. This variability keeps our brain engaged without overwhelming it, drawing our attention in a soft, restorative way. Think of it as a gentle mental massage—our minds become subtly entrained to these frequencies, helping us to enter a relaxed state that counters the effects of stress.

Story of John: Nature's Healing Power

To illustrate the effects of nature sounds, let's look at John's story. A high-powered executive, John suffered from chronic stress and insomnia, finding it nearly impossible to unwind even after work hours. His doctor recommended sound therapy, specifically incorporating nature sounds. Sceptical but willing to try anything, John began playing a recording of ocean waves as he prepared for bed. The rhythmic swell and retreat of the waves felt as though they were washing his stress away, calming his overactive mind.

After several weeks, John found that he could fall asleep faster and stay asleep longer. His mind, previously tense and strained, seemed to loosen as the sounds of nature filled his nights. The ocean waves not only helped him sleep but also began to carry over into his daytime routine, making him less reactive to stress. John's experience highlights how simple sounds from nature can create profound changes in our well-being, reducing stress, enhancing mental clarity, and even helping us reconnect with ourselves.

Why Nature Sounds Work: Psychological and Physiological Benefits

Nature sounds don't just feel soothing; they actually impact our body's physiology. Listening to the sounds of a gentle waterfall or rustling leaves can lower cortisol levels, the hormone responsible for stress. Additionally, exposure to these sounds helps activate the parasympathetic nervous system, which promotes rest and recovery. This activation opposes the sympathetic nervous system—responsible for the "fight or flight" response—which is frequently overstimulated by modern stressors.

One fascinating study showed that listening to natural sounds like water flowing or birds singing can significantly increase attention span and concentration. Nature sounds mimic the way our ancestors' environments sounded, creating a sense of safety and comfort. This alignment with our evolutionary past brings a unique form of mental peace, reducing anxiety and helping us to enter states of focus and relaxation.

Practical Tips: Integrating Nature Sounds into Daily Life

For anyone looking to tap into the healing power of nature sounds, here are some ways to make them part of your daily routine:

1. Morning Rituals with Birdsong: Start your day with the sounds of birds singing. Many apps and streaming services offer high-quality recordings of birdsong, which can be played as you get ready in the morning. Birdsongs in particular are linked to increased positive emotions, helping set a peaceful, optimistic tone for the day.

2. Midday Break with Water Sounds: During a break at work, listen to the sounds of a stream, waterfall, or ocean waves. Even a few minutes can refresh your mind, clearing away mental fatigue and improving focus for the rest of your day.

3. Sleep with Gentle Rain: Rain sounds are among the most popular for promoting sleep because of their constant and gentle nature. Try playing these sounds as part of your nighttime routine to create an environment conducive to falling asleep and staying asleep.

4. Outdoor Walks When Possible: While recordings are beneficial, nothing compares to the real experience. A walk in the park or by a river immerses you fully in natural sounds, combining the benefits of sound therapy with physical activity, fresh air, and sunlight.

5. Create a Nature Sanctuary at Home: If you can, set up a small fountain indoors or place wind chimes near an open window. These natural sounds provide a peaceful background to your home environment, grounding you in moments of calm throughout the day.

Looking Ahead: Nature Sounds in Modern Medicine

As researchers continue to explore the effects of nature sounds, they're finding new ways to integrate them into therapeutic settings. Hospitals are increasingly incorporating natural soundscapes to create calm atmospheres, reducing anxiety for patients before surgery or helping them relax during treatment. Some clinics even use virtual reality combined with nature sounds to transport patients into a forest or by a seaside, merging visual and auditory stimuli to amplify healing.

In the future, we may see nature sound therapy becoming a standard tool in mental health treatments, from managing anxiety to aiding in recovery from trauma. Imagine a world where nature sounds become an integral part of our healthcare system, from maternity wards to senior care homes, with every environment optimized for healing and tranquillity. These sounds connect us to a universal rhythm that modern life has often drowned out, reminding us of our roots and the peace found in nature's embrace.

Reflection: Reawakening Our Connection to Nature

In exploring the healing potential of nature sounds, we rediscover a part of ourselves that modern life often suppresses. These sounds speak to something ancient within us, inviting us to reconnect with a primal sense of peace and balance. For millennia, humanity found solace in nature's symphony—its songs offering reassurance and its rhythms guiding us toward well-being.

In a world filled with synthetic noise, we are reminded of the elegance in nature's simplicity. Water flowing, leaves rustling, birds singing—these sounds form the fabric of a healing experience that's available to us all. They invite us to pause, to listen, and to remember that healing can be as simple as returning to our natural state of harmony.

Conclusion: Embracing the Healing Power of Nature Sounds

Nature sounds offer more than a pleasant background; they serve as gateways to profound relaxation and holistic health. Whether you're seeking calm in the chaos of daily life, relief from stress, or a tool to enhance sleep, the sounds of nature are there to guide you.

As you incorporate these sounds into your life, may you experience not only relaxation but a deeper connection to the world around you. In this connection lies the secret of true healing: harmony with the natural rhythms of life. Embrace nature's symphony, and let its timeless sounds guide you toward a life of inner peace and balance.

PART VIII: PRACTICAL GUIDE TO SOUND THERAPY

CHAPTER 33: SELF-HEALING WITH SOUND

Imagine walking into a quiet room, tuning fork in hand, and striking it gently against a surface. The vibration hums through your hand and resonates in the air, creating a sound that seems to dissolve tension within seconds. You feel something powerful yet gentle happening within—a connection to something deeper as if your own cells are being reminded of a lost harmony. This is the essence of self-healing with sound: a practice that allows us to tune our bodies, much like tuning a musical instrument, to bring ourselves back into balance.

In this chapter, we delve into the transformative potential of using tuning forks and singing bowls for self-healing. By exploring the science, practical techniques, and ancient wisdom surrounding these tools, we open the door to a deeply personal journey of resonance and renewal.

The Science of Sound: Why Tuning Works

Just as every note in a song contributes to a harmonious melody, each cell in our body vibrates at a frequency that reflects health or dissonance. Cells under stress, illness, or fatigue emit frequencies that differ from those in a state of wellness. Sound healing, at its core, aims to restore harmony to these vibrations, guiding cells back to their natural state.

The science behind sound healing reveals that specific frequencies can encourage certain physiological responses. Research shows that sound frequencies can influence brain waves, heart rate, and even cellular repair. For instance, using a tuning fork at 528 Hz, known as the "miracle frequency," is believed to stimulate DNA repair and support healing on a cellular level.

Analogy: Think of the body like an orchestra where each cell resonates to its own note. If one instrument falls out of tune, the whole piece feels disjointed. Sound healing, with tools like tuning forks and singing bowls, "re-tunes" the cells, restoring harmony within the body.

The Art of Using Tuning Forks for Healing

Tuning forks, small and seemingly simple, are powerful tools that carry frequencies which resonate deeply within our bodies. When used mindfully, they can bring about calm, release stored tension, and foster healing.

How to Use Tuning Forks

1. Choosing the Right Fork: Begin by selecting a tuning fork that resonates with your goal. Common frequencies include:
    - 174 Hz: Pain relief and stress reduction.
    - 528 Hz: DNA repair and cell rejuvenation.
    - 639 Hz: Heart chakra alignment and improved relationships.

2. Activating the Fork: Hold the tuning fork by the handle and gently strike it against a rubber surface, like the sole of your shoe. The fork will begin to vibrate, creating a soft hum that lasts for several seconds.

3. Application: Once activated, bring the fork close to your body, focusing on specific areas of discomfort or stress. You can place the vibrating fork near your heart, abdomen, or any

area that feels "off." Let the frequency soak into your cells.

4. Breathing with Sound: As the vibration hums through you, take slow, deep breaths. Visualize the frequency moving through your body, clearing away blockages and restoring balance.

5. Consistency: For lasting effects, try using your tuning fork daily. Sound healing works best when it becomes a regular part of your routine, much like meditation or yoga.

Case Study: Maya's Journey with Tuning Forks
Maya, a 45-year-old architect, struggled with chronic neck pain and stress. After experimenting with physical therapy and massage, she turned to tuning forks as an alternative. By using a 174 Hz tuning fork daily, Maya reported a significant decrease in her pain and found herself less reactive to stress. She describes the experience as "tuning into her own harmony," feeling that her body and mind became better aligned.

The Healing Power of Singing Bowls

Singing bowls, particularly those crafted from metal or crystal, are another ancient instrument for healing. Originating from Tibetan and Himalayan cultures, these bowls produce sound by being struck or rubbed with a mallet, creating a unique vibrational frequency. Singing bowls are widely used for meditation, relaxation, and even physical healing, as the sound waves they emit can resonate deeply within the body, helping to release tension and balance energy.

Using Singing Bowls for Self-Healing

1. Creating the Space: Set up a calm environment where you can sit comfortably. Ideally, this should be a quiet room with minimal distractions.

2. Choosing the Right Bowl: Singing bowls are often tuned to specific frequencies associated with the chakras, the body's energy centres. For example:

- Root Chakra (432 Hz): Grounding and stability.
- Heart Chakra (639 Hz): Emotional healing and connection.
- Third Eye Chakra (852 Hz): Intuition and clarity.

3. Playing the Bowl: Hold the bowl in one hand and strike it gently with a mallet, or run the mallet around the rim to create a continuous hum. Feel the vibration extend from the bowl to your hand, and then radiate through your body.

4. Setting an Intention: Before you begin, set an intention for healing. Perhaps you want to release anxiety, ease physical pain, or simply relax. Let this intention guide your practice.

5. Listening and Absorbing: Close your eyes, breathe deeply, and let the sound of the bowl wash over you. Focus on the way the sound feels rather than simply how it sounds. Imagine each note resonating within you, balancing your cells and realigning your energy.

Practical Tip: If you're new to sound healing, start with a 10-minute session. Over time, you may wish to increase the duration as you become more attuned to the effects.

The Future of Personal Sound Healing

As we explore these tools, we may wonder: could sound healing one day replace traditional medicine in certain scenarios? Although we are still uncovering the full potential of sound, the stories of those who've experienced its healing effects suggest an inspiring future. A world where sound therapy becomes a natural part of wellness, where tuning forks and singing bowls share a place in our homes beside more conventional self-care items.

Imagine a world where sound healing is so mainstream that hospital recovery rooms are equipped with singing bowls, where tuning forks are used to relieve pain in physical therapy, and where every home has a small sanctuary for self-healing through sound.

Reflection: Tuning into Our Innate Harmony

As we embrace sound healing, we reconnect with a part of ourselves that remembers harmony. Each sound, each vibration reminds us that our bodies and minds are intricate, delicate instruments capable of resonating with life itself. When we fall out of balance, sound can guide us back— gently, intuitively, and naturally.

Philosophical Thought: If every cell in our body resonates, then we are, in essence, beings of sound. Every experience, thought, and emotion creates a unique frequency. Sound healing taps into our own symphony, restoring us to a state of natural, vibrant health.

Conclusion: Embarking on Your Journey with Sound

In exploring tuning forks and singing bowls, we gain not only new tools for healing but also a deeper awareness of our bodies. Through the resonance of sound, we learn to listen —not only with our ears but with our entire being. Embrace the journey, let these sounds guide you, and find your own harmony within.

Whether you are drawn to the simplicity of a tuning fork or the meditative hum of a singing bowl, remember: healing is not something that only comes from external sources. It is something that can be summoned from within. Sound is simply the key that helps us unlock this innate power.

CHAPTER 34: MEDITATIVE FREQUENCIES FOR STRESS RELIEF

Imagine the sound of a gentle stream, a soft hum resonating through your mind, gently washing away your worries. In a world where stress seems ever-present, meditative frequencies offer us a path to inner peace—a way to calm the mind, soothe the body, and restore balance.

In this chapter, we'll explore how meditative frequencies can be incorporated into daily routines for stress relief. We'll look at the science of how these frequencies work, share real-life examples of their impact, and offer practical steps to bring their healing power into your life.

The Science of Meditative Frequencies

Our brains constantly produce electrical impulses, measured as brain waves, which fluctuate depending on our state of mind. The frequencies of these waves range from high-energy beta waves, associated with active thinking, to low-energy delta waves, associated with deep, restorative sleep. Meditative frequencies work by gently guiding our brain waves into calmer states, like alpha (associated with relaxation and light meditation) or theta (linked to deep relaxation and creativity).

Research shows that listening to specific frequencies, such as 432 Hz and 528 Hz, can help induce these states, effectively reducing anxiety and stress. By syncing our brain waves with these soothing frequencies, we can activate the body's natural relaxation response, lowering heart rate, decreasing blood pressure, and reducing cortisol levels. In essence, meditative frequencies allow us to use sound as a bridge between the mind and body, creating harmony within.

Analogy: Think of meditative frequencies as a gentle wind guiding a leaf to the ground. Just as the wind helps the leaf fall slowly and peacefully, these frequencies guide our minds to a state of rest and tranquillity, softening the impact of stress.

Real-Life Transformations: Meditative Frequencies in Action

To better understand the power of meditative frequencies, consider the story of Sarah, a high-stress corporate executive who suffered from insomnia and anxiety. Desperate for relief, Sarah began incorporating daily sessions of 432 Hz music into her evening routine. Over time, she noticed profound changes—not only did she start sleeping better, but she also felt calmer throughout her day.

Case Study: Sarah's Healing Journey
Sarah's journey shows how frequency-based meditation can help the mind and body find a balanced state. By listening to these frequencies regularly, she trained her brain to associate certain sounds with relaxation, helping her switch from the fast-paced beta wave state of her workday to the calming alpha and theta states conducive to rest. After a few months, Sarah reported a notable reduction in anxiety and felt more present and peaceful.

Incorporating Meditative Frequencies into Your Daily

Routine

Whether you're new to meditative frequencies or familiar with their benefits, the following steps can help you integrate them seamlessly into your life for stress relief.

Morning Calm: Starting the Day with Clarity

- Choose a Frequency: Start with 528 Hz, known as the "miracle tone," which promotes clarity and transformation.
- Set a Time: Dedicate 10 minutes each morning to sit quietly with the frequency playing. Let the sound fill your space, closing your eyes and taking deep, mindful breaths.
- Visualization: As you breathe in, imagine the frequency gently cleansing your mind of any lingering worries or stress. Visualize a wave of calm washing over you, filling you with clarity for the day ahead.

Starting the day with calming frequencies can help you approach tasks with a centred mind, reducing stress before it even begins to build up.

Midday Reset: Easing Tension during Breaks

- Frequency Choice: Try 396 Hz, associated with releasing guilt and fear. This frequency is known to create a grounding effect, ideal for breaking up the day's tension.
- Mindful Break: Instead of grabbing your phone, put on headphones and listen to 396 Hz for 5-10 minutes. Take deep breaths, inhaling for a count of four, holding for two, and exhaling for four. Let your body relax with each breath.
- Reflect: Allow any stressful thoughts or situations to drift away with the sound, imagining the frequency dissolving tension with each exhale.

Even a short midday session can recharge you, giving you a mental "reset" to face the rest of the day with renewed calm and focus.

Evening Unwind: Preparing for Restful Sleep

- Use Soothing Frequencies: 432 Hz is particularly effective for evening relaxation as it promotes harmony and balance.
- Create a Relaxing Atmosphere: Dim the lights, close your eyes, and listen to 432 Hz music for 15-20 minutes before bed.
- Body Scan: As you listen, practice a body scan by focusing on each part of your body, from head to toe, consciously relaxing each area. Imagine the frequency as a soft wave, moving down your body, easing any lingering tension.
- Sleep Ritual: Make this part of your nightly ritual. By associating 432 Hz with sleep, you train your body to recognize it as a signal to relax fully and prepare for deep, restful sleep.

Practical Tips for Using Meditative Frequencies

To get the most out of meditative frequencies, consider these practical tips:

- Use Quality Headphones: Good-quality headphones or speakers help you experience the full depth of these frequencies, enhancing their effectiveness.
- Consistent Practice: As with any form of healing, consistency is key. Try to incorporate these practices daily for the most powerful results.
- Create a Dedicated Space: If possible, designate a quiet, comfortable area in your home where you can regularly listen to meditative frequencies without distraction.
- Experiment and Find What Works: Different frequencies affect people in unique ways. Experiment with various frequencies (396 Hz, 432 Hz, 528 Hz) to find the ones that resonate best with you.

Reflective Insight: A Journey into Inner Harmony

As you incorporate these meditative frequencies into your

life, remember that each session is more than just relaxation; it's an act of self-restoration. Just as an instrument goes out of tune, so do we, gradually becoming misaligned by the demands of modern life. Meditative frequencies are a way to re-tune ourselves, to restore our natural state of harmony.

By embracing sound as a healing tool, we honour the profound connection between our bodies and the vibrational frequencies surrounding us. The beauty of meditative frequencies lies in their simplicity—no side effects, no complex equipment—just sound and presence. With each listening session, we nurture a quieter mind, a peaceful heart, and a balanced body.

Philosophical Reflection: Consider that we, as beings, are made up of vibrations, each cell and thought resonating with its unique frequency. Sound is the bridge between the physical and the ethereal, the tangible and the intangible. By bringing meditative frequencies into our lives, we connect with a deep inner resonance, aligning ourselves with the rhythms of the universe.

Future Vision: The Role of Meditative Frequencies in Wellness

Imagine a future where meditative frequencies are a common part of mental health and wellness practices. Perhaps schools will use them to help students manage stress, workplaces to boost productivity and well-being, and healthcare providers to complement traditional treatments. The potential is vast, and as more people embrace this form of healing, we edge closer to a world where sound and wellness are inseparable.

Conclusion: Embrace the Healing Power of Frequency

Meditative frequencies invite us to journey inward, exploring realms of tranquillity and balance. They remind us that healing is a harmonious process, one that brings body and mind together in a state of peace. By integrating these sounds into our daily routines, we tap into a powerful source of calm that is always accessible—one that leads us back to ourselves.

Allow yourself the space and time to embrace meditative frequencies. Let each note and tone guide you, offering peace, clarity, and harmony. Healing is not a distant goal; it is a journey, and through sound, we find a path toward inner balance and serenity.

.

CHAPTER 35: CREATING A HEALING SOUND ENVIRONMENT AT HOME

Imagine stepping into a space where every sound soothes, every frequency heals, and every tone resonates with your body and soul. Creating a healing sound environment at home doesn't require complex equipment or a vast space —only an understanding of how to harness the power of frequencies and sound in a way that brings peace, balance, and rejuvenation into your daily life.

In this chapter, we'll guide you through the steps to set up a sound-based healing environment at home. By integrating scientific insights, practical tips, and inspiring ideas, we'll explore how to use frequencies to create a sanctuary of healing. You'll learn about the benefits of different frequencies, the equipment you might want to consider, and how to design a routine that supports both physical and mental well-being.

The Science Behind Healing Sound Environments

At the core of sound-based healing is the concept of

resonance. Think of every cell in your body as a part of a grand orchestra, each vibrating at its own unique frequency. When we experience stress or illness, it's as if certain cells fall out of tune, disrupting the body's natural harmony. Healing frequencies work like tuning forks, guiding these cells back into alignment.

Research shows that specific sound frequencies can influence brain wave activity, promote cellular repair, and reduce stress hormones. For instance, 432 Hz is associated with relaxation and well-being, while 528 Hz—often called the "miracle tone"—is known for its transformative effects, including DNA repair and emotional healing.

Designing Your Sound Sanctuary: Key Elements to Consider

Creating a healing sound environment requires a few essential elements: the right space, suitable sound sources, and a personalized routine. Here's a step-by-step guide to transforming your home into a healing haven.
Step 1: Choose Your Space

The ideal sound sanctuary is a place where you can be uninterrupted. It doesn't need to be an entire room—it could be a corner by a window, a section of your living room, or even a quiet nook in your bedroom. What matters is that the space feels safe, comfortable, and serene.

- Privacy: Choose a space with minimal outside noise to avoid distractions. If external sounds are unavoidable, consider noise-cancelling options or adding soundproofing materials.
- Ambiance: Use soft lighting, perhaps natural sunlight or warm, dimmable lights. Consider adding candles, essential oils, or calming decor elements like plants or tapestries that create a peaceful atmosphere.
Step 2: Select Sound Equipment

Your sound environment doesn't require high-end gear; even

basic equipment can be incredibly effective. Here are some options to consider based on your preferences and budget:

- Headphones: For immersive sound healing, invest in a pair of quality headphones, preferably noise-cancelling. This allows you to experience the frequencies deeply, blocking out external sounds.
- Speakers: If you prefer an open sound experience, a good set of speakers can fill the room with healing frequencies. Position them in a way that surrounds you with sound, creating an immersive atmosphere.
- Tuning Forks and Singing Bowls: For hands-on frequency healing, consider using tuning forks or singing bowls. These tools generate specific frequencies that can be directly applied to areas of your body or simply used to fill the room with calming resonance.
- Sound Apps and Playlists: Many sound therapy apps and streaming services offer playlists specifically designed for frequency-based healing. Search for playlists in the 432 Hz, 528 Hz, or 396 Hz ranges, which are popular for stress relief, transformation, and grounding.
Step 3: Choosing Your Healing Frequencies

Different frequencies offer distinct benefits. Here's a quick guide to help you choose the right frequencies for different aspects of healing:

- 432 Hz: Known for its relaxing and balancing properties, this frequency can help you destress and feel more grounded. It's a great option for winding down at the end of a busy day.
- 528 Hz: Often referred to as the "miracle tone," this frequency is associated with DNA repair, transformation, and love. Use it when you're looking to uplift your mood or release emotional blockages.
- 396 Hz: Known for promoting grounding and releasing fear, this frequency is ideal when you feel anxious or overwhelmed.

- 639 Hz: This frequency is often used for improving relationships, promoting harmony, and fostering emotional healing. It's perfect for meditative sessions focused on self-love and connection with others.

Step 4: Create a Routine

Consistency is key to experiencing the full benefits of a healing sound environment. Here's a simple routine that you can personalize to fit your lifestyle:

- Morning: Start the day with a 10-minute session at 528 Hz to energize and set a positive tone. You can use this time for visualization, setting intentions, or practicing gratitude.
- Afternoon: If you're feeling midday stress, take a 5-minute break with 396 Hz to help you reground and reset. Use headphones if you're in a public space or at work.
- Evening: Wind down with a 432 Hz session before bed. Combine this with a gentle body scan or deep breathing to fully relax and prepare for restful sleep.

Step 5: Mindful Listening and Intention Setting

Listening to healing frequencies isn't just about hearing the sound; it's about connecting with it. Practice mindful listening by closing your eyes and letting each note wash over you. Imagine the frequencies gently resonating with each cell in your body, bringing you into a state of harmony.

To deepen your practice, set a specific intention for each session. Whether you seek relaxation, emotional release, or mental clarity, infuse this intention into your listening. Letting your mind and body know your purpose can enhance the impact of the frequencies on your well-being.

Practical Tips for Enhancing Your Sound Environment

- Combine with Nature Sounds: Nature sounds like flowing water, birdsong, or rustling leaves can enhance your healing environment. Look for playlists that blend frequencies with these sounds to elevate your experience.

- Experiment with Essential Oils: Lavender, eucalyptus, and chamomile oils can heighten the relaxation effect. Try diffusing them in your space while listening to healing frequencies.

- Keep a Sound Journal: Track your experiences with different frequencies. Note how each one makes you feel, any shifts in mood or energy, and any personal insights you gain.

Reflective Insight: The Art of Healing Through Sound

Creating a healing sound environment at home is more than just arranging a space—it's an art that aligns your physical environment with your inner needs. Each session is a journey inward, where sound becomes a tool for self-discovery and self-care.

Philosophical Reflection: Consider how ancient cultures used sound in sacred rituals, healing ceremonies, and spiritual practices. By bringing this wisdom into our modern lives, we honour an age-old truth: sound and frequency have the power to heal not only our bodies but also our minds and spirits.

Future Vision: Sound Environments in Everyday Spaces

Imagine a future where sound-based healing environments are commonplace. Schools might integrate calming frequencies for students, workplaces may offer sound therapy breaks, and hospitals could incorporate sound healing into patient care routines. Sound environments could become as essential as lighting and air quality in promoting overall health.

Conclusion: Begin Your Journey

With your sound sanctuary in place, you are equipped to explore the profound benefits of frequency-based healing. Embrace each session as a sacred time of renewal, allowing sound to reconnect you to your natural state of balance and harmony. Healing is a journey, and each frequency is a step forward, leading you closer to peace, resilience, and well-being. Let your home resonate as a haven of sound, a place where healing harmonies gently guide you toward inner harmony.

PART IX: TECHNOLOGY-DRIVEN FREQUENCY HEALING

CHAPTER 36:
WEARABLE DEVICES
IN FREQUENCY
MEDICINE

In a world where stress, anxiety, and lifestyle diseases are more prevalent than ever, the integration of technology and frequency-based healing has become a beacon of hope. Wearable devices that harness the power of frequencies are now emerging as accessible, non-invasive tools that bring healing within arm's reach—literally. Imagine a device that, with a soft hum or gentle vibration, guides your body back to balance during a stressful day or helps you drift into restful sleep at night.

This chapter explores the revolutionary world of wearable frequency devices, breaking down their science, real-life applications, and future potential. We'll walk through how these devices work, present case studies of success, and offer tips on how to incorporate these tools into your life to enhance well-being.

The Science of Wearable Frequency Devices

The foundation of wearable frequency devices lies in bio-resonance—a principle suggesting that our bodies respond to certain frequencies. Just as each cell and organ vibrates at specific frequencies, imbalances or illnesses may cause parts of the body to "fall out of tune." By applying frequencies that resonate with the natural frequency of healthy cells, these devices aim to restore harmony at a cellular level.

Many wearable frequency devices employ low-level electromagnetic frequencies (EMFs) or microcurrent therapy to interact with the body. Some are based on pulsed electromagnetic field (PEMF) therapy, a technique that uses EMFs to encourage cellular repair. Others use infrared light or gentle electrical pulses to stimulate healing.

For example, devices targeting the alpha brainwave frequency range (8-14 Hz) can promote relaxation, creativity, and focus. Devices that emit frequencies in the beta range (13-30 Hz) can aid in alertness and mental clarity. Each frequency interacts uniquely with the body, allowing these devices to be tailored to meet specific needs, whether that's calming anxiety or relieving chronic pain.

The Everyday Benefits: Real-World Applications

Let's look at how wearable frequency devices are being used to address some of the most common health concerns of today.

1. Stress and Anxiety Management

One of the most compelling applications of wearable frequency devices is in managing stress and anxiety. Through devices that emit frequencies known to promote relaxation (such as 10 Hz), individuals can experience immediate calming effects. Research has shown that these devices can help decrease cortisol levels, the hormone associated with stress, while promoting the release of serotonin, our natural mood stabilizer.

Case Study: Sarah, a 35-year-old marketing professional, faced overwhelming work-related stress. After integrating a frequency-emitting wearable into her routine, she reported feeling more at ease and able to tackle her workload without feeling overwhelmed. Sarah's device pulsed at a steady 10 Hz during her workday, helping her maintain focus and calm in a high-stress environment.

2. Improving Sleep Quality

For those struggling with insomnia or poor sleep quality, wearable devices that emit frequencies in the delta range (1-4 Hz) are especially beneficial. The delta range is naturally produced by the brain during deep sleep, and introducing this frequency before bed can guide the body into a state of relaxation, priming it for restorative sleep.

Case Study: Tom, a 42-year-old engineer, battled with insomnia for years. After trying various medications and therapies, he turned to a frequency-based wearable that emitted a low-frequency pulse in the delta range. Over several weeks, Tom found that he could fall asleep faster and stay asleep longer. His wearable became a non-intrusive, safe alternative to sleep medications, allowing him to wake up feeling more refreshed.

3. Pain Relief and Inflammation Reduction

Devices using frequencies in the beta range are beneficial for managing pain and inflammation. This frequency range helps stimulate the body's natural pain-relief processes, including the release of endorphins. Wearable devices designed for pain relief can be worn directly over the area of discomfort, delivering targeted healing frequencies to the affected tissues.

Case Study: Janet, a 50-year-old with arthritis, found new freedom of movement after using a wearable device that emitted beta-range frequencies. Instead of relying solely on medication, Janet would wear her device around her knee, which helped reduce inflammation and relieve pain, giving her a natural, non-invasive way to manage her symptoms.

Practical Guide: How to Incorporate Wearable Frequency Devices into Your Life

Incorporating wearable devices into your daily routine is simple but requires a bit of practice to get the most benefit. Here's a guide to getting started:

- Choosing the Right Device: Select a device based on your specific needs. If stress management is your goal, look for wearables that offer low-frequency EMFs in the alpha range. For pain management, choose a device that targets the beta range.
- Set a Routine: Use your device at consistent times each day. For instance, stress-relieving devices can be worn during work hours, while sleep-enhancing devices work best 30 minutes before bed.
- Track Your Experience: Keep a journal of how you feel after each session. This will help you identify which frequencies and routines offer the most benefit, allowing you to customize your experience further.

Visionary Insight: The Future of Wearable Frequency Medicine

Wearable frequency devices represent the convergence of science, technology, and holistic health. Imagine a future where these devices become as ubiquitous as fitness trackers, seamlessly integrated into daily life. With advances in artificial intelligence, future wearables could adjust frequencies in real-time based on biofeedback, offering personalized healing sessions. Hospitals might use frequency wearables as part of post-surgical recovery, and workplaces could offer wearable devices to employees for mental clarity and stress management.

Reflective Thought: We are moving toward a world where healing frequencies are not confined to therapy sessions or clinics but become part of our lifestyle. This shift challenges traditional perspectives on medicine, emphasizing a preventative, harmonious approach that empowers individuals to take control of their health.

Navigating Scepticism with Openness and Inquiry

As with any emerging technology, scepticism toward wearable frequency devices is natural. Questions about long-term effects, consistency of outcomes, and scientific validation are valid and necessary for the field to advance responsibly. Yet, early studies and personal testimonials reveal a promising potential.

It's essential to maintain an open mind, acknowledging both the exciting possibilities and the need for rigorous research. Frequency medicine, after all, taps into the body's natural rhythms and vibrations—elements long recognized in traditional healing but only recently explored through modern science.

Closing Reflections: Empowering Healing at Your Fingertips

The journey toward wellness is deeply personal, and wearable frequency devices are but one path on that journey. For some, they may offer the missing link between modern living and inner peace, guiding them to a place of calm and resilience amidst life's demands.

Imagine a world where, in moments of stress, you can reach for a device that aligns you back to balance, or where a gentle pulse on your wrist supports your body's recovery from pain. Wearable devices represent a new age in health—a bridge between our body's natural frequencies and the healing harmonies waiting to be discovered within.

Through these devices, frequency medicine becomes a part of our everyday lives, putting the power of healing right at our fingertips. As we move forward, the dream of a world where frequency-based wellness becomes as routine as morning exercise or meditation draws closer, promising a future where health is truly harmonized.

CHAPTER 37: THE FUTURE OF FREQUENCY APPS AND PLATFORMS

As technology and wellness converge, our smartphones, tablets, and even smart speakers are becoming gateways to frequency-based healing. While the journey of frequency medicine started in dedicated therapy rooms and research labs, it's now available at our fingertips, anywhere and anytime. With this shift, frequency apps and platforms are quickly emerging as essential tools in promoting mental, physical, and emotional well-being.

This chapter explores how digital applications are harnessing the power of sound frequencies, vibrations, and electromagnetic fields to bring the benefits of frequency medicine to the masses. We'll look at the science behind these apps, delve into real-life examples, and envision a future where sound-based healing is a staple in personal health routines.

The Science Behind Frequency Apps: Why It Works

Before diving into the specific applications, it's essential to understand the science driving these digital tools. The foundation of frequency apps lies in principles such as resonance and entrainment. Resonance is the idea that objects can influence each other's vibrational state, much like how a tuning fork can cause another nearby tuning fork to vibrate in harmony. Similarly, frequency apps aim to create an environment where the body and mind "sync" with healing frequencies.

Frequency apps utilize sound waves, often in the form of binaural beats and isochronic tones, to influence brain wave states. Binaural beats, for instance, use two slightly different frequencies in each ear, creating a "phantom" third tone in the brain that encourages specific mental states. For example, theta waves (4-8 Hz) promote relaxation and creativity, while alpha waves (8-14 Hz) encourage calm alertness.

Through careful design and precise control of these frequencies, developers are able to target different aspects of well-being, offering tools for relaxation, pain management, focus, sleep, and more.

Real-World Applications: How Frequency Apps Are Transforming Wellness

As these apps and platforms grow in popularity, they're quickly becoming a part of daily wellness routines for many people. Here are a few areas where frequency apps are already making an impact.

1. Stress and Anxiety Relief

One of the primary reasons people turn to frequency apps is to manage stress and anxiety. Apps like Calm and Brain.fm use carefully curated soundscapes and binaural beats to help users shift into relaxed states. By offering music or tones in the alpha and theta frequency ranges, these apps help users unwind, reduce cortisol levels, and increase serotonin, creating a feeling of ease and tranquillity.

Case Study: Emma, a 29-year-old software engineer, struggled with work-related anxiety. By setting aside 10 minutes a day with an app that uses binaural beats in the 10 Hz alpha range, she found herself more focused, less anxious, and more resilient to stress. This small but regular practice made a noticeable difference in her mental health.

2. Enhancing Focus and Productivity

Frequency apps are also becoming popular tools for boosting productivity. With tracks that stimulate beta waves (13-30 Hz), some apps, like Focus@Will and Brainwave, aim to help users stay sharp and attentive. Research suggests that these frequencies can heighten alertness, making them ideal for tasks that require sustained focus, such as studying or work projects.

Case Study: Alex, a college student, found himself easily distracted while studying. After discovering an app that offered beta-range sounds, he began using it during his study sessions. The subtle background frequency helped him stay on track, turning study sessions into immersive experiences.

3. Improving Sleep Quality

Apps that promote delta wave frequencies (1-4 Hz), which are associated with deep sleep, are helping people achieve restful, rejuvenating sleep. By gradually guiding the brain into delta states, these apps can assist in falling asleep faster and staying asleep longer. Apps like Pzizz and Sleepytime incorporate these frequencies to create soothing sounds that act as digital lullabies.

Case Study: Sarah, a 41-year-old mother of two, had difficulty falling asleep due to her busy lifestyle. After trying various remedies, she incorporated a sleep app into her nightly routine. Using an audio track with a delta wave emphasis, she found herself able to relax and drift into sleep more naturally.

DIY Guide: Incorporating Frequency Apps Into Your Routine

Using frequency apps can be as simple as pressing play, but to get the best results, it's essential to establish a routine and choose frequencies suited to your goals. Here's a quick guide:

1. Define Your Goal: Do you need help with stress, focus, sleep, or pain relief? Choose an app or frequency range that aligns with your objective.

2. Consistency is Key: Try to use the app daily at the same time to maximize effectiveness. For example, listen to relaxation frequencies during your lunch break or focus frequencies before tackling a big task.

3. Use Headphones for Binaural Beats: Many apps, especially those that use binaural beats, require headphones for the best effect. Ensure you're in a quiet space to enhance the experience.

4. Start with Short Sessions: Begin with 5-10 minute sessions to get comfortable, then gradually increase the time as you become accustomed to the frequencies.

5. Combine with Other Practices: Frequency apps can be a powerful addition to meditation, yoga, or mindfulness practices. Try playing focus tracks during yoga sessions or sleep tracks while meditating before bed.

Looking Forward: The Future of Frequency Apps

The potential for frequency apps is immense. Imagine a future where your device doesn't just play sounds but actively measures your heart rate, stress levels, and brain activity to tailor frequencies in real-time. Using biofeedback and AI, the next generation of frequency apps could provide hyper-personalized healing, creating soundscapes designed to address your unique needs at any given moment.

As VR and AR technology advance, we may also see virtual reality environments that combine visual stimuli with sound frequencies to deepen healing experiences. Imagine meditating in a virtual forest, with ambient nature sounds tuned to relaxing frequencies, or working in a virtual study with tones optimized for focus.

Addressing Scepticism: A Balanced Approach

While the benefits of frequency-based apps are supported by preliminary research, they are still an emerging area of science. It's natural to approach these technologies with some level of scepticism. Not all claims in the app marketplace are fully backed by scientific studies, so it's essential to choose apps with transparency about their methods and a commitment to ongoing research.

Embrace these apps with an open but discerning mind. Test different apps and notice how they affect your mental state, but remember that frequency apps are supplements to well-being, not replacements for professional medical advice.

Reflective Thoughts: A New Era of Accessible Healing

In a world where digital devices often fuel stress, the emergence of frequency apps offers a refreshing alternative. These platforms remind us that our devices can be allies in our journey toward better health and harmony, not just sources of information and entertainment. With the simple act of listening, we can transform our smartphones into tools of healing, opening the door to a future where sound and frequency are as essential to self-care as a healthy diet or regular exercise.

As technology advances, the line between wellness and convenience will continue to blur, bringing the wisdom of frequency medicine into the palm of our hands. It's an invitation to explore the healing power of sound—a reminder that amidst the noise of the digital age, harmony is just a frequency away.

PART X: THE FUTURE OF FREQUENCY-BASED MEDICINE

CHAPTER 38:
FREQUENCIES IN
PREVENTATIVE
MEDICINE

Imagine a world where we could prevent chronic diseases, not just treat them. A world where technology tunes the body like an orchestra, ensuring that each cell resonates in harmony, preventing illness before it takes root. This chapter delves into the potential of frequency-based healing as a powerful tool in preventative medicine, exploring the possibilities of a future where frequencies are the foundation of health.

Introduction: Prevention as the New Frontier

Modern medicine has achieved remarkable feats, but much of it is reactive—addressing disease after it has already disrupted the body's natural harmony. Frequency-based medicine, however, offers a revolutionary vision of healthcare that targets the underlying vibrations and patterns that may contribute to disease long before symptoms appear. Picture your body as a vast musical instrument, with each organ and cell vibrating to a specific frequency. When one part falls out of tune, the entire symphony is affected, and eventually, illness may emerge. But what if we could continuously retune ourselves, creating an internal harmony that wards off disease?

The Science of Frequencies and Cellular Health

At a cellular level, the body operates in frequencies, a fact that science has only begun to explore deeply. Cells communicate through bioelectric signals, transmitting information and energy. Research shows that certain frequencies can influence cellular processes such as growth, repair, and immune function. When cells resonate at optimal frequencies, they're better equipped to fend off stressors, whether they're toxins, pathogens, or emotional stress.

Preventative frequency medicine thus seeks to establish a baseline of health by continuously engaging the body's cells in "tuning" practices that keep them in their ideal energetic states. Techniques like pulsed electromagnetic fields (PEMF), infrared light therapy, and sound wave exposure could become core components of wellness protocols, helping to fortify the body against chronic conditions such as inflammation, autoimmune disorders, and even metabolic imbalances.

Case Study: Frequency Therapy as a Shield Against Illness

Consider the case of Mark, a 45-year-old professional who, like many of us, faced relentless stress and fatigue. After a series of minor health issues and rising blood pressure, he decided to adopt a preventative approach using frequency therapy. Under the guidance of a frequency-based practitioner, Mark started a regimen of PEMF sessions twice a week, combined with daily listening sessions of healing frequencies designed for cellular repair and stress relief.

Over six months, not only did Mark's blood pressure stabilize, but he also reported improved sleep, higher energy, and mental clarity. He described feeling as if his body was "lighter," as if the weight of stress had dissipated.

Remarkably, regular screenings showed reduced markers of inflammation, which is often linked to chronic conditions like heart disease and diabetes. Mark's experience reflects how frequency therapy may not only heal but also serve as a preventative "shield," fortifying the body against the strains of modern life.

Practical Tips: Building Frequency Therapy Into Preventative Routines

1. Daily Sound Baths: A simple but powerful preventative tool is to incorporate daily sound baths with frequencies known for relaxation and cellular health, such as 528 Hz, which is thought to encourage DNA repair. Spend 10-15 minutes a day in a relaxed environment listening to these frequencies to foster resilience and reduce stress.

2. PEMF Devices: Small, portable PEMF devices are now available for home use, offering targeted electromagnetic fields to stimulate cellular health. For a holistic approach, use a PEMF device on key areas like the abdomen and spine, as these regions are often stress-sensitive and can benefit from regular "tuning."

3. Mindful Breathing With Frequencies: Combine frequency sessions with mindful breathing exercises to promote heart coherence—a state where the heart, mind, and emotions synchronize, creating physiological resilience. Apps like HeartMath can be used alongside sound frequencies, helping align breathing with the body's natural rhythms.

4. Routine Light Therapy: Low-level laser or LED therapy sessions a few times a week can offer a preventative "boost" by penetrating cells with beneficial wavelengths. Red light, in particular, promotes mitochondrial function, which is essential for energy production and cellular repair.

The Future of Preventative Frequency Medicine: A

Personalized Approach

Advances in technology could soon make it possible to monitor and adjust our vibrational health in real time, detecting subtle imbalances before they manifest as illness. Imagine a wearable device that continuously monitors your body's frequencies, identifying "misalignments" and suggesting corrective actions. Such devices could trigger a soothing frequency when you're stressed, or a healing wave when early signs of inflammation appear.

Artificial intelligence could also play a significant role in tailoring frequency therapy to individual needs. An AI-based health platform might analyse a user's unique energetic profile, factoring in genetics, lifestyle, and environment to create a personalized frequency regimen. This could include customized soundscapes, PEMF sequences, and light therapy patterns—all dynamically adapted to optimize cellular health and prevent disease.

A Reflective View: The Balance Between Science and Spirit

While frequency-based preventative medicine offers exciting potential, it's essential to view it as part of a holistic approach to health. True wellness doesn't come solely from external interventions; it also requires mindfulness, intention, and an inner commitment to harmony. Frequencies are a medium through which we can interact with the body's innate healing intelligence, but they're most effective when integrated with other wellness practices like balanced nutrition, physical activity, and emotional well-being.

Addressing Scepticism: A Rational Perspective

As with any emerging field, it's natural for frequency medicine to face scepticism, particularly in preventative applications. However, research is growing, with promising studies exploring its impacts on stress, inflammation, and even longevity. Although more rigorous clinical trials are needed, early findings suggest that frequencies have genuine, measurable effects on cellular and systemic health. For those sceptical , consider frequency therapy as one layer of a preventative approach, complementary to other evidence-based practices.

Closing Thoughts: A Symphony of Preventative Health

Preventative medicine's future may well be as simple as listening—to the body, to frequencies, and to the world around us. When we harmonize our cells through resonance, we're not just avoiding disease; we're cultivating a vibrant, resilient state of being. Think of frequency medicine as the symphony that tunes each cell, each system, into a state of readiness, strength, and peace. As science and spirit continue to bridge in this field, we can look forward to a world where the tones of preventative health guide us towards longevity and well-being.

Frequency therapy, in its essence, reminds us that we are part of a universal rhythm—a melody that, when aligned with intention and science, can be a profound force for life and vitality. In this symphony of frequencies, our health becomes the harmony, inviting us to embrace a future where illness prevention is not a reactive struggle but a proactive dance with life's most fundamental vibrations.

CHAPTER 39: INTEGRATING AI WITH FREQUENCY MEDICINE

Imagine a future where each sound, vibration, and frequency that touches your body is meticulously tailored to your unique biological rhythm. No more generic treatments or one-size-fits-all therapies. Instead, every sound bath, PEMF session, and therapeutic vibration is personalized, guided by artificial intelligence to support your body's specific needs in real-time. This chapter delves into how AI has the potential to transform frequency medicine, making treatments more precise, adaptable, and effective than ever before.

Introduction: A Personalized Symphony

Today's medicine is largely reactive, addressing symptoms after they emerge. But what if, instead, we could use frequency-based therapies to prevent disease, maintaining health by continuously fine-tuning our bodies like a symphony that's always in harmony? This vision is no longer far-fetched. The integration of AI with frequency medicine promises an era where the body's frequencies are constantly monitored and adjusted, creating a personalized preventative healthcare experience.

Imagine an AI system that learns and adapts to your body, analysing your biofeedback—heart rate variability, skin temperature, stress markers, and even brainwave activity—to detect shifts in your physical or emotional state. Based on these insights, the system might recommend specific frequencies to help balance, heal, or stimulate. With the marriage of AI and frequency medicine, the boundaries of healing extend beyond today's comprehension, moving us closer to a world where illness is prevented before it manifests.

How AI Tailors Frequency-Based Healing

AI in frequency medicine relies on biofeedback—biological data that tells the story of how your body is responding in real time. By gathering and interpreting this data, AI can identify the frequencies that align best with your current needs. Here's a breakdown of how this works:

1. Data Collection and Analysis: Advanced AI systems gather data through wearable devices that measure heart rate, brainwave activity, skin conductance, and more. This continuous data feed gives the AI insight into your physiological state, stress levels, and energy balance.

2. Pattern Recognition: Using machine learning, AI systems can recognize patterns in your biofeedback. For instance, if certain frequencies consistently lower your stress markers, the AI will learn to recommend those frequencies when it detects signs of stress.

3. Real-Time Adjustment: As your body changes moment to moment, so do its needs. AI allows for dynamic adjustments, modifying the frequency, intensity, and duration of therapy in real-time based on your ongoing biofeedback.

4. Predictive Capabilities: One of AI's most powerful features is its ability to anticipate. By recognizing early warning

signs, AI can suggest preventative treatments to maintain equilibrium and prevent issues before they arise.

Case Study: Sarah's Journey with AI-Guided Frequency Therapy

Sarah, a 38-year-old mother of two, struggled with chronic anxiety and fatigue that standard treatments couldn't fully alleviate. Her doctor recommended trying AI-assisted frequency medicine, where a wearable device would monitor her biofeedback and send this data to an AI platform. This AI, trained specifically for stress and energy management, analysed her body's rhythms, adjusting frequency recommendations on the fly.

On days when Sarah's biofeedback showed high stress and elevated cortisol, the AI would automatically suggest low-frequency sounds and PEMF sessions designed to calm her nervous system. When her biofeedback indicated a low energy state, the AI switched to higher frequencies aimed at stimulating cellular activity and energy production.

Over the course of several months, Sarah noticed remarkable improvements. Her anxiety diminished, her energy levels stabilized, and she reported a greater sense of peace. Her body, continuously tuned like a musical instrument, began to heal in ways she hadn't experienced before. Sarah's experience is a testament to the power of AI-guided frequency therapy, highlighting the profound potential for personalized, responsive healing.

Practical Tips: Introducing AI-Assisted Frequency Healing

While fully AI-integrated systems are still on the horizon, you can start exploring biofeedback-driven frequency therapy with these accessible methods:

1. Wearable Technology: Consider devices that measure heart rate variability (HRV) and skin conductance, as these

can provide valuable biofeedback. Some apps can help you match these metrics with frequency suggestions, offering a glimpse into AI's potential for real-time personalization.

2. Biofeedback Apps: A range of mobile apps can track stress levels, sleep patterns, and other wellness markers, allowing you to experiment with frequency therapy based on your body's changing needs.

3. Mindful Frequency Sessions: Start a journal that records how you feel before and after exposure to certain frequencies. By tracking patterns, you can begin to personalize your own frequency treatments, making note of what works best under different conditions.

The Future of AI in Frequency Medicine: A Harmonized World

In the near future, AI systems could merge with frequency therapy in ways that create a seamless healthcare experience. Imagine a world where:

- Home AI Systems Monitor Health: Equipped with sensors, your home environment could analyse your stress levels, energy, and mood. Based on this analysis, it might play specific frequencies through your speakers to help maintain your balance.

- Predictive Health: By detecting subtle shifts in biofeedback over time, AI systems could predict potential health issues long before they appear, recommending preventive therapies.

- Fully Personalized Frequency Therapy: Just as personalized medicine tailors drugs to your genetic profile, AI could tailor frequencies to your unique physiological and emotional landscape, creating an entirely personalized treatment plan.

Reflective Insight: The Symphony of Healing

As humans, we are intricate compositions of energy and matter, vibrating in complex rhythms that reflect our health, emotions, and essence. Frequency medicine taps into these rhythms, reminding us that healing isn't just about eliminating symptoms—it's about harmonizing every aspect of our being. AI, when integrated with frequency medicine, doesn't replace human intuition but enhances it, giving us the tools to better understand and nurture ourselves.

Addressing Scepticism: Balancing Hope with Science

The concept of AI-integrated frequency medicine might sound futuristic, but it's grounded in emerging science. Studies on biofeedback and frequency-based therapies show promising results, and while AI's full integration is still in development, current technology already hints at what's possible. As with all innovations, scepticism is natural. However, this fusion of AI and frequency therapy represents a balanced, science-backed approach that honours both the potential of new technology and the wisdom of ancient healing practices.

Final Reflection: A Vision for Tomorrow

Imagine waking up in a world where health is not a battle to fight but a state to maintain. With AI and frequency medicine as part of our daily lives, we may one day create a society where wellness is woven into the very fabric of our existence. This journey toward health through harmony —where technology and the body's natural resonance work in concert—offers a vision that is both revolutionary and deeply human.

In this future, AI will guide us not as a rigid system but as a dynamic partner, adapting to our needs and helping us sustain a state of vibrant health. As we tune into the wisdom of frequencies with the precision of AI, we open ourselves to a new realm of preventative, personalized, and profoundly resonant healing.

PART XI: FREQUENCY HEALING FOR SPECIFIC PHYSICAL CONDITIONS

CHAPTER 40: FREQUENCIES FOR MIGRAINE RELIEF

Imagine being in the grip of a migraine. The world feels distorted, sounds are too sharp, lights too intense, and pain throbs behind your eyes, refusing to relent. Now, picture slipping on a pair of headphones or lying in a room filled with soothing, low-frequency vibrations. As the frequencies wash over you, the pain gradually loosens its grip, your muscles relax, and your mind enters a state of calm. This isn't a fantasy—it's the potential of frequency-based therapy in migraine relief.

This chapter explores the transformative potential of sound and vibration therapies in managing migraines and chronic headaches, merging scientific insights with accessible, actionable steps. We'll explore how specific frequencies can impact the body and mind, soothing the intensity of migraines, and delve into real-life stories of relief and renewal.

Understanding Migraine: An Overwhelmed Symphony

The human brain is like an orchestra, with neurons firing and interacting to create harmonious mental states. But when migraines strike, it's as if the orchestra falls out of sync, with neurons firing too intensely, leading to a storm of over-activity that manifests as pain. Research shows that migraines are often triggered by neurological hypersensitivity and inflammation. Factors like stress, light sensitivity, hormonal changes, and even certain sounds can set off this imbalance, resulting in intense, pulsating pain.

Sound and vibration therapies offer a unique approach to restore harmony to this overwhelmed system. By introducing controlled frequencies, these therapies act as "conductors" to the brain's orchestra, gently nudging it back into rhythm. Frequencies help to quiet overactive neural pathways, reduce inflammation, and shift brainwaves into states that foster relaxation and relief.

The Science Behind Frequency Therapy for Migraines

1. Brainwave Entrainment: One of the most compelling methods to reduce migraines is brainwave entrainment. This involves using specific sound frequencies to guide the brain into desired states, like deep relaxation or sleep. For example, alpha waves (8-12 Hz) are associated with calmness, while delta waves (0.5-4 Hz) correspond with deep restorative sleep. Listening to alpha or delta frequencies during the onset of a migraine can help alleviate pain by promoting relaxation and interrupting the brain's overactive pathways.

2. Vibration Therapy: Low-frequency vibrations can reduce muscle tension, which is often exacerbated during migraines. By applying vibration to areas like the temples or neck, muscle contractions are minimized, leading to decreased pain. Vibration also increases blood flow to affected areas, helping to reduce the intensity of headaches.

3. Sound Baths and Binaural Beats: Sound baths—sessions in which individuals are "bathed" in sounds from instruments like Tibetan singing bowls or gongs—create a powerful sense of tranquillity. The resonant tones of these instruments have been shown to lower cortisol (stress hormone) levels, a common trigger for migraines. Binaural beats, where two

slightly different frequencies are played in each ear, have a similar calming effect, gently nudging the brain into a more relaxed state, relieving tension and pain.

Case Study: Emma's Path to Relief

Emma, a 42-year-old teacher, had suffered from chronic migraines for over a decade. Her migraines were often triggered by the stresses of work, and traditional medications brought only temporary relief with numerous side effects. Feeling trapped by the limitations of conventional treatments, she decided to explore alternative therapies, eventually discovering sound frequency therapy.

Emma began using a combination of binaural beats and sound baths during her migraine episodes. She found that listening to binaural beats in the alpha and delta range helped to shorten the duration of her migraines, and regular sound baths helped prevent them by maintaining her stress levels. Over time, she noticed a reduction in the frequency and intensity of her migraines, and her reliance on medication diminished significantly.

Emma's experience highlights how frequency-based therapies can offer a viable alternative or complement to conventional treatment, reducing both symptoms and dependency on medication.

Practical Tips for Using Frequency Therapy for Migraine Relief

For readers seeking to incorporate frequency therapy into their migraine management, here are some actionable steps:

1. Explore Binaural Beats: Search for binaural beats designed for relaxation, focusing on alpha (8-12 Hz) and theta (4-7 Hz) ranges. Try listening to these during the onset of a migraine, with comfortable headphones, in a quiet space.

2. Try Sound Baths: Look for a local sound bath session or explore online recordings featuring Tibetan bowls, gongs, or chimes. Aim to relax in a dark room and let the resonant frequencies wash over you. Many migraine sufferers find this experience deeply soothing, as it lowers stress and reduces muscle tension.

3. Experiment with Vibration Therapy: Devices like handheld massagers can be set to a low frequency and applied gently to the temples or base of the skull. Be cautious with intensity—only use low, gentle vibrations for migraine relief. Alternatively, some people find relief by lying on a vibration mat or bed, which offers whole-body relaxation.

4. Create a Calm Sound Environment: Ambient sounds in your living space can also influence migraine susceptibility. Use white noise machines, nature sound playlists, or calming music in your environment to maintain a peaceful backdrop that can prevent stress from accumulating.

Visionary Insights: The Future of Frequency Medicine in Migraine Care

As technology evolves, so too does the potential for frequency-based migraine treatments. Imagine a future where wearable devices monitor your neurological and physiological states in real time. Upon detecting early signs of a migraine, these devices could automatically play personalized frequencies tailored to disrupt the migraine's onset, even before the pain fully manifests.

AI-driven biofeedback systems may also play a role, adjusting frequencies dynamically based on real-time feedback from your body. This would allow for highly personalized treatments that adapt to each individual's unique needs, offering targeted relief and helping prevent migraines more effectively.

Reflecting on Healing Beyond Pain

Frequency-based therapies challenge us to rethink pain management and wellness. They invite us to view healing as a process that aligns mind and body rather than merely masking symptoms. When we experience pain, it's not simply a sensation to eliminate but a signal that our body's harmony has been disrupted. By tuning into frequencies, we honour the body's call for balance, promoting healing in a way that resonates with our entire being.

Migraines, though debilitating, are also a reminder of the sensitivity and interconnectedness of our brain and nervous system. Sound and vibration therapies remind us that healing can be gentle, immersive, and profoundly resonant, encouraging us to find harmony in even the most chaotic moments of pain.

Balanced Scepticism: Addressing Doubts and Encouraging Exploration

As with all emerging therapies, it's natural to approach frequency-based migraine treatments with a sense of curiosity and caution. Research into these therapies is growing but still in the early stages. For many, the relief experienced through frequency therapies is undeniable, but individual responses can vary.

A balanced approach involves integrating these therapies as part of a broader self-care and treatment plan, perhaps in conjunction with medical advice. Scepticism doesn't mean dismissing new possibilities; rather, it's about exploring with an open, informed mind.

Final Reflection: Harmonizing Healing

Frequency-based therapies offer more than just pain relief— they offer an invitation to connect with the body's rhythms, acknowledging that we are dynamic, resonant beings. In addressing migraines with sound and vibration, we engage with a form of healing that resonates on multiple levels—physical, mental, and emotional. This approach represents a shift from managing symptoms to embracing a holistic path of healing and self-care. In the symphony of healing, frequency medicine offers a powerful, harmonizing note, empowering each of us to find relief, resilience, and harmony within ourselves.

CHAPTER 41: HEALING DIGESTIVE DISORDERS WITH FREQUENCIES

In our fast-paced world, digestive disorders have become increasingly common, from irritable bowel syndrome (IBS) and acid reflux to chronic issues like Crohn's disease. Often linked to stress, lifestyle, and even our emotional states, these conditions challenge traditional medicine, pushing us to explore more integrative approaches. This chapter uncovers the fascinating role of frequency-based therapies in digestive health, addressing the gut-brain connection, stress management, and cellular regeneration.

Imagine, if you will, the digestive system as an intricate symphony. Each organ plays its part, each cell contributes its note, and together, they create the harmony that is our body's digestive process. But when stress, inflammation, or lifestyle imbalances disrupt this harmony, frequencies can act as a "tuner," helping to bring the body's natural rhythm back into balance.

The Gut-Brain Connection: Frequencies and the "Second Brain"

Science has long shown that the gut and brain are deeply

interconnected. The gut is often called the "second brain" due to its vast network of neurons, neurotransmitters, and its ability to influence mood and mental health. This gut-brain axis is a two-way communication system where the health of one influences the other.

When we experience stress or anxiety, the brain sends signals to the gut, often leading to issues like cramping, bloating, or even nausea. Conversely, digestive discomfort can influence mood, creating a cycle of tension that exacerbates both mental and physical symptoms. Frequency-based therapies, particularly those using sound and electromagnetic frequencies, offer a unique approach to breaking this cycle.

Example and Analogy: Imagine the gut and brain as two musicians in an orchestra. If one falls out of tune, the other will struggle to stay harmonious. Frequencies can help "tune" the gut-brain axis, restoring balance between these two systems.

How Frequencies Aid Digestive Health

Frequency therapies act on multiple levels to support digestive health. Let's explore some of the key ways frequencies interact with our body's digestive processes:

1. Reducing Stress and Anxiety with Sound Frequencies
Studies show that chronic stress is one of the biggest culprits behind digestive disorders. Sound frequencies, particularly those in the range of 3-8 Hz (delta waves) and 8-14 Hz (alpha waves), are known to reduce anxiety, calm the mind, and induce relaxation. When listening to these frequencies, the brain's activity synchronizes with the sound waves, naturally guiding the listener into a more relaxed state. In turn, this relaxation reduces stress on the gut, allowing it to perform its functions more smoothly.

Case Study: Sara, a 35-year-old mother of two, struggled with IBS for years, largely triggered by her stressful lifestyle. After integrating 20 minutes of daily sound therapy using alpha wave frequencies, she noticed a significant improvement in her symptoms. The consistent use of these frequencies reduced her anxiety levels, which had a direct, positive impact on her digestive health.

2. Supporting Cellular Regeneration with Pulsed Electromagnetic Field (PEMF) Therapy
PEMF therapy uses low-frequency electromagnetic fields to stimulate cell repair and regeneration. When directed at the abdominal area, PEMF has shown promise in enhancing blood circulation, reducing inflammation, and promoting the healing of tissues within the digestive tract. This can be particularly beneficial for conditions like Crohn's disease and ulcerative colitis, where inflammation and tissue damage are significant factors.

3. Binaural Beats for Gut-Brain Synchronization
Binaural beats work by playing two slightly different frequencies in each ear, resulting in a perceived third frequency that the brain aligns with. For gut health, binaural beats in the theta range (4-8 Hz) can be particularly effective, as they help synchronize the gut and brain, promoting relaxation and reducing stress. Theta waves have been linked to deep relaxation, reduced anxiety, and improved emotional resilience—all essential for a healthy digestive system.

DIY Tip: To try binaural beats at home, find a quiet space, put on headphones, and listen to theta-range binaural beats for 10-20 minutes daily. This practice can help calm the nervous system and, over time, support healthier gut function.

Real-World Application: A Journey of Relief

Let's meet Daniel, a 48-year-old software developer who had

battled chronic acid reflux and stomach pain for nearly a decade. Frustrated by the limited relief he found in medications, he turned to frequency-based therapies. Daniel began with daily sessions of sound therapy, listening to calming frequencies designed to reduce inflammation and promote relaxation. He also incorporated PEMF therapy once a week, using a device on his abdominal area.

Within a month, Daniel noticed that his acid reflux episodes were less frequent and less intense. The holistic approach of frequency therapy allowed him to reduce his medication and, in time, experience a significant improvement in his quality of life. Daniel's story highlights the power of frequencies to support the body's natural healing processes, particularly when traditional treatments fall short.

Practical Guide: Using Frequencies for Digestive Health

For readers looking to incorporate frequency therapies into their digestive health regimen, here are a few practical approaches:

1. Sound Therapy: Seek out sound therapy recordings specifically tailored to digestion. These might include sounds in the delta and theta ranges, which promote relaxation. You can find these frequencies in apps or playlists focused on healing sounds. Set aside a few minutes daily to listen, especially after meals or during stressful moments.

2. PEMF Therapy: For those with access to a PEMF device, applying it to the abdominal area can help reduce inflammation and support cellular repair. Aim for 10-20 minutes a few times a week, adjusting based on how your body responds. Be sure to consult a practitioner to understand proper usage and frequency.

3. Mindful Breathing with Binaural Beats: Combine binaural beats in the theta range with mindful breathing. Breathe

deeply and slowly as you listen, allowing your body to enter a state of calm. This practice can alleviate stress and create a positive feedback loop between the brain and gut.

Looking Forward: Frequencies in Preventive Digestive Health

Imagine a future where individuals use frequency-based therapies as a preventive measure for digestive health. As research advances, frequency-based wellness practices could become as common as yoga or meditation, helping to maintain digestive harmony before issues even arise. Wearable devices may eventually detect stress or digestive imbalance and play personalized frequencies to prevent discomfort, merging technology with the body's natural rhythms in a seamless dance of balance and wellness.

Reflective Thoughts on Healing Harmony

Our digestive system is more than just a series of organs—it's a mirror reflecting our mental and emotional well-being. By turning to frequency-based therapies, we honour the mind-body connection, nurturing the gut and, by extension, our overall health. These therapies invite us to embrace a deeper understanding of health, one that moves beyond simply addressing symptoms and seeks to restore harmony across all levels of our being.

Balanced Scepticism and Embracing New Modalities

As with any emerging field, it's essential to approach frequency-based therapies with open-minded curiosity tempered by rational scepticism. While anecdotal evidence and preliminary studies are promising, frequency-based digestive therapies are still under exploration. We encourage readers to integrate these approaches thoughtfully and consult healthcare providers as part of a comprehensive approach to digestive health.

Conclusion: Rediscovering Balance

Frequency-based therapies offer a profound yet gentle approach to restoring digestive health. By engaging the body's natural rhythms and supporting the gut-brain connection, they provide a path to healing that is as resonant as it is restorative. In this journey toward health, may we remember that harmony within ourselves is both a science and an art, one that frequencies are uniquely suited to help us rediscover.

CHAPTER 42: MANAGING INFLAMMATION THROUGH VIBRATIONAL THERAPY

In our pursuit of health and healing, few discoveries have been as profound as the understanding of inflammation. Often called the "silent killer," inflammation lies at the root of numerous chronic diseases, including arthritis, cardiovascular issues, autoimmune disorders, and even cancer. The revelation that inflammation can be managed, and even reduced, through vibrational therapy unveils a new horizon for non-invasive, holistic medicine.

Imagine your body as a complex symphony where every cell, tissue, and organ contributes to the harmony of life. When inflammation strikes, it's as though certain sections of the orchestra are out of sync, creating dissonance that affects the whole. Vibrational therapy, by using specific frequencies to "retune" the body, aims to restore that harmony and bring about balance.

Unveiling the Science of Vibrational Therapy and Inflammation

At its core, vibrational therapy operates on a simple yet profound concept: every cell in the human body vibrates at a particular frequency. Inflammation, however, disrupts this natural rhythm, causing cellular distress that leads to pain, swelling, and other chronic symptoms. By introducing external frequencies to the affected area, vibrational therapy encourages cells to "reset" to their healthy resonance, promoting cellular repair and reducing inflammatory responses.

Research has shown that certain frequencies can directly impact inflammatory markers in the body. For example, studies on Pulsed Electromagnetic Field (PEMF) therapy —a form of vibrational therapy—have demonstrated that it reduces pro-inflammatory cytokines, proteins that play a significant role in the inflammation process. As these cytokines decrease, inflammation diminishes, helping the body return to a state of equilibrium.

A Case Study in Healing: Anna's Journey with Arthritis

Anna, a 57-year-old artist, spent years battling rheumatoid arthritis. Conventional treatments had helped her manage the pain, but she was eager to explore alternatives to reduce her reliance on medication. After learning about vibrational therapy, she started using PEMF therapy on her hands and knees daily, combined with occasional sound therapy sessions tuned to 174 Hz—a frequency believed to promote pain relief.

Within a month, Anna noticed a decrease in stiffness and swelling. Her pain had reduced to the point where she could paint for hours without discomfort, something she hadn't experienced in years. For Anna, vibrational therapy didn't just relieve her pain; it empowered her to take control of her health.

How Specific Frequencies Target Inflammation

Vibrational therapy includes various methods, each with unique benefits. Here are some key approaches and the frequencies used to address inflammation:

1. Pulsed Electromagnetic Field (PEMF) Therapy

   - PEMF therapy involves the application of low-frequency electromagnetic waves to target inflamed tissues. Frequencies ranging from 1-30 Hz are particularly effective, as they stimulate cellular repair, improve circulation, and promote anti-inflammatory responses. PEMF devices are now widely available for home use, allowing people to incorporate this therapy into their daily routines.

2. 174 Hz and Sound Therapy

   - Known as a "natural anaesthetic," 174 Hz is one of the Solfeggio frequencies, a set of tones thought to have therapeutic effects on the body. This frequency is associated with pain relief and relaxation, making it valuable for reducing tension in inflamed areas. Many sound therapists recommend this frequency for chronic pain patients, especially those with conditions like fibromyalgia and arthritis.

3. Vibrational Massage Therapy

   - Vibrational massage devices combine mechanical vibrations with therapeutic frequencies, usually around 50-100 Hz, which are known to penetrate deep tissue layers. These devices can target inflamed areas directly, helping to reduce swelling and improve circulation. Vibrational massage is often used by athletes to manage inflammation and speed up recovery after intense workouts.

4. Infrared and Light Frequencies

- Although not strictly vibrational, light frequencies —particularly in the infrared spectrum—are known to penetrate skin and muscle, reducing inflammation at the cellular level. Devices emitting near-infrared light around 660 nm have become popular for treating joint and muscle pain, as they improve blood flow and decrease oxidative stress.

Practical Guide to Using Vibrational Therapy for Inflammation

Integrating vibrational therapy into your life doesn't require a large time commitment or expensive equipment. Here are some simple methods to explore:

1. Daily PEMF Sessions: Many PEMF devices are portable and can be placed on inflamed areas for 10-20 minutes. Start with a frequency between 5-15 Hz to see how your body responds, gradually increasing based on comfort and effectiveness.

2. Listening to Healing Frequencies: Incorporate sound therapy into your daily routine. Find recordings tuned to 174 Hz or similar therapeutic frequencies. Use headphones to immerse yourself for 10-15 minutes, especially if stress contributes to your inflammation. Apps and online platforms now offer playlists specifically designed for reducing inflammation.

3. Using Vibrational Massage: Apply a vibrational massage device directly to affected areas for a few minutes each day. This can help release muscle tension, decrease swelling, and improve blood flow to reduce pain.

4. Exploring Infrared Therapy: Infrared lamps or wearable devices targeting inflamed areas can be effective. Aim for sessions of 10-15 minutes, especially on joints or muscles. Combining infrared with vibrational therapy can have a

synergistic effect, amplifying the benefits of each modality.

A Future Vision: Inflammation Management with Personalized Frequencies

Imagine a future where wearable devices monitor your body's inflammation levels in real-time, using sensors to detect early signs of cellular distress. These devices could then emit targeted frequencies directly to affected areas, preventing inflammation from escalating into chronic disease. As artificial intelligence integrates with vibrational medicine, personalized frequency protocols may be developed based on an individual's biofeedback, offering highly tailored anti-inflammatory treatments.

Such advances could transform our approach to healthcare, shifting the focus from treating symptoms to preventing disease at its root—by maintaining cellular harmony before imbalance occurs. This would represent a paradigm shift in medicine, one that honours the body's natural rhythms and its ability to self-regulate with minimal intervention.

Reflective Thought: Embracing the Power of Frequencies

While vibrational therapy is still emerging as a mainstream treatment, its ancient roots and promising scientific findings invite us to rethink the nature of healing. Inflammation, once seen purely as a response to injury or infection, is now recognized as a complex, systemic issue affected by our emotional, physical, and environmental states. By embracing the power of frequencies, we honour a more holistic view of health—one that seeks harmony within, rather than simply managing symptoms.

Balanced Scepticism and Continued Exploration

As with all complementary therapies, it's essential to approach vibrational medicine with curiosity and discernment. While the science is growing, not all claims are validated, and responses to therapy can vary widely. Some may experience significant relief, while others may find it to be only part of a broader treatment plan. Consulting healthcare providers and seeking evidence-based sources can enhance your journey with vibrational therapy, ensuring it complements your unique needs.

Conclusion: Rediscovering Health Through Vibrational Harmony

Managing inflammation through vibrational therapy is more than a treatment; it's an invitation to return to balance. In our fast-paced world, where stress and environmental factors constantly challenge our well-being, these therapies remind us of our body's innate capacity to heal and thrive when supported by the right frequencies. Through the careful application of sound, electromagnetic waves, and light, we can foster a healing environment within ourselves, empowering our cells to work in unison. Embracing vibrational therapy is not just about addressing physical symptoms—it's a step toward a deeper harmony, one that nourishes both body and soul.

CHAPTER 43: FREQUENCY APPLICATIONS FOR DIABETES MANAGEMENT

The modern world has witnessed a surge in chronic conditions, with diabetes at the forefront. The management of diabetes, particularly Type 2, requires a blend of medical intervention, lifestyle adjustments, and stress management. While traditional methods focus on dietary control, medication, and exercise, an emerging body of research suggests that frequency therapies—using specific sound and electromagnetic waves—may offer a complementary approach to managing diabetes.

Imagine the body as a finely tuned instrument, each system vibrating at a specific frequency to maintain harmony. When blood sugar levels are out of balance, the body's natural rhythms become disrupted. Frequency-based therapies can help restore balance, alleviating stress and supporting the body's natural regulatory functions. Though still in its nascent stages, the potential of frequency therapy in diabetes care is promising and worth exploring for those seeking holistic approaches.

The Science Behind Frequencies and Blood Sugar Regulation

To understand how frequency therapy might aid diabetes management, it's essential to recognize the physiological interplay between stress, blood sugar, and the nervous system. Stress activates the sympathetic nervous system, releasing cortisol and other hormones that can cause blood sugar levels to spike. Over time, chronic stress impairs the body's ability to regulate glucose, exacerbating diabetic symptoms.

Sound frequencies, electromagnetic fields, and even light-based therapies have shown potential to calm the nervous system, reducing stress and indirectly influencing blood sugar levels. Some studies indicate that specific frequencies may enhance cellular insulin sensitivity and promote glucose metabolism, creating a pathway to help the body manage blood sugar more effectively.

Case Study: John's Journey with Frequency Therapy and Diabetes

John, a 52-year-old accountant, had managed his Type 2 diabetes for years with medication and diet. Yet, he found that stress would often lead to sudden spikes in his blood sugar, throwing his well-laid plans into disarray. Intrigued by frequency therapy, he began incorporating sound sessions focused on 528 Hz, a frequency often associated with cellular repair and transformation, alongside low-frequency PEMF (Pulsed Electromagnetic Field) therapy for stress relief.

After three months, John noticed more stable blood sugar readings and an increased sense of calm. While he continued with his prescribed treatment, he felt that frequency therapy offered him a supportive layer that traditional methods alone hadn't achieved.

How Frequency Therapies Work in Diabetes Management

Several types of frequency-based therapies are currently being explored for diabetes management, each targeting different aspects of the condition.

1. Sound Therapy (528 Hz and 639 Hz)
   - Known for its reputed healing properties, 528 Hz is thought to encourage cellular regeneration and support overall wellness. When combined with 639 Hz—a frequency linked to emotional balance—it can create a calming environment, reducing stress and potentially helping blood sugar control. Listening to these frequencies for 10-15 minutes a day may benefit individuals struggling with stress-related blood sugar fluctuations.

2. PEMF Therapy
   - PEMF therapy involves applying electromagnetic fields at specific low frequencies to improve circulation and cellular metabolism. For diabetes, frequencies between 5-15 Hz have been found to help in cellular repair, enhancing insulin sensitivity. Regular sessions of PEMF therapy can be a valuable addition to the routines of those managing diabetes, as it may improve the body's ability to utilize insulin.

3. Binaural Beats for Stress Management
   - Stress is a major factor that exacerbates diabetes. Binaural beats—two slightly different frequencies played in each ear —help the brain achieve a calm state. Listening to binaural beats tuned to the theta frequency (around 4-8 Hz) can reduce anxiety and foster relaxation, potentially helping to stabilize blood sugar levels by curbing cortisol spikes.

Practical Tips for Integrating Frequency Therapy in Diabetes Care

While frequency therapy should not replace traditional

diabetes management, it can serve as a complementary tool. Here are some practical ways to incorporate it into daily life:

1. Daily Sound Therapy: Set aside 10-15 minutes to listen to frequencies like 528 Hz and 639 Hz. You can find these sounds on many streaming platforms or apps designed for therapeutic sound. Headphones can enhance the effect, allowing you to focus solely on the frequencies.

2. PEMF Devices for Home Use: Portable PEMF devices are increasingly available and user-friendly. Place the device on your abdomen or lower back (areas associated with insulin sensitivity) and use a low-frequency setting for 10 minutes, twice daily. As with any new therapy, start with short sessions to see how your body responds.

3. Using Binaural Beats Before Bed: Many people with diabetes struggle with restful sleep, often linked to stress. Listening to theta-tuned binaural beats before bed can promote deep relaxation, helping reduce cortisol levels and preparing the body for balanced glucose levels upon waking.

4. Mindful Breaks with Frequency Playlists: Stress management is essential in diabetes care. Use short breaks throughout your day to listen to calming frequencies or a playlist designed for relaxation. These micro-moments of stress relief can have a cumulative effect, supporting better blood sugar control over time.

A Vision for the Future: Personalized Frequency Protocols for Diabetes

As frequency medicine advances, one can envision a future where healthcare providers can tailor treatments based on an individual's unique biofeedback. Imagine wearing a device that measures real-time blood sugar, cortisol levels, and heart rate variability, instantly adjusting frequencies to maintain balance.

With the integration of AI, personalized sound and PEMF protocols could be developed, adapting to your body's needs and providing frequency interventions before blood sugar spikes even occur. Such innovation would make diabetes management more proactive, minimizing reliance on medication and reducing complications.

Reflective Thought: Tuning the Body for Harmony

At its core, frequency medicine invites us to consider health as a state of harmony rather than mere absence of disease. In managing diabetes, this approach encourages us to see the body as a dynamic system, responding to sound and frequencies that support its intrinsic regulatory mechanisms. Rather than viewing diabetes as a purely biochemical issue, frequency therapy allows us to explore the energetic dimensions of healing, embracing a holistic approach that considers mind, body, and spirit.

Addressing Scepticism with Openness

It's natural to question new approaches to chronic conditions like diabetes, especially those involving non-conventional methods. However, the emerging research on frequency medicine is promising, and numerous anecdotal cases reflect positive outcomes. For those managing diabetes, exploring frequency therapy with an open mind offers an additional tool—one that aligns with the body's innate rhythms and its ability to heal.

Conclusion: The Power of Frequency in Diabetes Care

In a world where diabetes continues to rise, frequency medicine offers a fresh perspective. Through sound, electromagnetic pulses, and an understanding of the body's natural vibrations, we gain access to a new dimension of health. By tuning into our body's frequencies, we support not just physical wellness, but also emotional resilience—a critical aspect of managing any chronic condition.

While frequency therapy may not be a standalone solution, it has the potential to work alongside traditional methods to offer a more balanced and empowered approach to diabetes care. In embracing this harmony, we not only manage our conditions but also reconnect with the deeper rhythms of health that resonate within us all.

CHAPTER 44: FREQUENCIES IN CARDIOVASCULAR HEALTH

Imagine the human body as an intricate orchestra, where each cell, each heartbeat, plays a part in creating a harmonious melody of health. Just like a single out-of-tune instrument can disrupt the flow of an orchestra, imbalances in the body, especially in the cardiovascular system, can interrupt this harmony, leading to conditions like high blood pressure, poor circulation, and heart disease. Could frequencies and vibrations offer a way to restore balance? Emerging research suggests that vibrational therapy may indeed support cardiovascular health, offering a gentle yet profound way to impact blood pressure, circulation, and overall heart health.

The Science Behind Frequencies and the Heart

The heart operates rhythmically, beating around 100,000 times a day to maintain the flow of blood and sustain life. This rhythm is not isolated but resonates with the rhythms of every cell and organ, creating a resonance that promotes harmony across the body. Vibrational therapy leverages sound, electromagnetic frequencies, and even light to influence the heart's rhythmic functions.

At its core, frequency-based healing recognizes that the heart's electromagnetic field is the most powerful in the body. Research has shown that frequencies in the range of 0.5 to 30 Hz can stimulate the body's autonomic nervous system, which regulates blood pressure, heart rate, and stress levels. By working at these frequencies, vibrational therapy can create a calming, balancing effect, helping the cardiovascular system to perform optimally.

Case Study: Sarah's Journey with Frequency Therapy and Blood Pressure

Sarah, a 60-year-old retired teacher, had battled high blood pressure for years. Despite her medications, she still experienced sudden spikes whenever she was stressed. Curious about non-invasive therapies, she decided to try vibrational therapy, specifically binaural beats designed for cardiovascular health.

Within a month, Sarah noticed something remarkable: not only were her blood pressure levels more stable, but she felt a sense of calm and relaxation she hadn't experienced in years. By incorporating 10 minutes of daily listening to 10 Hz and 15 Hz binaural beats, Sarah's blood pressure gradually stabilized, and she found herself less affected by daily stresses.

How Frequencies Affect Cardiovascular Health

Several frequency-based therapies have shown promise for enhancing heart health, each working through slightly different mechanisms.

1. Binaural Beats for Blood Pressure and Stress Reduction
   - Binaural beats are sound frequencies that produce a calming effect by stimulating specific brainwave patterns. For blood pressure and stress management, binaural beats in the alpha (8-14 Hz) and theta (4-8 Hz) ranges encourage

relaxation and decrease anxiety, which is essential for stabilizing blood pressure.

2. Pulsed Electromagnetic Field (PEMF) Therapy for Circulation

- PEMF therapy uses electromagnetic fields to improve blood flow and enhance the oxygenation of tissues. For cardiovascular health, PEMF therapy is often applied at frequencies between 1 and 15 Hz to stimulate vasodilation —the widening of blood vessels, which improves blood circulation and lowers blood pressure.

3. Sound Therapy for Heart Rate Variability

- Heart Rate Variability (HRV) is an important indicator of heart health. Higher HRV reflects a resilient and adaptable cardiovascular system, while low HRV is linked to stress and cardiac issues. Sound therapy, particularly with Tibetan singing bowls tuned to 432 Hz, has been associated with improved HRV. Listening to these sounds for 15-20 minutes a day has shown potential to increase HRV, promoting relaxation and heart resilience.

Practical Tips for Incorporating Frequency Therapy for Cardiovascular Health

If you're interested in using frequency therapy as part of your heart health routine, here are a few simple ways to get started:

1. Daily Binaural Beat Sessions: Use headphones to listen to alpha or theta-tuned binaural beats for 10 minutes in the morning or evening. Regular sessions can help reduce stress and regulate blood pressure.

2. Try PEMF Therapy: Many portable PEMF devices are now available, allowing you to use the therapy in the comfort of your home. Focus on areas where blood flow is essential, such as the legs and lower back. Start with sessions of 10-15

minutes, twice daily, and monitor any improvements in circulation and blood pressure.

3. Heart-Centric Meditation with Sound: Integrate Tibetan singing bowls or tuning forks into a heart-centred meditation practice. Spend 5-10 minutes a day in a quiet space, focusing on the vibrations and sounds as they resonate within you. This practice not only enhances relaxation but may also improve HRV and emotional balance.

4. Frequency Playlists for Relaxation: Many music streaming services offer frequency-based playlists specifically designed to improve cardiovascular health. Consider incorporating these playlists during your commute or while unwinding in the evening.

A Vision for the Future: Personalized Cardiovascular Frequency Protocols

As frequency medicine continues to evolve, we can envision a future where healthcare becomes more personalized. Imagine a device that could assess your real-time HRV, blood pressure, and stress levels, and then deliver tailored frequencies to maintain balance. AI could analyse your body's responses to specific frequencies and automatically adjust the therapy to match your needs, ensuring that your cardiovascular system remains resilient and balanced.

Such innovations would mean that cardiovascular therapy could extend beyond medications and invasive procedures, focusing on restoring balance at the cellular and energetic levels. This vision aligns with the future of holistic medicine, where prevention and non-invasive therapies are prioritized.

Reflective Thought: Listening to the Heart's Symphony

The concept of using frequencies to improve heart health invites us to consider health in a new light. Rather than viewing the heart as just a muscle that pumps blood, frequency therapy encourages us to see it as an organ of connection, resonance, and harmony. As we tune into frequencies that resonate with our heart's rhythms, we can nurture a deeper relationship with our body's innate wisdom.

Consider this: if each heartbeat echoes as a note within the body's larger symphony, then frequency therapy becomes the process of tuning the heart's music, ensuring that it flows harmoniously through every cell, tissue, and organ.

Embracing Scepticism with Openness

For those new to the concept of frequency medicine, the idea of using sound and electromagnetic fields to manage cardiovascular health may seem unconventional. However, it is important to acknowledge that many therapies we accept today—like ultrasound and MRI—were once seen as radical ideas. Frequency-based therapy invites us to keep an open mind while exploring its potential alongside proven medical practices.

Conclusion: Frequencies for a Heart in Harmony

Frequency-based therapies offer a new dimension to cardiovascular care, addressing not just the physical heart but also the subtle rhythms that regulate our body's harmony. By tapping into specific frequencies, we support the body's natural abilities to regulate blood pressure, improve circulation, and enhance resilience against stress.

In this journey of healing, the heart emerges as more than an organ; it becomes a guide, leading us toward a state of health that is both dynamic and harmonious. Embracing frequency medicine invites us to listen deeply to our body's rhythms, trust in the body's capacity to heal, and, ultimately, to honour the music within that sustains us all.

CHAPTER 45: LIVER AND KIDNEY HEALTH THROUGH FREQUENCY MEDICINE

The liver and kidneys, the unsung heroes of the human body, work tirelessly to cleanse, detoxify, and balance our systems. These organs operate like a finely tuned filtration system, removing toxins and waste products that, if left unchecked, could disrupt our well-being. But what if, beyond diet and medicine, we could support these essential organs through vibrational frequencies? Recent advancements in frequency medicine suggest that specific sound waves and electromagnetic frequencies may aid in detoxification, cellular repair, and overall organ health, offering a non-invasive and complementary approach to supporting the liver and kidneys.

The Science of Frequency Medicine in Organ Health

At the core of frequency medicine is the idea that every cell, tissue, and organ in our body vibrates at its own unique frequency. By applying specific frequencies to the body, we can stimulate cells to function optimally, aiding in detoxification and repair processes. This concept is especially relevant for the liver and kidneys, which manage a continuous workload of processing and eliminating waste.

The liver is responsible for filtering blood, processing nutrients, and neutralizing toxins, while the kidneys regulate fluids, remove waste, and maintain essential electrolytes. Both organs are made up of millions of cells that communicate and function in rhythmic harmony. Frequency-based therapies aim to enhance this natural rhythm, helping cells in these organs to repair, regenerate, and detoxify more effectively.

A Personal Story: John's Path to Recovery

John, a 45-year-old mechanic, had long struggled with high blood pressure and liver enzyme imbalances due to years of poor lifestyle choices. His doctor recommended medication, but John was eager to explore complementary approaches. When he heard about frequency therapy, he was intrigued and decided to try it alongside traditional treatments.

John began regular sessions with low-level pulsed electromagnetic field (PEMF) therapy, which was set at frequencies specifically aligned with liver and kidney function. Over the course of a few months, his liver enzyme levels showed marked improvement, and he felt an increase in energy and overall well-being. While he continued to follow his doctor's advice, John credited frequency therapy with helping his liver and kidneys regain a level of balance he hadn't experienced in years.

How Frequencies Support Liver and Kidney Health

Frequency-based therapies target the liver and kidneys in a few specific ways:

1. Detoxification Through Vibrational Stimulation
   - Vibrational therapies, including PEMF and sound frequencies, can stimulate cellular activity within the liver

and kidneys, promoting efficient detoxification. By applying frequencies between 1-10 Hz, which align with natural cellular rhythms, the liver and kidneys can operate more effectively in flushing out toxins.

2. Cellular Repair and Regeneration
   - Frequency therapy also shows promise in supporting cellular repair and regeneration, a critical process for organs that endure heavy workloads. Low-frequency electromagnetic fields and sound waves can enhance mitochondrial function—the powerhouse of the cell—boosting energy production and aiding in cellular repair. For the liver, which has a high capacity for regeneration, these frequencies can expedite recovery after damage from alcohol or medication.

3. Improving Blood Flow and Reducing Inflammation
   - Good blood circulation is essential for liver and kidney function. PEMF therapy and ultrasound frequencies encourage vasodilation, or the widening of blood vessels, which enhances blood flow to these organs, bringing essential nutrients and oxygen. Certain frequencies have anti-inflammatory properties as well, helping to alleviate conditions such as fatty liver disease or kidney inflammation.

Practical Tips for Using Frequencies to Support Liver and Kidney Health

If you're interested in exploring frequency therapy for liver and kidney health, here are some practical ways to incorporate these therapies into your routine:

1. PEMF Therapy for Daily Detoxification: Many home-use PEMF devices offer settings specifically for detoxification. Starting with low frequencies around 5-10 Hz, you can use PEMF therapy for 10-20 minutes daily, ideally in the morning, to support daily detox and cellular repair.

2. Sound Healing with Tuning Forks or Singing Bowls: Instruments like tuning forks and Tibetan singing bowls tuned to frequencies between 120 Hz and 528 Hz can stimulate relaxation and enhance cellular function. By focusing on these specific frequencies, you can create a soothing environment for your body's detox processes. Spend about 10 minutes with these sounds, focusing on the abdomen, where the liver and kidneys are located.

3. Binaural Beats for Cellular Regeneration: Binaural beats are another accessible way to incorporate frequency therapy into your routine. Frequencies in the theta (4-8 Hz) and delta (0.5-4 Hz) ranges are thought to promote cellular healing and regeneration. Listening to these beats while resting can support the liver and kidneys, especially if you're recovering from an illness or detoxing.

4. Guided Frequency Meditation: Many frequency therapy apps and guided meditation services offer sessions specifically designed for organ health. Set aside 10-15 minutes in a quiet space, and let the frequencies guide you into a meditative state where your body can engage in healing.

Envisioning the Future: Tailored Frequency Therapies for Organ Health

As we look to the future, advancements in AI and biofeedback may enable highly personalized frequency treatments. Imagine a device that can scan your body, identify the unique frequency imbalances within your liver and kidneys, and then generate specific vibrational protocols to target those needs. In this vision of the future, frequency therapy could become a preventive tool, detecting early signs of stress on organs before they manifest as illness.

Reflective Insights: Honouring the Body's Rhythm

When we think about supporting liver and kidney health, frequency therapy offers a profound reminder of the importance of balance and rhythm. These organs operate tirelessly, often taken for granted until they start to falter. Frequency medicine encourages us to reconnect with these rhythms, honouring the body's natural capacity to heal.

This approach to health invites us to consider our organs not just as biological machines but as parts of a harmonious system in tune with the broader rhythm of life. By tapping into these frequencies, we are reminded of the interconnectivity between body, mind, and the vibrations that govern the universe.

Balancing Scepticism and Openness

The notion of healing organs through frequencies may seem unconventional to some. However, many modern medical practices, like ultrasound and MRI technology, were once seen as futuristic concepts. As frequency medicine continues to develop, it's important to approach these therapies with both curiosity and critical thinking, combining them with evidence-based approaches to ensure safe and effective use.

Conclusion: Resonating with Health

Frequency-based therapies for liver and kidney health offer a powerful, non-invasive way to support two of the body's hardest-working organs. By embracing this form of healing, we engage in a more holistic view of health—one that sees the body as an interconnected, resonant system.

In the end, frequency therapy for liver and kidney health is not just about healing; it's about reconnecting with our bodies, tuning into their natural rhythms, and respecting the vibrational patterns that sustain us. As we learn to listen and respond to these frequencies, we open ourselves to a new world of possibilities for health, healing, and harmony.

CHAPTER 46: IMPROVING LUNG HEALTH WITH FREQUENCIES

The human body is a symphony of vibrations and resonances, each organ humming to its own frequency, yet harmonizing in a complex orchestra of life. Nowhere is this musicality more vital than in our lungs, the organs that set the rhythm of our existence, breathing life into every cell. As chronic respiratory conditions like asthma and COPD (chronic obstructive pulmonary disease) become increasingly prevalent, the potential of frequency-based therapies to support lung health is both timely and revolutionary. In this chapter, we delve into how vibrational healing may open a new frontier for those who struggle to breathe freely, offering hope and tangible results.

Understanding Frequency Medicine for Respiratory Health

At its essence, frequency medicine taps into the vibrations at which our cells and tissues naturally operate. When applied to lung health, specific frequencies can target cellular functions, enhance tissue repair, reduce inflammation, and improve oxygen exchange. Much like a tuning fork can adjust the pitch of a musical note, frequency therapy can help recalibrate the cells within the lungs to function optimally.

This approach includes sound therapy, such as the use of tuning forks and specific sound frequencies, as well as electromagnetic frequencies applied through devices like pulsed electromagnetic field (PEMF) systems. In the case of respiratory health, frequencies between 3 Hz and 10 Hz have shown promising results in encouraging relaxation of bronchial airways and enhancing cellular oxygenation.

A Real-World Story: Sarah's Journey with Asthma

Sarah had struggled with asthma since childhood. In her mid-30s, her condition worsened to the point where she was constantly using inhalers and felt trapped by her limited lung capacity. Conventional treatments provided temporary relief but did little to address her long-term discomfort. Then, Sarah heard about frequency therapy. With some hesitation, she decided to try a series of PEMF sessions targeting her lungs.

Within weeks, Sarah noticed subtle changes. Her breathing felt smoother, her need for inhalers reduced, and she felt an unexpected boost in energy. Over time, her respiratory function improved, and she felt empowered, regaining a sense of control over her own health. While she continued to follow her doctor's advice, she attributed much of her improvement to the supportive effects of frequency-based therapy.

How Frequencies Impact Lung Function and Health

Frequency-based therapies have demonstrated a range of benefits in supporting lung health, particularly for individuals with chronic respiratory conditions:

1. Relaxation of Bronchial Muscles
   - PEMF therapy and low-frequency sound waves can promote the relaxation of bronchial muscles, easing the constriction that occurs in conditions like asthma. By

applying frequencies between 5-10 Hz, the lungs' airways become more receptive to airflow, which can reduce the sensation of tightness in the chest.

2. Enhanced Oxygen Uptake and Circulation

- Certain frequencies help stimulate cellular oxygen uptake, allowing for more efficient oxygen exchange within the lungs. This not only improves overall lung capacity but also enhances circulation, delivering vital oxygen more effectively throughout the body. Frequencies around 7 Hz, in particular, support oxygenation, creating an environment where cells can thrive.

3. Reduction of Inflammation

- Inflammation is at the root of many respiratory issues, from asthma to bronchitis to COPD. Frequency therapy—specifically, low-level PEMF and sound therapy—has shown anti-inflammatory effects, reducing oxidative stress within lung tissue and easing the chronic inflammation that limits lung capacity.

4. Support for Mucus Clearance

- Frequencies can also play a role in stimulating cilia, the tiny hair-like structures in the respiratory tract responsible for clearing mucus. By aiding mucus clearance, frequency therapy can reduce respiratory infections and prevent the buildup of congestion, a common issue for those with chronic respiratory conditions.

Practical Tips for Using Frequencies to Support Lung Health

For those interested in incorporating frequency-based practices into their respiratory health routine, here are some methods and tips:

1. PEMF Therapy for Bronchial Relaxation: Begin with daily PEMF therapy sessions using frequencies between 5-10 Hz, focusing on the chest area. Start with 10-15 minutes

per session, gradually increasing as your body adjusts. These sessions can help relax bronchial muscles and make breathing feel more natural.

2. Sound Therapy with Low-Frequency Tones: Create a relaxing environment with sound frequencies around 7 Hz, which supports cellular oxygenation. Devices like tuning forks, singing bowls, or specialized soundtracks can help. Take 10-15 minutes to relax in a quiet place and focus on slow, deep breathing while immersing yourself in the sounds.

3. Binaural Beats for Respiratory Health: Binaural beats in the delta and theta ranges (between 1-8 Hz) are ideal for enhancing relaxation and supporting cellular recovery. Incorporate these beats during meditation or before sleep to encourage restful, deep breathing, and ease respiratory inflammation.

4. Guided Breathwork with Frequency Apps: Many frequency-based wellness apps offer guided breathwork sessions that incorporate specific sound frequencies. Look for sessions aimed at respiratory relaxation and follow the prompts to synchronize your breath with the frequency. This not only enhances lung capacity but can also reduce anxiety, a common companion of respiratory issues.

Looking to the Future: Personalized Frequency-Based Respiratory Therapies

The future of frequency medicine for respiratory health lies in personalized treatments. Advances in AI and biofeedback technology will likely allow practitioners to design custom frequency protocols based on individual needs. Imagine a small wearable device that monitors your respiratory health and delivers targeted frequencies to the lungs to help manage symptoms before they escalate. This kind of tailored frequency therapy could bring transformative relief to those with chronic conditions.

Reflecting on Breath as Life's Rhythm

In a very real sense, each breath we take is a note in the grand symphony of life. When we struggle with respiratory issues, it is as if that note has fallen out of tune, creating discord within the body. Frequency medicine invites us to think of the body as a musical instrument, one that can be retuned to play in harmony with our natural rhythms. By attuning ourselves to these frequencies, we support our lungs' vital role, embracing the life force they bring to every cell.

Balancing Hope and Caution

While the potential of frequency medicine for lung health is exciting, it's essential to approach it as a complementary therapy, not a replacement for conventional treatments. The field is still evolving, and much research remains to be done. For now, integrating these therapies with established respiratory care offers a balanced path toward healing, one that bridges modern science and ancient principles of sound and vibration.

Conclusion: Breathing with New Possibilities

The power of frequency-based therapies for lung health represents more than just physical healing—it embodies the possibility of liberation for those who have long struggled to breathe freely. In tuning our bodies to frequencies that support respiratory health, we align ourselves with the natural rhythms that sustain life itself.

As you explore these techniques, remember that each breath is both a gift and a reminder of the harmony within. With frequency medicine, we take a step closer to a world where chronic conditions are met with resonance, where the simple act of breathing becomes a celebration of life, and where healing harmonies are available to all who seek them.

CHAPTER 47: FREQUENCY-BASED SUPPORT FOR IMMUNE FUNCTION

Imagine the body as an orchestra, with each cell vibrating at its own pitch, creating a symphony of life. When our immune system faces a threat, this symphony can become disrupted, like an orchestra suddenly out of tune. Frequency-based therapies offer a way to help the body "retune" itself, supporting the immune system in ways that conventional medicine is only beginning to understand. In this chapter, we'll explore how specific frequencies can enhance immune response, aid recovery, and even serve as a preventive approach to wellness.

The Science Behind Frequency and Immunity

The immune system, our body's natural defence mechanism, relies on countless cellular processes. Every immune cell—be it a white blood cell fighting off pathogens or a lymphocyte targeting viruses—functions based on electrical impulses and vibrations. Frequency medicine taps into these natural rhythms, delivering specific vibrations that can support immune cells and facilitate a faster, more efficient response.

Low-frequency electromagnetic fields (PEMF) and sound

frequencies have demonstrated significant effects on cellular health, improving circulation, reducing inflammation, and increasing cellular repair. Researchers have found that low frequencies can stimulate cell activity, effectively energizing cells to perform their roles more effectively. Much like providing a tune-up to an engine, frequency therapy helps the immune system work smoothly and efficiently.

Story: Mark's Recovery from Chronic Illness

Mark had lived with a chronic immune condition for years, feeling exhausted and frequently falling ill. Conventional treatments offered some relief but left him feeling dependent on medications and lifestyle restrictions. He heard about frequency-based therapies and decided to give them a try, hoping for a gentler way to support his body.

Within weeks of beginning PEMF sessions targeting his lymphatic system and immune cells, Mark noticed a shift. He had more energy, experienced fewer infections, and his chronic inflammation was significantly reduced. He felt as though his immune system had finally received the support it needed to function at its best. Mark continued with frequency-based therapy as part of his regular routine, giving his immune system the tools to naturally strengthen and protect his health.

How Frequency Medicine Boosts Immune Function

Research into frequency therapy reveals multiple pathways through which frequencies can benefit the immune system:

1. Enhanced Cellular Energy and Repair
   - Frequencies between 5 and 15 Hz have shown to boost ATP production, which provides cells with energy. Immune cells rely heavily on ATP to function optimally, and by increasing cellular energy, they can respond more effectively to pathogens.

2. Reduction of Inflammation
   - Chronic inflammation is often an obstacle for the immune system, draining resources that would otherwise be directed toward fighting infections. Frequencies around 10 Hz have demonstrated anti-inflammatory effects, helping reduce systemic inflammation and freeing up the immune system to focus on threats.

3. Increased Lymphatic Circulation
   - The lymphatic system is crucial to immune health, as it transports white blood cells and clears toxins from the body. PEMF frequencies between 3 and 10 Hz have shown to stimulate lymphatic flow, helping the body eliminate waste more efficiently and giving immune cells greater mobility.

4. Stimulating Immune Cell Activity
   - Some studies suggest that specific frequencies can stimulate immune cells like lymphocytes and macrophages, helping them target and neutralize invaders more effectively. These frequencies effectively "wake up" the immune system, making it more alert to potential threats.

Practical Tips: Incorporating Frequencies into Immune Support

For those looking to use frequency therapy to support their immune system, here are a few practical ways to get started:

1. PEMF Therapy for Lymphatic Health: Start with low-frequency PEMF sessions targeting the lymphatic system and immune cells. Sessions of 10-15 minutes at frequencies between 5-10 Hz can help stimulate lymphatic flow and immune cell activity.

2. Sound Baths for Inflammation Reduction: Sound baths featuring low-frequency vibrations can promote relaxation and reduce inflammation. Frequencies around 10 Hz, often found in therapeutic soundtracks, are ideal for relieving

stress and inflammation, which indirectly supports immune function.

3. Daily Binaural Beats for Energy and Recovery: Listen to binaural beats within the delta and theta ranges (between 1-8 Hz) during meditation or sleep. These frequencies can encourage cellular repair and immune function, especially when incorporated into a daily routine.

4. Cold and Flu Frequency Support: At the onset of cold or flu symptoms, frequencies around 8 Hz have shown to boost immune response and reduce the severity of symptoms. Listen to frequency-based music or engage in brief PEMF sessions to support the body's natural defences.

Looking Ahead: Personalized Immune Frequency Protocols

As research and technology advance, we can expect to see more personalized approaches to frequency-based immune support. With AI and biofeedback devices, practitioners could one day create tailored frequency protocols based on real-time analysis of immune activity and individual needs. Imagine a wearable device that senses immune function and delivers specific frequencies to boost immune response as soon as a pathogen is detected—offering protection before symptoms even appear.

Reflection: The Immune System's Quiet Harmony

Our immune system works tirelessly, largely unnoticed, defending our body against threats big and small. But like any other complex system, it sometimes needs support and balance. Frequency-based therapies remind us of the delicate vibrational dance within our cells and the interconnected nature of our body's systems.

Reflecting on immune health through the lens of frequency medicine allows us to think beyond treating symptoms, instead focusing on harmonizing the body's natural defences. This perspective encourages us to respect the body's innate intelligence, offering support in a way that honours its complex and subtle design.

Balancing Caution and Hope

While the potential of frequency-based therapies for immune support is promising, it's essential to view these approaches as complementary rather than a substitute for conventional immune care. Although studies continue to reveal exciting possibilities, frequency medicine remains a growing field, and much more research is needed. For now, combining frequency therapy with traditional treatments offers a balanced approach, one that combines scientific rigor with ancient principles of resonance and harmony.

Conclusion: Empowering the Body's Defences

Frequency-based immune support represents more than a novel treatment—it's a return to understanding health as a symphony, where each cell, tissue, and organ plays its part in harmony with the rest. By tuning into frequencies that support immune function, we offer our bodies not only protection but a sense of balance and resilience.

With each discovery, we edge closer to a world where healing is gentle, aligned with our body's rhythms, and accessible to all. Whether you're looking to prevent illness, recover faster, or simply strengthen your immune system, frequency therapy offers a path of resonance and empowerment, inviting us to participate in the delicate, wondrous dance of our body's natural defences.

PART XII: FREQUENCY HEALING ACROSS DEMOGRAPHICS

CHAPTER 48: FREQUENCY THERAPY FOR INFANTS AND CHILDREN

Imagine a lullaby—a gentle melody that soothes and comforts a baby into sleep. But what if this music did more than calm? What if the sound waves themselves nurtured the baby's body, enhancing cellular health, improving focus, or even aiding in development? Frequency therapy for children harnesses these possibilities, offering safe, non-invasive ways to support their well-being from infancy through adolescence.

In this chapter, we'll explore the science behind frequency therapies specifically tailored for young bodies and minds, discover real-life examples of families who've incorporated sound and frequency into their children's lives, and discuss practical, safe applications for parents interested in gentle, frequency-based healing methods.

Understanding How Frequencies Affect Young Minds and Bodies

Children's bodies are in a constant state of growth

and change. Their cells are rapidly dividing, and their neural pathways are being built and reinforced through experiences. Frequencies can affect their physiology differently than adults because children's systems are more responsive to subtle environmental cues.

Certain frequencies resonate with various parts of the body, promoting relaxation, enhancing cognitive function, and supporting emotional health. For instance, low-frequency sounds—around 4-8 Hz—can help with relaxation, making them ideal for sleep and calmness, while higher frequencies, such as 10-14 Hz, may aid concentration and cognitive alertness.

A Story: Sarah's Sleep Struggles

Sarah was a vibrant five-year-old with boundless energy. However, her parents struggled with her sleep routine; bedtime often turned into an hour-long process, with Sarah finding it hard to wind down and relax. After reading about sound therapy, they decided to try a bedtime routine that included low-frequency sounds played softly in her room.

Within days, they noticed a change. Sarah seemed to relax faster, and her energy transitioned more smoothly from playful to calm. Bedtime became a shorter, more pleasant experience for everyone involved. The gentle sounds, calibrated to a calming frequency, helped align Sarah's mind and body with a state conducive to rest, a true revelation for her parents.

Safe and Effective Frequencies for Children

Parents often ask if frequency therapy is safe for children, and the answer is yes—when used appropriately. Frequency-based therapies can be a gentle way to nurture the natural rhythms in children, helping with sleep, focus, and emotional balance. Here are some safe, age-appropriate ways to introduce frequencies to young ones:

1. Lullaby Frequencies for Sleep and Calm
  - Frequencies in the range of 4-8 Hz promote relaxation and deep sleep. These frequencies are often embedded within lullabies or ambient soundtracks.
  - DIY Tip: Try playing calming music with binaural beats set within the delta or theta range during bedtime. Just 10-15 minutes can create a calming environment that helps children drift into restful sleep.

2. Focus and Learning with Higher Frequencies
  - Frequencies between 10-14 Hz, which align with alpha brainwave states, are ideal for improving concentration and focus. These can be helpful for children during study time or other activities requiring attention.
  - Practical Application: Try using focus-enhancing music or binaural beats during homework time. Studies show these frequencies can aid in maintaining attention and reducing distractions, especially in children with attention challenges.

3. Emotional Balance through Sound Baths
  - Sound baths, featuring instruments like singing bowls or chimes, emit frequencies that resonate throughout the body, bringing balance to the nervous system. These sessions can be particularly helpful in reducing anxiety and enhancing emotional resilience in children.
  - Case Study: A therapist used weekly sound baths with a group of children at a wellness centre. The parents reported that their children became calmer, with a notable reduction in anxiety levels and an increased ability to handle emotional upsets.

Practical Tips: How Parents Can Use Frequency Therapy at Home

Parents interested in incorporating frequency therapy can do so easily, thanks to accessible tools and resources. Here are some practical steps for making frequency-based therapies a part of your child's routine:

1. Create a Bedtime Ritual with Calming Sounds

Integrate calming frequencies into your child's bedtime ritual. You can use a playlist that incorporates delta and theta waves to promote relaxation. Keep the volume low and let the soothing sounds ease your child into a restful sleep.

2. Incorporate Focus Frequencies During Study Time

Set up a learning environment with background music containing alpha frequencies to improve focus. Many streaming platforms offer music designed for concentration, making it easy to find playlists suited for studying and mental clarity.

3. Introduce Weekly Sound Bath Sessions

If possible, consider incorporating a family "sound bath" time, where you use singing bowls or ambient music with resonant frequencies. This can be a bonding activity that also nurtures emotional and physical well-being.

4. Practice Mindful Listening with Nature Sounds

Frequencies in nature sounds—like flowing water or birds chirping—have been found to positively affect mental health. Taking your child to a natural setting or playing nature sounds indoors can provide grounding, peace, and balance.

The Future of Frequency Therapy for Children

As technology advances, we can expect more sophisticated, child-friendly devices that allow parents to customize and tailor frequency-based therapies to individual needs. Imagine wearable devices for children that monitor stress levels and emit calming frequencies when anxiety spikes. Or picture classrooms equipped with sound panels emitting focus-enhancing frequencies, helping children stay engaged and calm.

Reflecting on the Potential of Frequency Therapy for Our Youngest Generation

Children today are growing up in a world filled with stimuli, from constant digital engagement to the fast-paced rhythm of modern life. Frequency therapy offers a natural, calming counterbalance. It supports their development in a way that honours their body's inherent wisdom, providing a gentle way to nurture their minds, bodies, and spirits.

Integrating frequencies into a child's environment acknowledges that health is more than a physical state— it's an emotional and energetic harmony. By incorporating frequency-based practices, parents can empower their children with tools for resilience, focus, and inner peace, setting the stage for a balanced, healthy life.

A Balanced Perspective: Safety, Research, and the Path Forward

While frequency therapy shows great promise, it's essential to approach these methods with balance and awareness. As frequency medicine continues to be researched, we recommend parents choose reputable sources, consult experts when needed, and be attentive to their child's responses. Frequency therapy should complement, not replace, conventional health care, and its integration into daily routines should be thoughtful and gentle.

Conclusion: A Harmonious Start to Life

Children, with their open minds and sensitive systems, are beautifully receptive to the benefits of frequency-based therapies. These methods offer an innovative way to support development, reduce stress, and foster emotional resilience. Imagine the profound impact of growing up in a world where well-being is achieved not through invasive interventions, but through the gentle resonance of sound and frequency.

As frequency-based medicine grows, we see the possibility of a future where children's health is nurtured with harmonious vibrations, shaping not just their physical growth, but their overall well-being. By embracing these methods, we lay the foundation for a generation that is balanced, resilient, and connected to the subtle yet powerful rhythms of life.

CHAPTER 49:
ADOLESCENT
MENTAL HEALTH
AND FREQUENCIES

The teenage years mark a profound transition—a time of self-discovery, identity formation, and emotional growth. But adolescence is also when mental health challenges often emerge, from anxiety and depression to issues with self-esteem and focus. Frequency-based therapies offer a non-invasive, supportive tool for helping teenagers navigate this critical period, promoting emotional resilience, mental clarity, and even physical well-being.

In this chapter, we'll explore how frequencies interact with the adolescent brain, introduce case studies showing the potential impact of these therapies, and provide practical guidance for incorporating frequency healing into daily life. We'll also imagine a future where frequency medicine is fully integrated into adolescent mental health care.

Understanding Frequency and the Adolescent Brain

During adolescence, the brain undergoes significant restructuring. The prefrontal cortex, which manages decision-making and impulse control, is still maturing, while the amygdala, involved in emotional processing, often dominates, making teenagers more susceptible to emotional highs and lows. Neuroplasticity is at its peak, meaning that both positive and negative experiences can strongly influence development.

Frequencies and sound-based therapies can support the adolescent brain during this period of growth by gently influencing brain wave states and, therefore, mood and cognition. For example:
- Delta (1-4 Hz): Associated with deep sleep, essential for growth and emotional balance.
- Theta (4-8 Hz): Linked with creativity, relaxation, and emotional processing.
- Alpha (8-14 Hz): Promotes calmness, focus, and reduced anxiety.
- Beta (14-30 Hz): Stimulates active thinking and can enhance focus and learning.

Using frequencies aligned with these brainwave states can help address common adolescent challenges, from improving focus and reducing anxiety to supporting emotional stability.

A Story: Liam's Journey with Anxiety

Liam was 15 when he began experiencing anxiety attacks. His parents noticed he became more withdrawn, often skipping social events and struggling with sleep. After researching non-invasive therapies, they decided to try frequency-based healing, introducing calming theta frequencies before bedtime and alpha waves for focus during study sessions.

Within weeks, Liam experienced fewer anxiety episodes. He reported feeling more "in control" and less reactive to stress. The frequencies didn't eliminate his anxiety but helped create a foundation of calm that made managing his emotions easier. For Liam, frequency therapy provided a supportive tool, helping him navigate the intense emotions that came with adolescence.

Practical Applications for Supporting Adolescents

Frequency-based therapies are easy to integrate into daily routines and can provide adolescents with tools to manage stress, improve focus, and stabilize mood. Here are some ways to use frequency therapy for supporting teenagers:

1. Calming Anxiety and Promoting Emotional Balance
  - Theta waves (4-8 Hz) are associated with relaxation and introspection, helping adolescents with emotional processing.
  - DIY Tip: Set up a playlist with ambient music embedded with theta waves. Encourage teens to listen during quiet moments, especially when feeling overwhelmed, or before bed to promote restful sleep.

2. Enhancing Focus and Study Skills
  - Alpha frequencies (8-14 Hz) can improve focus, making them ideal for study sessions or homework time. Beta frequencies (14-30 Hz), although higher energy, can also aid in tasks requiring sustained concentration.
  - Practical Application: During study hours, play background music featuring alpha and beta frequencies. Many teenagers find that frequency-enhanced music helps reduce distractions and improves concentration, especially when they struggle with focus.

3. Improving Sleep and Recovery
  - Delta waves (1-4 Hz) are linked with deep sleep, essential for growth, mental clarity, and emotional regulation in teenagers.
  - Case Study: A small group of teenagers struggling with insomnia participated in a sound therapy program, using delta frequencies before bed. After a month, the majority reported not only improved sleep but also feeling more

balanced emotionally and mentally the following day.

Frequency-Based Tools for Parents and Teens

There are many ways families can incorporate frequency-based support into their daily routines. These tools require minimal effort, making them accessible even to busy teenagers:

1. Sound Baths: Sound baths using singing bowls, chimes, or other harmonic instruments can create a calm, balanced environment. Regular sessions can help with emotional regulation, making them ideal for teens who struggle with mood swings.

2. Binaural Beats Apps: Binaural beats are sound-based technologies designed to induce specific brainwave states. Many apps provide frequencies for focus, relaxation, and sleep, making them convenient options for teens with smartphones.

3. Personalized Frequency Playlists: Parents and teens can create playlists with specific frequencies for different needs —like focus during homework, relaxation during stress, or calming during social situations. Integrating these playlists into routines can subtly support emotional resilience and well-being.

4. Meditation with Frequency-Enhanced Tracks: Meditation can be challenging for teens, but adding frequency-enhanced tracks helps them reach a calm state more easily. Even a few minutes of listening can make a difference.

The Future of Adolescent Mental Health and Frequency Medicine

Imagine a future where frequency-based therapies are embedded into schools and mental health programs. Teens could enter a "calm room" during moments of stress, where frequencies tailored to relaxation are played. Counselling sessions might incorporate frequency therapy, offering a non-verbal support tool alongside traditional talk therapy.

Schools might even integrate specific frequencies into the classroom environment to enhance focus during testing or foster creativity during projects. As our understanding of brainwave states and mental health deepens, the potential for personalized, frequency-based interventions could revolutionize adolescent mental health care.

Reflecting on the Role of Frequencies in Supporting Adolescents

Adolescence is a turbulent yet formative time, a period where habits, emotions, and self-concept are woven together, forming the foundation for adulthood. By integrating frequency-based therapies, we're offering more than just a way to manage emotions—we're nurturing a balanced mind, encouraging self-regulation, and promoting resilience.

Frequency therapy aligns with the natural rhythms of the brain, offering gentle support without the invasive approaches that can sometimes be too intense for young people. As teens use these methods, they learn that mental health isn't just about fixing problems but about tuning into their inner harmony, developing a toolkit they can carry through life.

Embracing Balanced Scepticism and Safety

While frequency-based therapies show great promise, it's important to approach them with balanced scepticism. Not every approach works for everyone, and frequency therapy should be a supportive tool rather than a standalone solution. Parents should research reputable sources, consult with experts if needed, and monitor their teen's response to different frequencies.

As with any therapy, safety comes first. Frequency therapy for teens should always be non-invasive, age-appropriate, and used as part of a broader approach to well-being.

Conclusion: A Resonant Path Forward

Frequency therapy provides a hopeful, harmonious addition to the landscape of adolescent mental health care. It speaks to the unique needs of teenagers, who are still discovering who they are and how to manage the waves of their inner world. By embracing frequencies, we offer teens a gentle yet powerful means to navigate this challenging phase, building a foundation of resilience, calm, and clarity.

As we tune into the potential of frequency-based healing, we're not just addressing symptoms—we're nurturing the whole person, helping young minds find harmony in a world that can often feel overwhelming. This approach plants seeds of self-awareness and well-being, empowering our teens to step confidently into adulthood, balanced and strong.

CHAPTER 50: WOMEN'S HEALTH AND FREQUENCIES

From hormonal shifts to the transformative experiences of pregnancy, birth, and menopause, women's health encompasses a wide spectrum of physiological and emotional changes. Frequency-based medicine is emerging as a promising tool in this field, offering non-invasive support for hormone regulation, menstrual health, pregnancy, and postpartum recovery. This chapter delves into how frequencies resonate with the unique rhythms of women's bodies, supporting health and healing at every stage.

Introduction: The Unique Symphony of Women's Health

Imagine the body as an orchestra, where each organ, cell, and process contributes its own unique note. In women, this symphony changes over time, with hormonal shifts and life phases creating cycles of growth, transformation, and renewal. By using specific frequencies, we can support and balance this natural rhythm, helping the body maintain harmony.

One such story involves Emily, a 35-year-old woman who struggled with severe menstrual cramps. Traditional treatments only offered temporary relief, so she explored frequency therapy. Through regular sessions with specific frequencies, Emily experienced reduced pain, and over time, her cycle became more manageable. Emily's journey shows how frequency medicine can provide gentle, effective support for women's health issues that often go untreated or inadequately addressed.

The Science of Frequencies and Hormone Balance

Hormonal health is central to many aspects of women's wellness. Estrogen, progesterone, and other hormones fluctuate monthly and across life stages, affecting everything from mood to metabolism. Research shows that frequencies can influence the body's neuroendocrine system —the system that connects the brain and hormone-releasing glands.

How Frequencies Work: Specific frequencies stimulate the brain's hypothalamus and pituitary gland, which play a role in regulating hormones. Studies in bioenergetics and frequency medicine suggest that regular exposure to these frequencies can help the body maintain hormonal balance.

Example Frequencies for Hormone Balance:
- 7.83 Hz (Schumann Resonance): Often used to promote general wellness, this frequency is thought to synchronize the body's rhythms with the Earth's electromagnetic field.
- 10 Hz (Alpha Brainwave State): Encourages relaxation and can help manage stress, which is closely linked to hormonal fluctuations.

Frequency Medicine for Menstrual Health

Menstrual health is a vital aspect of well-being for many women, yet it's often overlooked or under-discussed. Frequencies can help alleviate menstrual discomfort, regulate cycles, and support emotional balance.

Case Study: Alleviating Menstrual Pain with Frequencies

Consider Sarah, a 28-year-old dealing with painful cramps that disrupted her life every month. After trying multiple treatments, she experimented with frequency therapy. By using binaural beats and PEMF (Pulsed Electromagnetic Field) devices tuned to lower frequencies, Sarah noticed her cramps diminished significantly. The relaxation induced by these frequencies helped her reduce stress, which further contributed to her overall wellness.

Practical Tips for Menstrual Health
- Frequency Playlists: Create a calming playlist that includes frequencies in the alpha range (8-14 Hz) to support relaxation and pain management. Listen during the days leading up to and during menstruation.
- Binaural Beats: Use binaural beats tuned to 7.83 Hz for 10-15 minutes each day to foster hormonal balance.

Pregnancy and Frequency Medicine

Pregnancy is a time of profound physical and emotional change. Frequencies can offer gentle, supportive care for pregnant women by reducing stress, improving sleep, and supporting relaxation.

Important Considerations: Frequency therapies should always be gentle during pregnancy, as the developing fetus is highly sensitive to external stimuli. Frequencies in the theta (4-8 Hz) and low alpha (8-10 Hz) range are generally safe and beneficial for relaxation and reducing anxiety.

Benefits of Frequency Therapy During Pregnancy
- Reduced Stress: Pregnancy can bring anxiety, and frequencies that promote relaxation can be invaluable. Listening to calming frequencies or sound baths with lower frequencies can reduce cortisol levels and create a sense of calm.
- Enhanced Sleep: Theta frequencies, associated with deep

relaxation, can help improve sleep, which is crucial for pregnant women.

DIY Frequency Therapy for Expectant Mothers
- Gentle Sound Baths: Set aside 15-20 minutes each day for a sound bath using low-frequency instruments like singing bowls, which emit frequencies that promote calm.
- Guided Meditation with Frequencies: Listen to guided meditation tracks with embedded theta waves to promote relaxation and connection with the baby.

Frequency Medicine for Postpartum Recovery

The postpartum period, or "fourth trimester," can be one of the most challenging stages, marked by physical recovery, hormonal shifts, and emotional adjustments. Frequencies can provide a subtle yet powerful support system for new mothers during this time, helping with recovery, emotional balance, and sleep.

Supporting Postpartum Healing
- Physical Recovery: Frequencies that support tissue repair and circulation, like PEMF therapy at 10 Hz, may aid in recovery from childbirth.
- Emotional Balance: The stress of adjusting to life with a newborn can trigger mood fluctuations. Frequencies in the alpha range can foster a state of calm, helping new mothers manage the emotional ups and downs.

A Story of Postpartum Healing
After giving birth, Marie struggled with fatigue and mood swings. Introduced to frequency-based therapies by her doula, she incorporated PEMF sessions with low frequencies and began listening to binaural beats designed for relaxation. Over the next few months, Marie found herself feeling more balanced and resilient.

Menopause and Beyond: Frequency Medicine for Mature

Women

Menopause brings about another phase of hormonal transition, with shifts that can lead to symptoms like hot flashes, mood swings, and sleep disturbances. Frequencies can provide relief by promoting relaxation and supporting hormonal equilibrium.

How Frequencies Help During Menopause
- Reducing Hot Flashes: Frequencies in the alpha range (8-14 Hz) can help manage the nervous system's response, which may reduce the intensity of hot flashes.
- Improving Sleep Quality: Theta frequencies (4-8 Hz) are effective for relaxation and may alleviate insomnia, a common issue during menopause.

Practical Tips for Menopause Support
- Daily Frequency Practice: Incorporate alpha frequencies in the evening to promote calm and theta frequencies before bed for better sleep.
- PEMF Therapy: Consider low-frequency PEMF sessions for overall wellness and to support bone health, which becomes a concern post-menopause.

Reflecting on Women's Health and Frequency Medicine

Frequency medicine offers a natural, gentle means of supporting the diverse health needs of women. By aligning with the body's rhythms, these therapies provide non-invasive options for managing pain, regulating hormones, and supporting mental and emotional well-being through every phase of life. It is not a cure-all but rather a complement that respects the wisdom of the female body.

Envisioning the Future of Women's Health with Frequency Therapy

Imagine a future where frequency therapy is available as a routine part of women's health care, integrated into prenatal visits, menopause counselling, and menstrual health support programs. Women would have access to frequency playlists for pain relief, sleep, and emotional balance, customized to their unique hormonal profiles.

In this future, frequency medicine isn't just an alternative —it's a primary method of support, empowering women to embrace natural, non-invasive healing. As research continues to deepen, frequency-based treatments may become mainstream, enhancing the quality of life for women of all ages.

Conclusion: Embracing Harmony

Women's health is a journey, a lifelong interplay of cycles and phases that deserves compassionate, holistic care. Frequency medicine aligns with this journey, offering harmony and balance through the language of sound and vibration. By tuning into these supportive frequencies, women can embrace their health with greater awareness, resilience, and peace, allowing the natural rhythms of life to guide their wellness journey.

CHAPTER 51: MEN'S HEALTH AND FREQUENCY HEALING

Men's health, like women's, encompasses a unique set of physiological needs and challenges. From prostate health and cardiovascular wellness to managing stress and mental clarity, frequency-based therapies offer men a powerful, non-invasive tool to support overall well-being. As frequency medicine continues to evolve, it's becoming clearer how specific frequencies resonate with the unique biology of men, aiding them in achieving optimal health at every stage of life.

Introduction: Frequency Healing for the Modern Man

Imagine the body as an intricate machine, with each part resonating to its own specific frequency. For men, maintaining this machine requires balance and attention to the unique needs of male physiology. Think of frequency healing as a finely-tuned maintenance tool that helps keep every aspect of this machine running smoothly.

To illustrate, let's consider Tom, a 45-year-old businessman who faced stress and mild prostate issues. Traditional therapies provided some relief, but Tom turned to frequency healing as a supplementary practice. Through tailored frequency sessions, Tom noticed improvement in his stress levels, and after several months, his prostate discomfort had lessened. Tom's experience exemplifies how frequency healing can provide holistic support for men's health.

Prostate Health: Targeting the Core of Male Well-being

Prostate health is a priority for men, particularly as they age. The prostate gland, often taken for granted, plays a crucial role in men's overall health, and maintaining it is essential. Frequency therapies can support prostate health by reducing inflammation and enhancing blood flow.

The Science Behind Frequency Therapy and Prostate Health

Studies in bioelectromagnetics show that certain frequencies can stimulate cellular repair and reduce inflammation in targeted tissues. In the case of the prostate, frequencies between 8-12 Hz, which help reduce stress and inflammation, may promote better health by improving blood flow to the area and supporting cellular rejuvenation.

Case Study: Alleviating Prostate Discomfort

Consider James, a 50-year-old who experienced discomfort due to an enlarged prostate. He incorporated frequency therapy into his routine, using a device designed to emit specific low-frequency pulses. Over the course of three months, James reported a noticeable reduction in discomfort and greater ease in his daily life, allowing him to maintain his active lifestyle.

Practical Tips for Prostate Health
- Daily Frequency Sessions: Use a device tuned to frequencies between 8-12 Hz for 10-15 minutes daily to support prostate health.
- Sound Meditation: Low-frequency sound baths can also provide relaxation, reducing stress that may exacerbate

prostate issues.

Cardiovascular Health: Tuning the Heart's Rhythm

Cardiovascular health is central to overall wellness, impacting energy levels, mental clarity, and longevity. Frequency therapy offers a promising approach for maintaining cardiovascular health by helping to regulate blood flow, manage blood pressure, and even reduce cholesterol levels.

How Frequencies Affect the Heart

The heart functions in rhythm, and certain frequencies can influence this rhythm to support healthy circulation. Frequencies in the range of 1-10 Hz, particularly when used in pulse therapy like PEMF (Pulsed Electromagnetic Field therapy), have shown potential in improving heart rate variability and circulation.

Example: Supporting Cardiovascular Health

Consider Mark, a 60-year-old man who was advised by his doctor to lower his blood pressure. Along with lifestyle changes, Mark began using a PEMF device at 7.83 Hz for 20 minutes each day. Over a period of six months, his blood pressure decreased significantly, and his doctor noted improvement in his heart rate variability, an indicator of cardiovascular resilience.

Practical Tips for Cardiovascular Support
- PEMF Therapy: Daily PEMF sessions at 7-10 Hz can support cardiovascular health by enhancing circulation and heart rate variability.
- Relaxing Frequency Playlists: Listening to calming frequencies, such as those in the alpha range (8-14 Hz), can reduce stress, a common factor in cardiovascular strain.

Managing Stress: Finding Balance with Frequency Therapy

Stress affects all aspects of men's health, from heart health to mental clarity and hormone balance. Frequency therapy can be a powerful ally in managing stress by promoting relaxation, improving sleep, and fostering mental clarity.

The Role of Frequencies in Reducing Stress

Stress management through frequency therapy taps into brainwave entrainment, where specific sound frequencies guide the brain into more relaxed states. Alpha waves (8-14 Hz) and theta waves (4-8 Hz) are particularly effective for relaxation and stress reduction.

Story of Stress Relief

John, a 35-year-old with a high-stress job, struggled with burnout. Traditional methods like exercise and meditation helped, but he found deeper relaxation through frequency therapy. Listening to binaural beats tuned to alpha and theta frequencies, John experienced a notable decrease in stress and reported better sleep quality and focus.

Practical Tips for Stress Management
- Daily Binaural Beats: Spend 10-15 minutes listening to binaural beats in the alpha range to promote relaxation.
- Evening Frequency Meditation: Wind down with a guided meditation that includes theta waves to support restful sleep and reduce stress.

Frequency Therapy for Cognitive Function and Mental Clarity

Mental clarity and focus are essential for men, particularly in demanding work environments. Frequency therapy can support cognitive function, enhancing focus, memory, and creativity.

Brainwave Entrainment and Focus

Beta waves (13-30 Hz) are associated with active thinking

and concentration, making them ideal for boosting cognitive function. Brainwave entrainment techniques, like binaural beats and PEMF, can guide the brain into beta frequencies, improving mental clarity.

Case Study: Enhancing Focus with Frequencies

Paul, a 40-year-old who struggled with focus in his demanding job, integrated frequency therapy into his routine. Using binaural beats at 15 Hz while working, Paul found that his concentration improved, and he felt more alert and productive throughout the day.

Practical Tips for Mental Clarity
- Beta Frequency Binaural Beats: Listen to beta-frequency binaural beats (13-18 Hz) to enhance focus and concentration.
- Midday Frequency Breaks: Take a short break midday to listen to low-frequency pulses for mental refreshment.

Visionary Reflections: The Future of Men's Health and Frequency Healing

The potential of frequency therapy in men's health is only beginning to unfold. As we deepen our understanding, imagine a future where frequency therapy becomes a routine part of healthcare. Men may one day be able to access devices specifically designed for prostate health, cardiovascular support, and cognitive enhancement, allowing them to manage health proactively and non-invasively.

In this envisioned future, men would have frequency playlists tailored to each stage of life, from enhancing focus and reducing stress in their working years to supporting heart health and prostate wellness as they age. With frequency therapy, the healthcare of tomorrow could empower men to take control of their well-being through personalized, accessible, and side-effect-free methods.

Conclusion: Embracing Balance and Wellness

Men's health requires a holistic approach, one that respects the unique physiological needs of men while offering non-invasive, proactive solutions. Frequency medicine provides a gentle yet effective tool for achieving this balance, whether for supporting prostate health, maintaining cardiovascular fitness, managing stress, or enhancing mental clarity.

By tuning into these frequencies, men can embrace a path of wellness that goes beyond traditional medicine, harmonizing with the body's natural rhythms and fostering a life of resilience, vitality, and peace. Frequency medicine allows men to take an empowered approach to health—one that resonates at every level, supporting the body, mind, and spirit as they navigate life's journey.

CHAPTER 52: FREQUENCY HEALING IN SENIOR CARE

Aging is a journey that comes with its own set of challenges and joys. As people grow older, physical and cognitive changes can affect daily life, from mobility issues and joint pain to memory lapses and mental clarity. Frequency healing has emerged as a promising approach to support senior health, providing non-invasive therapies that resonate with the body's natural systems, aiding in the management of pain, improving mobility, and enhancing cognitive function. Imagine frequency-based therapies as gentle waves that tune the body, restoring balance and harmony to areas in need.

In this chapter, we'll explore how frequency healing can empower seniors to maintain a vibrant and fulfilling life, with a closer look at scientific evidence, practical applications, and inspiring success stories.

Introduction: Embracing Aging with Frequency Healing

Meet Eleanor, a retired teacher in her seventies who started experiencing joint pain and stiffness, which affected her ability to enjoy her favourite activities. Frustrated with the side effects of traditional pain medications, Eleanor turned to frequency therapy after a friend recommended it. She began using low-frequency PEMF (Pulsed Electromagnetic Field) therapy daily, targeting her knees and lower back. After a few weeks, Eleanor found her pain significantly reduced, and her mobility improved. This newfound freedom allowed her to return to gardening and long walks.

Eleanor's story demonstrates the potential of frequency healing for aging individuals. With the right techniques, seniors can reclaim aspects of their health, allowing them to engage in the activities they love.

The Science Behind Frequency Healing in Senior Care

Aging bodies undergo cellular changes, including reduced blood flow, a slowdown in cell regeneration, and an increase in inflammation. Frequency-based therapies address these challenges at the cellular level, providing targeted support where it's needed most. Here are some key ways that frequency healing can benefit senior health:

- Pain Reduction: PEMF and low-level laser therapy (LLLT) have shown promise in reducing inflammation and pain in joints and muscles, helping seniors manage chronic conditions like arthritis without relying on medication.
- Enhanced Mobility: By improving blood flow and supporting cellular regeneration, frequencies between 5-20 Hz can promote joint flexibility and reduce stiffness.
- Cognitive Health Support: Brainwave entrainment techniques using alpha (8-14 Hz) and theta (4-8 Hz) waves can support mental clarity, memory, and relaxation, essential for healthy cognitive function as people age.

Case Studies and Real-World Applications

Case Study: Frequency Therapy for Arthritis

Arthritis affects millions of seniors worldwide, often limiting mobility and reducing quality of life. Fred, an 80-

year-old former athlete, struggled with severe arthritis in his hands, which made it difficult to perform even simple tasks. After learning about PEMF therapy, he decided to try a portable device specifically designed for joint pain. Over three months of daily 10-minute sessions at 10 Hz, Fred reported reduced pain and improved flexibility in his fingers, allowing him to regain independence in daily tasks.

Case Study: Cognitive Clarity and Memory

Memory challenges are common with age, and many seniors experience mild cognitive decline. Patricia, a retired professor, noticed she had trouble remembering names and organizing her thoughts. She began using brainwave entrainment with alpha frequencies (10 Hz) through binaural beats. Patricia listened to these sessions for 20 minutes daily, and within a few weeks, she observed a marked improvement in her focus and memory. This simple practice became an anchor in her daily routine, enhancing her mental clarity and helping her feel more engaged with her surroundings.

Practical Tips: Incorporating Frequency Healing into Senior Care

1. Daily Pain Management: Use a PEMF or low-level laser device targeted at joints for 10-15 minutes per session to reduce inflammation and pain.
2. Cognitive Health: Seniors can listen to alpha or theta frequency binaural beats for 15-20 minutes each day to support memory and mental clarity.
3. Improve Circulation: Sessions with frequencies around 8-12 Hz can enhance blood flow, benefiting both physical mobility and overall health.

For those new to frequency therapy, it's essential to start slow and incorporate these practices gradually. Even short, consistent sessions can produce noticeable benefits over time.

Visionary Reflections: The Future of Frequency Healing for Aging Populations

As technology advances, the potential of frequency healing in senior care is becoming clearer. Imagine a world where every senior has access to a customized frequency healing program, tailored to their unique needs, from pain relief to cognitive support. In the near future, we could see frequency-based "health pods" in assisted living facilities, where residents can enjoy non-invasive therapies that support vitality and longevity.

One emerging area is the development of wearable frequency devices that adjust in real-time to the body's feedback. For seniors, these could be revolutionary, allowing for continuous, on-the-go therapy that adapts to their needs throughout the day. This vision holds the promise of a healthcare system that prioritizes preventive, holistic care, empowering seniors to live independently and vibrantly for as long as possible.

Reflections: Beyond Physical Health

Frequency healing offers seniors more than just physical relief; it can also support emotional well-being and social engagement. Many seniors face feelings of isolation or anxiety, and frequency-based therapies can foster a sense of calm and connectedness. By aligning the mind and body, these therapies create a space for seniors to reconnect with themselves and others, enhancing their overall quality of life.

As Eleanor discovered, frequency healing goes beyond physical wellness. It fosters a holistic experience, reminding seniors that aging doesn't have to mean losing vitality. With frequency therapy, they can continue to grow, adapt, and embrace every chapter of life with joy and resilience.

Conclusion: A New Path to Aging Gracefully

Incorporating frequency healing into senior care isn't just about managing symptoms; it's about enhancing quality of life. By addressing the specific challenges of aging with non-invasive techniques, seniors can feel empowered and supported on their journey. Frequency healing offers a bridge between scientific rigor and the intuitive wisdom that our bodies are, at their core, vibrational beings capable of harmony and healing.

Through the stories of Eleanor, Fred, Patricia, and countless others, we see the transformative power of frequency medicine in supporting aging individuals. As more seniors and caregivers discover these tools, frequency healing will continue to redefine what it means to age gracefully, reminding us that true wellness is about fostering harmony within.

Frequency medicine has the potential to empower seniors to embrace aging with dignity, strength, and joy—a path to wellness that resonates deeply, not just with the body, but with the spirit.

CHAPTER 53: FREQUENCY THERAPY FOR ATHLETES AND FITNESS ENTHUSIASTS

Imagine an orchestra, each instrument resonating in harmony to create a powerful, moving sound. Now, imagine your body as that orchestra, where each cell, muscle, and joint plays its own note, collectively creating the symphony of human movement. When our bodies perform physical feats—whether sprinting across a finish line, lifting weights, or manoeuvring through yoga poses—this symphony can fall out of tune. Muscles tire, joints ache, and energy wanes. For athletes and fitness enthusiasts, maintaining the balance between performance and recovery is essential. This is where frequency therapy comes into play, offering a powerful, non-invasive way to restore harmony, promote faster healing, and enhance performance.

In this chapter, we'll explore how frequency therapy can revolutionize the world of athleticism. From scientific insights to real-life stories and practical applications, this chapter will illuminate how frequencies can accelerate muscle recovery, reduce injury risk, and take performance to the next level.

The Science Behind Frequency Therapy and Athletic Recovery

Athletes often push their bodies to the edge, which can result in microtears in muscles, inflammation, and fatigue. Frequency therapy, whether through pulsed electromagnetic fields (PEMF), sound waves, or vibration plates, works by realigning cellular frequencies to promote healing and cellular repair.

Key Mechanisms of Action:
1. Cellular Repair: PEMF therapy enhances cellular energy production and communication, accelerating the repair of damaged tissues. Frequencies in the range of 8-15 Hz —similar to the Earth's natural frequency (Schumann resonance)—are known to support cellular recovery and tissue repair.
2. Reduced Inflammation: Low-level laser therapy (LLLT) and PEMF can reduce inflammation at the cellular level, alleviating pain and promoting healing by increasing blood flow to injured areas.
3. Enhanced Circulation: Frequencies in the range of 5-20 Hz improve blood flow, delivering oxygen and essential nutrients to muscles and tissues. This increased circulation aids in endurance, speeds up recovery, and reduces soreness.

Case Studies: Real-Life Transformations Through Frequency Therapy

Case Study 1: A Runner's Return to Form

Lily, a marathon enthusiast, experienced chronic calf pain after years of rigorous training. Searching for a solution, she began using a PEMF device targeting her calves at 10 Hz for

20 minutes post-run. Within weeks, she felt less soreness, and her post-race recovery time improved significantly. Frequency therapy restored her strength, allowing her to compete pain-free.

Case Study 2: Injury Prevention in a Professional Soccer Player

Carlos, a pro soccer player, faced frequent ankle sprains, which hindered his game. After incorporating vibration therapy at 30 Hz to strengthen his ankle, he noticed increased flexibility and stability. This proactive use of frequency therapy reduced his injury risk, enabling him to play more confidently.

Practical Applications of Frequency Therapy for Fitness Enthusiasts

For athletes at all levels, frequency therapy offers practical solutions for performance and recovery. Here are some key ways to incorporate it into your routine:

1. Post-Workout Recovery: Use PEMF mats or devices targeting sore muscles for 10-15 minutes after exercise to speed up recovery and reduce soreness.
2. Enhanced Warm-Up: Vibration plates or low-frequency devices can stimulate muscles before workouts, potentially improving performance and reducing the risk of strains.
3. Injury Rehabilitation: Frequency therapies like LLLT or PEMF offer targeted relief for strains and sprains, promoting quicker and more efficient recovery.

DIY Frequency Therapy Techniques for Athletes

If you're interested in exploring frequency therapy at home, several accessible tools and techniques can support your fitness journey:

- Binaural Beats for Mental Focus: Listening to specific

binaural beats during warm-ups can boost concentration. Theta waves (4-8 Hz) create a calm but focused state, ideal for maintaining steady energy in endurance sports.

- Vibration Plates for Muscle Activation: Standing on a vibration plate for 5-10 minutes pre-workout can activate and warm up muscles, making them more responsive to exercise.

- Handheld PEMF Devices: These portable devices can be used on sore or strained areas post-workout, providing an on-the-go solution for muscle recovery and tension relief.

Visionary Future of Frequency Therapy in Sports

As research in frequency therapy continues, its applications for athletes are bound to expand. Envision a future where every professional sports team has a customized frequency therapy suite, where AI-driven frequency devices analyse real-time recovery data and adjust treatments based on the athlete's needs. Imagine trainers using frequency-based therapies to target specific muscle groups without the need for high-impact weights, allowing athletes to train effectively while minimizing injury risks.

In this potential future, athletes might even use frequency therapy to enhance their mind-body connection, achieving states of peak concentration and resilience. By tailoring therapy to both the body and the mind, frequency technology could redefine what it means to be "in the zone."

Reflections: Frequency Therapy as a Pathway to Balance

Frequency therapy does more than enhance physical performance; it helps athletes cultivate mental resilience. In sports, mental focus can be as critical as physical strength. Frequency therapies like binaural beats or sound frequencies can help athletes manage stress and maintain composure under pressure. By tuning both mind and body, frequency therapy allows athletes to unlock their full potential in a balanced, holistic way.

Practical Guide to Implementing Frequency Therapy

If you're ready to try frequency therapy, here are a few steps to start:

1. Identify Your Needs: Consider whether you need help with recovery, injury prevention, or mental focus.
2. Choose Your Tool: PEMF devices, vibration plates, and binaural beats each offer unique benefits. Experiment to find what works best for your needs.
3. Consistency Is Key: For optimal results, integrate frequency therapy into your regular routine. For example, try PEMF therapy after every workout for a month and track your recovery and performance improvements.

Conclusion: Elevate Your Performance with Healing Frequencies

As athletes, fitness enthusiasts, or simply individuals committed to a healthier life, frequency therapy provides a revolutionary approach to achieving peak performance. Through scientifically backed, non-invasive techniques, frequency therapy can harmonize the body's rhythms, promoting recovery, resilience, and a new level of physical and mental harmony.

Imagine a future where frequency-based devices are as common in gyms as treadmills or dumbbells, where athletes intuitively use frequencies to reset and recharge. This journey toward self-discovery through frequency therapy is not just about strengthening the body—it's about orchestrating a symphony of balance, resilience, and excellence.

In this quest for harmony, athletes like Lily and Carlos demonstrate that frequency therapy is more than just a tool—it's a partner in the journey to peak performance. As you explore this realm, may you discover your own healing harmonies, resonating with the profound potential within. The journey toward peak performance, where every step, stretch, and stride resonates with possibility, is a symphony awaiting your unique rhythm.

PART XIII: ADVANCED THERAPIES AND TECHNIQUES

CHAPTER 54: TRANSCRANIAL MAGNETIC STIMULATION (TMS) FOR MENTAL HEALTH

Imagine if, by harnessing the power of magnetic fields, we could reshape the brain's activity and alleviate mental suffering. This is not a distant dream but a present-day reality made possible by Transcranial Magnetic Stimulation, or TMS. In a world where millions struggle with mental health conditions like depression and anxiety, TMS offers a powerful, non-invasive alternative to medications, which often come with a host of side effects and inconsistent results. TMS taps into the body's natural ability to respond to frequency-based treatments, providing hope to those who haven't found relief through traditional means.

This chapter explores the science, application, and profound potential of TMS, offering both an accessible understanding of how it works and an inspirational look at its transformative impact on people's lives.

How TMS Works: A Gentle Reset for the Mind

At its core, TMS works by stimulating specific regions of the brain with targeted magnetic pulses. These pulses are delivered through a coil placed near the scalp, generating a magnetic field that reaches approximately 2-3 centimetres into the brain's outer layers. This field creates tiny electrical currents that "reset" neural pathways and modulate brain activity in areas linked to mood, particularly the prefrontal cortex.

To make this clearer, think of the brain like a network of highways. In people with depression, the "traffic" of neural signals slows down or even stalls in certain regions. TMS acts like a traffic conductor, signalling when and where to direct energy flow, allowing these neural pathways to resume normal function and activity.

The Science of TMS: Evidence and Effectiveness

TMS has undergone extensive clinical trials, particularly as a treatment for depression. Studies indicate that it is particularly effective for those with treatment-resistant depression, a condition in which conventional therapies, including medication, have proven insufficient. Research shows that approximately 60% of patients respond to TMS, with many experiencing a significant reduction in depressive symptoms, and around a third achieving full remission.

Key Mechanisms at Play
1. Enhanced Neuroplasticity: TMS promotes neuroplasticity, the brain's natural ability to adapt and reorganize by forming new neural connections. By stimulating specific areas, TMS encourages the brain to "retrain" itself, potentially altering maladaptive thought and behaviour patterns.
2. Regulation of Neurotransmitters: TMS has been shown to

influence the release and uptake of neurotransmitters, such as serotonin and dopamine, which play crucial roles in mood regulation.

3. Targeted Impact: Unlike medication, which affects the entire brain, TMS focuses solely on the targeted areas associated with mood, thus reducing the risk of side effects.

Real-Life Stories: TMS in Action

Sarah's Journey to Relief
Sarah, a 32-year-old teacher, had been battling depression for most of her adult life. After years of trying various medications and therapies with limited success, she felt trapped in a cycle of sadness and fatigue. Her psychiatrist suggested TMS, and though hesitant, she decided to try it.

After her third week of TMS sessions, Sarah began noticing subtle shifts. Tasks that once felt overwhelming became manageable, and her persistent feelings of hopelessness started to fade. By the end of her treatment cycle, Sarah felt more in control of her life, experiencing clarity and energy she hadn't felt in years. For her, TMS was more than a treatment—it was a lifeline.

David's Battle with OCD
David, a 28-year-old programmer, struggled with obsessive-compulsive disorder (OCD). He'd tried medications, behavioural therapies, and even alternative treatments with little success. When he heard about TMS, he was intrigued by its non-invasive approach. Though initially sceptical , he gave it a chance.

During his treatment, David experienced a gradual reduction in the intensity of his obsessive thoughts. By the end of his sessions, he reported feeling more liberated, with his compulsions drastically reduced. TMS provided him with a new lease on life, offering relief from the constant mental burden of OCD.

What to Expect from a TMS Treatment

For anyone considering TMS, knowing what to expect can alleviate some of the anxiety surrounding the procedure. Here's a breakdown:

1. Preparation: The process is straightforward. Patients sit in a comfortable chair, and a technician positions the TMS device close to the head.
2. The Treatment: Magnetic pulses are delivered in quick bursts. The sensation is often described as a light tapping, which some people find unusual but not painful.
3. Duration and Frequency: Sessions usually last between 20-40 minutes, typically administered five days a week for four to six weeks.
4. Minimal Side Effects: Unlike medications, TMS's side effects are minimal, most commonly including mild scalp discomfort or headaches.

DIY Mindfulness Techniques to Complement TMS

While TMS is powerful on its own, integrating it with mindfulness practices can amplify its benefits. Here are a few techniques that patients undergoing TMS might find helpful:

- Breathing Exercises: Slow, deep breaths can help calm the mind and prepare it to receive TMS treatments. Try inhaling for a count of four, holding for four, and exhaling for six.
- Visualization: Picture your brain's neural pathways resetting, like dust settling after a storm. Visualizing healing can create a supportive mental environment that complements TMS's physical effects.
- Journaling: Track your progress, noting any changes in mood or thought patterns. Journaling can provide a reflective space, helping you recognize subtle shifts that may otherwise go unnoticed.

The Future of TMS and Mental Health Treatment

As technology advances, TMS is likely to become even more precise and personalized. Imagine devices that analyse brain activity in real-time, adjusting magnetic pulses to the unique neural patterns of each individual. With ongoing research, TMS could also expand to treat a broader range of conditions, from anxiety disorders to PTSD.

Moreover, the future may hold collaborative treatments where TMS is combined with AI-driven insights. AI could track individual responses to frequency, customizing treatment in ways that deepen and extend therapeutic outcomes. This vision isn't far off, and it could make TMS an even more accessible, widely used mental health tool.

Reflecting on TMS: A Gentle Invitation to Healing

TMS offers more than symptom relief; it opens doors to mental clarity and emotional resilience. Unlike traditional methods that may feel invasive or come with side effects, TMS feels like a gentle nudge to the brain, reminding it of its innate capacity to heal and adapt.

In a world where mental health conditions are on the rise, TMS offers hope. It's a reminder that science and technology can work in harmony with the body's own frequencies to support wellness. For anyone feeling lost in their mental health journey, TMS serves as a testament to the power of innovative medicine to bring light into dark spaces.

Summary and Key Takeaways

- Non-Invasive and Precise: TMS provides targeted relief for mental health conditions, specifically treatment-resistant depression, with minimal side effects.
- Proven Results: Clinical trials show a success rate of approximately 60% in relieving depressive symptoms, with around a third achieving full remission.
- Complementary Practices: Techniques like breathing

exercises, visualization, and journaling can enhance TMS's effectiveness.

- A Bright Future: TMS's future holds potential for more personalized and even more effective treatments.

As TMS continues to evolve, it shines as a beacon of hope for those seeking new pathways to healing, an invitation to experience mental health recovery in ways once thought unimaginable. Through TMS, we see the profound potential of frequency medicine in the quest for mental clarity, resilience, and peace.

CHAPTER 55:
CYMATICS AND
VISUAL FREQUENCY
THERAPY

Imagine watching as sound comes to life, as vibrations create patterns in water or particles on a metal plate, forming shapes so precise and harmonious that they seem almost magical. This is the world of cymatics—a field that transforms invisible sound frequencies into visible, often breathtakingly beautiful, forms. In essence, cymatics allows us to "see" sound, and with it, gain deeper insight into the harmony that exists between frequency and form, and ultimately, between energy and health. This chapter will explore cymatics' role in frequency medicine, offering a window into how sound and vibration can transform our understanding of healing and wellness.

The Basics of Cymatics: Making the Invisible Visible

The word "cymatics" comes from the Greek word kyma, meaning wave. At its heart, cymatics involves using different media, like water or sand, to make sound vibrations visible. Each frequency produces a unique geometric pattern, like a fingerprint for that specific sound. Imagine striking a note on a piano and watching particles of sand or drops of water arrange themselves into symmetrical, almost mandala-like patterns. These forms reveal the natural resonance and inherent order in sound frequencies, creating an awe-inspiring connection between sound and matter.

One of the foundational figures in cymatics was Swiss physician and natural scientist Dr. Hans Jenny, who, in the 1960s, extensively studied the effects of sound waves on various materials. His experiments revealed that certain frequencies create intricate, stable patterns, while others lead to chaos and disorder. Jenny's work demonstrated that sound not only influences physical matter but also has the potential to guide it into harmonious states.

Cymatics and the Body: An Orchestra of Cells

If we consider the human body, it too is constantly resonating with frequencies. Our cells communicate with each other, often through electrical impulses, creating their own vibratory language. Just as a cymatic pattern responds to sound, our cells and tissues respond to frequencies. In a healthy body, cells "vibrate" at specific, optimal frequencies, working together like an orchestra, each with its own unique tone yet harmonizing with the whole. When these vibrations fall out of harmony due to illness, stress, or injury, we experience physical symptoms. Cymatics in medicine aims to use frequency to restore these harmonious states.

A compelling example is how cymatic patterns have been applied to visualize the heartbeat or brainwaves, offering real-time glimpses into the frequencies that dictate our physiological state. By translating these frequencies into visible forms, researchers and practitioners can study and even manipulate them to encourage healing and balance.

Therapeutic Applications of Cymatics: Sound, Frequency, and Healing

As cymatics has gained traction in scientific and

alternative medicine communities, researchers have explored therapeutic applications. Here are some of the most promising ways cymatics is being used:

1. Cymatherapy: Healing with Sound Waves
 - Cymatherapy, inspired by cymatics, uses sound frequencies to stimulate the body's natural healing processes. By applying specific sound frequencies through devices that create cymatic patterns, practitioners aim to bring the body back into resonance. These devices deliver precise frequencies to targeted areas of the body, encouraging cellular repair and balance.
 - In practice, cymatherapy has shown potential in treating chronic pain, inflammation, and even in accelerating wound healing. Many practitioners believe that by "tuning" the body, they can reduce stress, boost immunity, and improve overall wellness.

2. Water and Vibration Therapy
 - Water is highly responsive to sound vibrations, and researchers have found that exposing water to certain frequencies can "structure" it, forming organized patterns. Given that our bodies are about 60% water, this discovery is significant. Structured water within our cells is thought to be more bioavailable, supporting cell function and hydration.
 - In frequency-based healing, practitioners use cymatics to infuse water with frequencies that promote cellular health. When this structured water is consumed, it may aid in detoxification and cellular regeneration, creating a ripple effect of health throughout the body.

3. Visual Sound Therapy for Mental Health
 - Cymatics is also finding a role in mental health therapy through "visual sound baths." In these sessions, patients are exposed to soothing sound frequencies while watching the corresponding cymatic patterns projected on screens. This combination engages both auditory and visual senses,

deepening relaxation and helping individuals achieve meditative states.

- For people with anxiety, depression, or PTSD, visual sound therapy offers a gentle way to reprogram the mind and body by fostering inner calm and emotional release. By experiencing sound visually, patients may connect more deeply with their own inner harmony.

Case Studies: Healing Stories Through Cymatics

Maria's Journey with Chronic Pain
Maria, a 50-year-old artist, had struggled with chronic back pain for years. Traditional therapies offered little relief, so she began exploring alternative options, including cymatherapy. During her sessions, she listened to specific frequencies while watching cymatic visuals corresponding to those sounds. She described the experience as "hypnotic," feeling as though her cells were realigning.

Over a few months, Maria reported a gradual reduction in her pain levels, finding relief that lasted longer after each session. Today, she uses cymatherapy as a regular part of her wellness routine and has seen profound improvements not only in her pain levels but in her overall vitality.

Healing with Structured Water
In another case, a naturopathic clinic experimented with structured water infused with cymatic frequencies for clients with inflammatory conditions. Patients who consumed the water daily reported enhanced hydration, reduced inflammation, and a greater sense of energy. While the science behind structured water is still evolving, these anecdotal reports suggest promising potential for cymatics in cellular health.

DIY Cymatics: Exploring Healing Frequencies at Home

If you're curious about experiencing the principles of

cymatics yourself, there are simple ways to explore this transformative world of sound and vibration:

1. Experiment with Frequency Videos: Many online videos display cymatic patterns for various frequencies. Find a frequency that resonates with you, sit comfortably, and watch the patterns while allowing yourself to connect with the sound's vibration. This can be a deeply calming practice, ideal for stress relief or meditative focus.

2. Sound Baths with Cymatic Art: You can create a mini sound bath at home by playing calming frequencies (such as 432 Hz or 528 Hz) while watching cymatic patterns or artwork inspired by these frequencies. Many people find these visuals comforting, bringing a sense of inner peace.

3. Water Experiment: Fill a glass with water, play soft frequencies nearby, and observe any subtle ripples or movements. While it's not as elaborate as a scientific cymatic setup, this simple experiment can remind you of how sound affects water—and, by extension, our bodies.

Reflecting on Cymatics: A Gateway to Inner Harmony

Cymatics offers more than a visual spectacle; it unveils the hidden geometry of sound, a reminder of the intricate connection between frequency and form. When we witness sound creating order from chaos, we are reminded of our own potential for healing. Cymatics shows that harmony is not just an abstract concept—it's a tangible, visible phenomenon that can bring real change to our health and well-being.

Looking to the Future: The Expanding Potential of Cymatics in Medicine

As cymatics gains recognition, we can anticipate its growing role in personalized medicine. Imagine practitioners using cymatic analysis to diagnose imbalances, applying precise frequencies to guide the body back to health. With advances in technology, it's not far-fetched to envision a world where hospitals and clinics routinely incorporate cymatics as a diagnostic and therapeutic tool, bridging the gap between science and art in the pursuit of health.

In witnessing the beauty of cymatics, we see that harmony is not just a concept but a state we can nurture within ourselves. By tuning into the healing frequencies of sound, we step closer to understanding the profound potential of frequency medicine in our journey toward balance and wholeness.

CHAPTER 56: PHOTOTHERAPY: LIGHT FREQUENCIES FOR HEALING

The healing potential of light has been revered for centuries, with sunlight once celebrated as a source of vitality and health. Today, science has uncovered the profound ways that specific wavelengths of light interact with our cells, prompting regeneration, relief, and balance. This chapter will delve into phototherapy—a field that uses the power of light frequencies to treat a variety of conditions, from skin ailments to mood disorders.

Light as Medicine: The Foundation of Phototherapy

Imagine that every cell in your body is like a solar panel, absorbing energy from light to fuel its processes. In a way, that's exactly how phototherapy works. By exposing the body to specific wavelengths, phototherapy can trigger cellular responses that lead to healing. This is no longer a mystical concept but one firmly rooted in biophysics: our cells contain light-sensitive receptors that respond to different frequencies in remarkable ways.

At the heart of phototherapy lies the concept of using specific wavelengths—visible light, infrared, or ultraviolet

—to target particular cells and tissues. Each wavelength penetrates the skin to a different depth, stimulating a variety of biochemical processes. For example, red light reaches the mitochondria, the powerhouses of the cell, enhancing energy production, while blue light, with its shorter wavelength, targets the skin's surface, killing bacteria and reducing inflammation.

The Science Behind Light Frequencies and Cellular Healing

Phototherapy taps into the body's inherent ability to respond to light, a field known as photo biomodulation. In photo biomodulation, light is absorbed by chromophores, light-sensitive molecules in cells. When these molecules absorb light, they initiate a cascade of biological processes that can lead to healing and restoration.

Consider how plants undergo photosynthesis, using light to produce energy. In our cells, phototherapy promotes a similar process where specific wavelengths stimulate the production of ATP (adenosine triphosphate), the energy currency of cells. This boost in cellular energy allows cells to repair damage, regenerate tissue, and reduce inflammation.

One compelling example of this is the use of red and near-infrared light to support muscle recovery and reduce joint pain. By promoting blood flow and reducing inflammation, red light has become a popular choice among athletes and those with chronic pain conditions.

Applications of Phototherapy: Skin Health, Mood Enhancement, and More

Phototherapy is applied in various forms, from high-tech devices to home-based light therapy lamps. Here are some of the most common therapeutic applications:

1. Skin Health and Acne Treatment
   - Blue Light Therapy: Blue light therapy has gained

popularity as an effective, non-invasive treatment for acne. Blue light penetrates the skin's surface, targeting the bacteria that cause acne without harming the surrounding tissues. By killing acne-causing bacteria and reducing inflammation, blue light helps improve skin health with minimal side effects.

- Red Light Therapy: Red light, with its deeper penetration, promotes collagen production and improves skin elasticity, making it an effective tool in anti-aging treatments. Dermatologists often use red light therapy to reduce wrinkles, improve skin tone, and accelerate wound healing, especially after procedures like laser resurfacing.

2. Mood Disorders and Seasonal Affective Disorder (SAD)

- Bright Light Therapy: For individuals struggling with seasonal affective disorder (SAD), a type of depression that typically occurs during the darker months, bright light therapy is a transformative tool. Exposure to a bright light box for 20-30 minutes each morning can mimic natural sunlight, triggering the release of serotonin, the "feel-good" hormone, and helping regulate circadian rhythms. This therapy is highly effective in improving mood, energy levels, and sleep patterns.

- Full-Spectrum Light Therapy: Full-spectrum light, which simulates natural daylight, is used to regulate melatonin production, the hormone responsible for sleep. Many people find that full-spectrum light therapy helps combat fatigue, depression, and other mood disorders associated with lack of sunlight.

3. Pain Relief and Muscle Recovery

- Infrared Therapy: Infrared light, invisible to the human eye, penetrates even deeper than red light, reaching muscles, nerves, and bones. Infrared therapy has shown promise in reducing inflammation, easing chronic pain, and accelerating muscle recovery, making it a popular choice for athletes and those suffering from chronic pain conditions like arthritis.

- Low-Level Laser Therapy (LLLT): LLLT uses low-intensity lasers to stimulate cellular function, reduce inflammation, and promote tissue repair. Often used in physical therapy, LLLT is helpful for those with muscle injuries, joint pain, and even neuropathy. It's a non-invasive, drug-free alternative that provides relief by enhancing the body's natural healing processes.

Case Studies: Real-Life Healing with Phototherapy

A New Outlook for Sarah's Skin Condition
Sarah, a 28-year-old with severe cystic acne, had tried countless topical treatments without success. Her dermatologist recommended a combination of blue and red light therapy sessions. Within weeks, she noticed a reduction in inflammation and fewer new breakouts. Over the course of several months, her skin continued to improve, restoring her confidence and helping her find a natural, non-invasive solution to a lifelong struggle.

James's Battle with Seasonal Depression
James, a marketing executive, dreaded winter every year due to seasonal affective disorder. When his therapist suggested bright light therapy, he was sceptical but willing to try. After setting up a light box at home and using it every morning, James experienced a profound shift. His energy improved, he felt more focused, and the once-daunting winter months became manageable. Light therapy had given him a new sense of control over his mood and well-being.

DIY Light Therapy: Bringing Healing Frequencies into Your Daily Life

If you're curious about trying phototherapy at home, there are simple, accessible ways to incorporate healing light into your routine:

1. Bright Light Therapy for Mood: If you struggle with low energy or mood during darker months, consider a light therapy box with 10,000 lux intensity. Place it 16-24 inches from your face and use it in the morning for 20-30 minutes to start your day on a positive note.

2. Red Light for Skin and Muscle Recovery: Red light panels, now available for home use, can be used on areas where you want to enhance skin health or relieve muscle soreness. Follow the manufacturer's guidelines for duration and frequency of use to ensure safe and effective results.

3. Full-Spectrum Lighting for Workspace: If you work indoors or away from windows, a full-spectrum light can mimic daylight, supporting a natural circadian rhythm. Many people find it boosts focus, productivity, and mood throughout the day.

Reflecting on Phototherapy: Light as a Path to Inner and Outer Health

Light has always symbolized hope, renewal, and clarity. With phototherapy, we're discovering that light is not only a metaphor but a powerful, tangible force for healing. By harnessing specific wavelengths, phototherapy taps into the body's natural rhythms, offering gentle, non-invasive treatments that empower the body's own healing mechanisms.

The Future of Light Frequencies in Medicine

As research advances, we may see phototherapy expand into even more areas of medicine. Scientists are exploring light therapy for neurological conditions, wound healing, and even cancer treatment, with early results showing promise. With continued exploration, phototherapy could become an essential component of mainstream medicine, providing healing solutions that align with the body's natural processes.

In a world dominated by artificial lights and screen glare, phototherapy brings us back to the pure essence of natural light, connecting us with rhythms that have sustained life for millennia. Embracing light as a healing frequency allows us to experience the harmony and balance that nature offers, illuminating our path to well-being and offering a glimpse into the limitless potential of frequency-based medicine.

CHAPTER 57: LASER ACUPUNCTURE AND FREQUENCY STIMULATION

In recent decades, a powerful synergy has emerged at the intersection of ancient healing practices and cutting-edge technology. This union is laser acupuncture, a modern evolution of traditional acupuncture that uses specific frequencies of light instead of needles to stimulate healing within the body. By combining the time-honoured knowledge of acupuncture points with the precision of laser technology, laser acupuncture brings a new dimension to healing—offering a needle-free, non-invasive approach that is both effective and gentle.

A New Twist on Ancient Wisdom

Acupuncture has a history that spans thousands of years, with roots in Traditional Chinese Medicine (TCM). Central to this practice is the belief that the body's energy, or "Qi," flows through pathways called meridians. According to TCM, illness and discomfort occur when this energy becomes blocked or unbalanced. Acupuncture, by targeting specific points along these meridians, is thought to restore the body's natural energy flow, promoting healing and harmony.

Laser acupuncture maintains this philosophical foundation but replaces the traditional needles with focused light. By using lasers tuned to specific wavelengths, practitioners can stimulate these same energy points on the body, encouraging a natural healing response without piercing the skin. Think of it as tuning an instrument. Each point on the body responds to specific "notes" or frequencies. The laser acts as a tuning fork, helping the body recalibrate and return to balance.

How Laser Acupuncture Works

To understand laser acupuncture, it's helpful to explore the science behind laser technology. Lasers produce focused beams of light at specific wavelengths, which can penetrate the skin and interact with cells. When applied to acupuncture points, these laser frequencies stimulate cellular activity, particularly in mitochondria, the cell's energy centres. This process, known as photo biomodulation, leads to increased production of ATP (adenosine triphosphate), the molecule responsible for cellular energy.

The advantage of using lasers over needles lies in their precision. Each acupuncture point can be targeted with specific light frequencies, allowing practitioners to customize treatments according to the needs of each patient. For example, lower-energy lasers can be used for a calming effect, suitable for anxiety and stress, while higher-energy lasers may be directed to areas of pain or inflammation to provide relief and promote healing.

Laser acupuncture is particularly advantageous for individuals who may be hesitant about needles, including children and those with a low pain tolerance. The method is non-invasive, painless, and often more accessible to those who are wary of traditional acupuncture.

The Benefits and Applications of Laser Acupuncture

Laser acupuncture has shown promise across a wide range of health conditions, from chronic pain and inflammation to stress and fatigue. Here are some of the primary areas where it has proven effective:

1. Pain Management

- Chronic Pain Relief: Many studies have documented the efficacy of laser acupuncture for chronic pain management. The light stimulates endorphin release, the body's natural painkillers, offering relief without the need for medication. Conditions like arthritis, fibromyalgia, and neuropathic pain have responded particularly well to laser acupuncture.

- Sports Injuries: Athletes and fitness enthusiasts are also turning to laser acupuncture for faster recovery and pain relief. By targeting injured muscles or tendons, laser acupuncture can reduce inflammation and stimulate healing, making it a popular choice for those seeking non-invasive recovery methods.

2. Mental Health and Stress Reduction

- Anxiety and Depression: Laser acupuncture has a calming effect on the nervous system, helping regulate the body's response to stress. Many practitioners have observed that patients with anxiety and depression find relief through regular laser acupuncture sessions, as the light frequencies help calm the body and mind.

- Sleep Disorders: Certain acupuncture points are associated with promoting relaxation and restful sleep. Laser acupuncture, by gently stimulating these points, can help regulate sleep cycles, offering a natural remedy for insomnia and disrupted sleep patterns.

3. Digestive Health

- Irritable Bowel Syndrome (IBS): Laser acupuncture can be used to target points associated with digestive health.

Research has shown that regular sessions can alleviate symptoms of IBS, such as abdominal pain, bloating, and irregular bowel movements, by calming the nervous system and reducing inflammation in the gut.

- Improved Metabolism: Laser acupuncture is also associated with improved metabolism and digestion. By stimulating points linked to metabolic functions, it can help balance digestive processes, supporting those with sluggish digestion or metabolic disorders.

4. Immune System Support

- Immune Modulation: Laser acupuncture has shown potential in modulating the immune system, enhancing its ability to fight infections. By targeting specific points, practitioners can stimulate immune function, which can be especially beneficial for individuals with autoimmune disorders or weakened immunity.

Case Studies: Real-World Success with Laser Acupuncture

Maria's Journey with Chronic Pain
Maria, a 50-year-old teacher, had struggled with chronic knee pain for years due to arthritis. Traditional acupuncture had provided some relief, but she was seeking a more convenient option that wouldn't involve needles. Her acupuncturist suggested laser acupuncture, and within weeks of starting treatment, Maria noticed a significant reduction in pain. The sessions were painless and easy, allowing her to go about her day without interruption. Laser acupuncture not only alleviated her pain but also improved her overall mobility.

Liam's Battle with Insomnia
Liam, a 35-year-old tech entrepreneur, had been plagued by insomnia due to work-related stress. After trying medications, which left him groggy, he turned to laser acupuncture. Targeting points associated with relaxation

and sleep, Liam's therapist helped him restore his natural sleep patterns. Within a month, Liam's sleep improved drastically. He reported feeling more rested and focused, and his insomnia became a distant memory.

DIY Tips: Integrating Laser Acupuncture at Home

For those interested in exploring the benefits of laser acupuncture at home, there are now devices designed for safe, at-home use. While professional guidance is recommended, here are some general tips for getting started:

1. Choose the Right Device: Look for a low-level laser therapy (LLLT) device that is specifically intended for acupuncture or phototherapy. Many devices come with instructions and recommendations for various conditions.

2. Start with Relaxation Points: If you're using laser acupuncture for stress relief or sleep, try applying the laser to the space between the eyebrows (known as the "third eye" point) or the wrist's inner crease.

3. Follow Recommended Dosages: Consistency is key with laser acupuncture. Follow the manufacturer's guidelines on duration and frequency to ensure safe and effective results.

Reflecting on Laser Acupuncture: A Glimpse into the Future of Healing

Laser acupuncture represents a profound merging of traditional wisdom and modern science. It respects the philosophy of acupuncture while addressing the needs and preferences of modern patients who seek non-invasive treatments. By offering a gentle yet powerful alternative to needles, laser acupuncture has expanded the possibilities of frequency-based medicine, inviting more people to explore its benefits.

The Future Potential of Laser Acupuncture in Mainstream

Medicine

With ongoing research and technological advancements, laser acupuncture is likely to gain broader acceptance within mainstream medicine. As more studies affirm its efficacy and safety, healthcare providers may integrate laser acupuncture into treatments for chronic pain, stress, immune support, and more. By continuing to explore the potential of laser-based therapies, we are not only honouring ancient healing traditions but also paving the way for a future where healing is as precise as it is profound.

Laser acupuncture stands as a testament to the endless potential of frequency-based healing—revealing how light, like sound, can harmonize the body, restore balance, and connect us to the deep currents of healing energy that have always been within reach.

CHAPTER 58: HYPERBARIC OXYGEN THERAPY (HBOT) AND FREQUENCY

Hyperbaric Oxygen Therapy (HBOT) is an established medical treatment that involves breathing pure oxygen in a pressurized environment. The process, by saturating the body with oxygen, stimulates cellular repair and accelerates healing from various ailments, from chronic wounds to brain injuries. But what happens when we introduce sound frequencies into this process? The combination of HBOT and sound frequencies opens up exciting possibilities, presenting a unique approach to health and recovery. This chapter explores how this synergy works, grounded in science but with an inspiring look toward the potential future of frequency-based healing.

Oxygen: The Essential Fuel for Healing

Oxygen is vital for our survival, but it's more than just an element in the air we breathe. Every cell in the human body relies on oxygen to produce energy, repair itself, and maintain normal function. When we are injured or fighting an illness, our body's demand for oxygen increases, especially in the affected tissues. However, in cases of severe injury, infection, or inflammation, oxygen supply to tissues can become limited, which slows down the healing process. This is where HBOT comes into play.

In an HBOT session, patients enter a hyperbaric chamber, where the atmospheric pressure is increased, allowing the lungs to take in more oxygen than would be possible at normal pressure. This enriched oxygen travels through the bloodstream, flooding cells and tissues, promoting faster repair and regeneration. Studies have shown that HBOT can reduce inflammation, encourage new blood vessel formation, and even aid in the healing of traumatic brain injuries.

The Science of Frequency and Cellular Resonance

To understand how sound frequencies enhance the effects of HBOT, we must first understand cellular resonance. Think of each cell as a mini-orchestra, vibrating at its unique "note" when healthy. However, when a cell is damaged, it can become "out of tune." Certain frequencies can help these cells regain their original resonance, essentially helping them "tune back" to their healthiest state.

The field of frequency medicine suggests that specific sound frequencies can stimulate cellular activity, aid in tissue repair, and reduce stress. When we apply these sound frequencies to the body, they interact with cells on a subtle level, similar to how a tuning fork can resonate and create harmony. Studies have shown that different frequencies affect different tissues, with some frequencies more effective at calming the nervous system while others promote circulation or immune response.

Merging HBOT and Sound Frequencies: A Powerful Synergy

When HBOT is combined with sound frequencies, we create

a therapeutic environment that not only supplies tissues with oxygen but also "tunes" them to optimize healing. Imagine each cell in your body as a musician in an orchestra, ready to play. Oxygen is the energy that fuels their performance, while sound frequencies provide the rhythm that keeps them in sync.

For example, frequencies between 40 Hz and 100 Hz have been shown to encourage blood flow, aiding in oxygen distribution throughout the body. When this is combined with the increased oxygen intake from HBOT, the effect is amplified. Increased blood flow and oxygen support a faster, more efficient healing response.

In studies with patients recovering from injuries, the combination of HBOT and specific frequencies has demonstrated enhanced cellular repair and recovery times. Athletes recovering from muscle injuries and patients with chronic wounds have particularly benefited from this combined therapy.

Case Study: Accelerated Recovery in Brain Injury

Consider the case of Jacob, a 30-year-old man who suffered a traumatic brain injury (TBI) after a car accident. Traditional therapies provided minimal improvement, leaving him with persistent cognitive issues. His doctors recommended HBOT, which he started alongside a specific protocol of sound frequencies tailored for brain health, focusing on low-frequency gamma waves (30-40 Hz).

Within weeks, Jacob noticed an improvement in mental clarity and memory. Over time, the combination of oxygen and frequency therapy helped reduce inflammation in his brain, supporting his cognitive function. This case suggests the promising potential of combining HBOT with frequency therapy for brain injuries.

Practical Tips for Using HBOT and Frequency Therapy

For those interested in exploring this combination therapy, here are some recommendations to consider:

1. Find a Certified HBOT Facility: HBOT requires a medical-grade hyperbaric chamber, so be sure to work with a licensed facility.
2. Work with Frequency Experts: Professionals in sound therapy can help select appropriate frequencies tailored to individual health needs. Frequencies between 40-100 Hz are often used for circulation and tissue repair, while lower frequencies (10-20 Hz) are more relaxing and anti-inflammatory.
3. Personalized Protocols: Frequency therapy should be personalized to the individual. Consult with both your HBOT provider and frequency therapist to establish a regimen suited to your specific health condition.

Future Vision: Personalized Frequency Protocols with AI

Looking forward, we can imagine a future where artificial intelligence (AI) helps determine the precise frequency and oxygen levels each individual needs. Imagine an AI-guided HBOT chamber that reads your body's current cellular state and selects frequencies in real time, "tuning" your body's cells while infusing them with oxygen.

This future of frequency therapy could open doors to treating ailments ranging from chronic illnesses to mental health conditions. With ongoing research, we may find that combining HBOT and frequency therapy not only heals but also rejuvenates the body, helping us stay vibrant as we age.

Reflecting on the Potential of Frequency and HBOT Therapy

The combination of HBOT and frequency therapy may initially seem unusual—yet, when we consider the common thread between these treatments, it makes profound sense. Both methods operate on principles that are as old as life itself: oxygen as a vital force and vibration as an inherent quality of all matter.

The possibilities for integrating HBOT and sound frequencies are as exciting as they are profound. These therapies speak to a future where we move beyond invasive methods and embrace more subtle, yet powerful approaches to healing. In this vision, healing becomes a harmonious experience that involves not only the physical body but also the mind and spirit, bringing us closer to a holistic understanding of health.

As the fields of HBOT and frequency medicine continue to evolve, we may be at the dawn of a new era in healthcare—one that is less about correction and more about resonance, alignment, and optimization.

CHAPTER 59: VIBRATIONAL THERAPY FOR POST-SURGICAL RECOVERY

After surgery, the body embarks on a complex journey of repair, a journey that demands immense cellular coordination, ample energy, and time. Traditionally, the process of recovery has relied on rest, physical therapy, and often medication to manage pain and inflammation. However, vibrational therapy, a groundbreaking approach that uses sound frequencies to aid in post-surgical recovery, is opening new doors for healing—non-invasively and without side effects. Imagine frequencies working alongside your body's natural rhythms, gently guiding cells toward a state of health and harmony. This chapter explores the profound potential of vibrational therapy in post-surgical recovery, bridging scientific insights with stories of healing and empowerment.

The Role of Frequencies in Cellular Communication

Think of your body as an orchestra, where every cell plays a specific role, resonating at a natural frequency when in optimal health. After surgery, this orchestration can become chaotic. Inflammation, pain, and tissue damage disrupt cellular communication, making it challenging for the body to coordinate the healing process. Vibrational therapy works by reintroducing harmony to this orchestration. Through specific frequencies tailored to address pain, inflammation, and tissue repair, cells are "reminded" of their natural rhythms, accelerating recovery and reducing discomfort.

The science behind this lies in cellular resonance. Each cell and tissue type responds to certain frequencies, similar to how musical notes resonate differently with different objects. By applying these frequencies, we can influence cellular processes, encouraging everything from collagen formation in tissues to the reduction of inflammatory cytokines. Research has shown that low-frequency sound waves, in particular, can increase blood flow, reduce pain, and promote the repair of soft tissue.

A Story of Recovery: Sarah's Journey

To illustrate vibrational therapy's potential, consider Sarah, a 45-year-old marathon runner who faced knee surgery after an injury. The surgery was successful, but her recovery was marked by persistent pain and swelling, which delayed her physical therapy and weighed on her mental health. Her physician recommended a vibrational therapy regimen, which she approached with curiosity and scepticism.

Her therapist used frequencies between 40 and 100 Hz, scientifically recognized to aid in tissue repair and reduce inflammation. The therapy involved short, 20-minute sessions where sound waves were directed at her knee using a specialized device. Within a week, Sarah noticed a shift. Her pain began to ease, and the swelling subsided. Over the next month, her physical therapist observed accelerated progress, noting that her muscle strength and joint mobility were improving faster than expected. Sarah returned to running in record time, attributing her recovery to the harmonious blend of traditional rehabilitation and vibrational therapy.

Understanding the Mechanisms of Vibrational Healing

The mechanisms by which vibrational therapy accelerates post-surgical recovery are still an active area of research, but several promising insights have emerged:

1. Pain Reduction: Frequencies around 90 Hz have been shown to interact with nerve pathways associated with pain perception. By targeting these pathways, vibrational therapy can effectively "distract" the nervous system from pain signals, offering relief without the need for pharmaceutical intervention.

2. Anti-Inflammatory Effects: Inflammation is both a healing process and a source of discomfort after surgery. Studies suggest that frequencies between 40-50 Hz reduce inflammatory markers, helping to limit swelling and reduce the risk of complications associated with excessive inflammation.

3. Enhanced Blood Flow and Tissue Repair: Specific frequencies stimulate microcirculation, delivering oxygen and nutrients more effectively to the healing tissue. This increase in blood flow not only accelerates recovery but also helps prevent scar tissue formation, an often-overlooked aspect of long-term recovery.

4. Mental Well-being: Surgery and recovery are taxing not only on the body but also on the mind. Certain low frequencies, like 10 Hz, are known to promote relaxation by influencing brain wave patterns, helping patients manage anxiety and stress during recovery.

Practical Tips for Integrating Vibrational Therapy Post-Surgery

For readers interested in using vibrational therapy as part of their recovery process, here are practical steps and tips:

- Consult with Your Healthcare Provider: Always consult with your physician or surgeon to ensure vibrational therapy is suitable for your specific recovery needs. While non-invasive, this therapy may have different protocols based on the type of surgery you've undergone.

- Work with a Trained Therapist: Look for a certified vibrational therapist or a rehabilitation clinic that offers frequency-based treatments. Trained therapists can customize frequencies to address your unique healing needs, maximizing the therapy's effectiveness.

- Frequency and Duration: Research suggests that short, consistent sessions (20-30 minutes, 2-3 times per week) are effective for post-surgical recovery. Low to mid-range frequencies (40-100 Hz) are often used for tissue repair and pain management.

- Consider At-Home Devices: Several portable vibrational devices are now available for home use, allowing patients to incorporate sound therapy into their recovery routine conveniently. While these devices may not replace professional therapy, they can be a supplementary tool to speed up healing.

Reflecting on the Potential of Vibrational Therapy

Reflecting on vibrational therapy's role in post-surgical recovery, one cannot help but marvel at the way science is rediscovering and repurposing ancient principles. Sound and vibration have been tools for healing for centuries, from Tibetan singing bowls to indigenous drumming practices. Today, these ancient principles are validated by modern science, showing us that sound and vibration are more than just tools for relaxation—they are powerful agents of healing.

In a world where we increasingly seek alternatives to

invasive procedures and medication, vibrational therapy offers a hopeful glimpse into a future where healing is natural, gentle, and holistic. Imagine a world where recovering from surgery involves not just physical therapy and medication but also a frequency regimen designed to harmonize the body, mind, and spirit.

A Vision for the Future

Looking forward, we can imagine a healthcare system where vibrational therapy is a standard part of post-surgical care. Perhaps each hospital recovery room will be equipped with personalized sound frequencies that match each patient's unique needs. Advances in biofeedback and AI could even allow us to "listen" to the body's needs in real time, adjusting frequencies dynamically to support optimal healing.

This vision may seem futuristic, but the seeds of this reality are already planted. The intersection of medicine and frequency therapy speaks to a shift in how we view health —not as a process of fixing broken parts but as a journey toward harmony within the body's intricate orchestra.

Closing Thoughts

In the same way that an orchestra needs all its instruments tuned to create harmony, our bodies require balance and resonance to heal. Vibrational therapy stands as a testament to the healing power of frequencies, reminding us that we are, at our core, beings of energy and rhythm. As we embrace vibrational therapy in post-surgical recovery, we are opening a door to a more harmonious approach to health—one that honours the natural rhythms within us and harmonizes the science of medicine with the art of healing.

Vibrational therapy is not just a tool for recovery; it's an invitation to experience healing on a deeper level, where science meets spirit, and where the body's journey to wellness is guided by the gentle power of sound.

CHAPTER 60: GUIDED SOUND MEDITATION FOR HEALING

In a world where stress often seems inescapable, the power of guided sound meditation offers a remarkable antidote. Imagine a form of healing that requires nothing more than closing your eyes, letting go of daily worries, and surrendering to the soothing embrace of sound. Guided sound meditation, using specific frequencies, is a practice that allows our minds and bodies to enter a state of deep relaxation and healing, harnessing the transformative potential of sound vibrations to soothe the nervous system, elevate mood, and even alleviate pain. This chapter explores the ways in which frequency-guided meditation can unlock inner peace, heal emotional wounds, and empower individuals to reach new levels of mental well-being.

The Science of Sound and Mind

At its core, sound meditation works by stimulating brainwave states associated with relaxation, focus, and healing. Our brains operate at different frequencies depending on our state of mind, measured in hertz (Hz). For instance, the beta frequency (13-30 Hz) is where we operate during alert, problem-solving states, while the alpha (8-12

Hz) and theta (4-7 Hz) states promote relaxation, creativity, and meditative awareness.

Guided sound meditations specifically designed with frequencies that resonate with the alpha and theta states can effortlessly shift our minds from a state of high tension to one of calm and receptivity. These frequencies encourage a shift from the fast-paced thinking of beta waves to the slower, deeper mental states where healing begins. When practiced consistently, guided sound meditation can help individuals achieve mental clarity, emotional resilience, and physical relaxation.

Case Study: Maria's Path to Healing

Maria, a 38-year-old teacher, had struggled for years with anxiety and chronic stress. As an educator in a high-pressure environment, she often felt overwhelmed by her responsibilities, which impacted her sleep and relationships. She began exploring various therapeutic options and found herself drawn to the idea of sound meditation.

With the help of a guided meditation program designed to use alpha and theta frequencies, Maria began meditating each evening, allowing the calming sounds to wash over her as she closed her eyes and breathed deeply. After a week, she noticed subtle changes: she felt more at ease during her days, her mind was clearer, and her sleep had improved. Over months, guided sound meditation became an essential part of her life, helping her manage stress and cultivate a sense of inner peace. Maria's journey is just one of many illustrating the profound impact sound meditation can have on mental well-being.

How Guided Sound Meditation Works

Guided sound meditation combines two powerful elements: intentional sound frequencies and structured guidance.

Here's how it facilitates healing:

1. Inducing a Relaxed State: Sound frequencies between 8-12 Hz resonate with alpha brainwaves, which are associated with relaxation without drowsiness. Listening to meditative tracks designed with these frequencies helps people reach this state more quickly, often within minutes.

2. Activating the Parasympathetic Nervous System: In deep meditation, the body's "rest and digest" system, or the parasympathetic nervous system, becomes active. This reduces heart rate, lowers blood pressure, and releases tension, which can promote healing, particularly for those who suffer from chronic stress or anxiety.

3. Enhancing Emotional Release and Processing: Sound vibrations can help unlock emotional blockages by reaching deep layers of the subconscious mind. Theta waves (4-7 Hz), often included in guided meditations, allow the mind to drift into a semi-dreamlike state, making it easier to process and release pent-up emotions.

4. Improving Focus and Mental Clarity: While meditation is often associated with relaxation, it also enhances mental focus. Meditative frequencies can improve concentration by quieting the constant chatter of the mind, allowing for clearer thinking and greater insight.

DIY Tips for Practicing Guided Sound Meditation

For those interested in trying guided sound meditation, here are some practical tips to make the most of each session:

- Choose the Right Environment: Find a quiet space where you won't be interrupted. Dim lighting or a candle can enhance relaxation. If you prefer to lie down, make sure you're comfortable.

- Select Suitable Frequencies: Alpha (8-12 Hz) and theta

(4-7 Hz) frequencies are ideal for relaxation and healing. Many meditation apps and online resources offer frequency-specific tracks for these states.

- Use Headphones: For optimal experience, especially with binaural beats, use high-quality headphones. This ensures that each ear receives the correct frequency, helping the brain create the intended response.

- Start Small and Build Up: Beginners can start with 10-15 minutes a day. Gradually increase as you feel comfortable. Consistency is key for long-term benefits.

- Practice Deep Breathing: Focus on your breath. Slow, deep breathing helps synchronize with the meditative frequencies and enhances relaxation.

Reflection: The Power of Inner Harmony

Guided sound meditation does more than simply alleviate stress. It opens a doorway to inner peace and alignment, connecting us to our deeper selves. As we dive into the healing vibrations, we begin to understand that mental well-being is not simply the absence of stress or negative thoughts; it is the cultivation of harmony within. Sound meditation gently reminds us that healing is both a physical and spiritual process—a return to our natural rhythms.

Incorporating sound meditation into one's routine can feel like discovering a hidden sanctuary within. As more people are drawn to these practices, the collective impact could be profound. Imagine a world where more of us are able to manage our stress, cultivate resilience, and share calm with those around us.

Looking Ahead: A World Embracing Frequency Healing

As we look to the future, guided sound meditation could become an integral part of mainstream mental health practices. With advances in neurofeedback and digital health technologies, it's possible we'll soon have personalized soundscapes designed for each individual's needs, helping people achieve the mental states most conducive to healing and wellness. Imagine wearable devices that track your stress levels and automatically play calming frequencies to help restore balance.

The potential for guided sound meditation in medical settings is also exciting. Some hospitals are already integrating sound therapy for post-surgical recovery and anxiety reduction. The future could see mental health centres, hospitals, and even schools using guided sound meditation to help individuals of all ages foster resilience, focus, and emotional well-being.

Final Thoughts: The Healing Journey

Guided sound meditation reminds us that healing is not always about curing or fixing; sometimes, it's about allowing. Allowing ourselves to pause, to release, and to reconnect with the quiet spaces within. It's a powerful reminder that our mental well-being is intricately connected to the vibrations we allow into our lives. Guided sound meditation serves as both a mirror and a guide, reflecting the calm within us while leading us to deeper, lasting peace.

In a world where we are constantly bombarded with noise, stress, and distraction, guided sound meditation offers a return to simplicity and presence. It calls us to embrace healing through harmony, showing us that sometimes, all we need is a gentle frequency to lead us back to ourselves. The journey of healing is, at its heart, a journey of returning to the profound peace that lies within—a journey that begins with a single sound.

CHAPTER 61:
FREQUENCY-DRIVEN DETOXIFICATION PROTOCOLS

In the realm of frequency medicine, the idea that certain frequencies can aid in detoxification opens new possibilities for maintaining wellness. Imagine that the body, much like a symphony orchestra, has distinct frequencies at which each organ or cell performs best. Tuning the body to these frequencies, especially when it comes to detoxification, can help cleanse cells, improve metabolic processes, and expel toxins. This chapter delves into the science and practice of frequency-driven detoxification, illustrating how vibrational energy may encourage the body's natural ability to cleanse and renew itself.

The Science of Frequency-Driven Detoxification

Detoxification is the body's natural process of removing toxins, involving organs like the liver, kidneys, skin, and lymphatic system. While conventional methods such as dietary adjustments and hydration can support detox, frequency-based approaches take it further by stimulating cellular functions through precise vibrational frequencies. Cells communicate using subtle electrical impulses, and research suggests that exposing cells to certain frequencies can enhance their detoxification processes.

Using frequency-based devices, such as Rife machines, bio resonance technology, and PEMF (Pulsed Electromagnetic Field) therapy, practitioners can stimulate detoxification on a cellular level. These devices emit specific frequencies that interact with the body's natural electrical systems, potentially improving cellular performance and flushing out toxins.

A Case Study: Revitalizing Through Frequency Detox

Consider John, a 52-year-old man who struggled with chronic fatigue, digestive issues, and sluggish metabolism. Traditional detox methods, including dietary cleanses, provided limited relief. Curious about alternative therapies, John discovered frequency-based detox protocols.

Over six weeks, John underwent sessions using a Rife machine, which was programmed with frequencies designed to support liver and lymphatic detox. Following each session, he felt noticeably more energized, and over time, his digestion improved. John's case illustrates the potential of frequency detox to enhance overall wellness, showing how detoxification can transcend conventional methods by aligning with the body's own vibrational needs.

How Frequency-Driven Detoxification Works

The concept of frequency detoxification revolves around the belief that each type of cell and organ has an optimal frequency at which it functions best. Exposure to these frequencies can enhance cellular performance, making detoxification more efficient. Here are key principles behind this process:

1. Stimulating Cellular Metabolism: Frequencies that match the vibrational patterns of cells can enhance cellular metabolism, promoting waste removal and nutrient uptake.

PEMF therapy, for example, is used to stimulate cellular energy production, aiding in the removal of toxins at the cellular level.

2. Activating Detoxification Organs: The liver, kidneys, and lymphatic system play central roles in detoxification. Frequencies targeting these organs can potentially improve their performance, assisting in the breakdown and elimination of harmful substances. Liver-focused frequencies may enhance enzyme production, while lymphatic frequencies help move stagnant lymph fluid, expelling toxins more efficiently.

3. Breaking Down Environmental Toxins: Environmental toxins, including heavy metals and chemicals, can accumulate in body tissues. Some frequency therapies are aimed specifically at detoxifying these compounds, using targeted vibrations that disrupt the bonds holding these toxins in place, enabling the body to remove them.

Practical Frequency-Driven Detox Protocols

For those intrigued by frequency-driven detox, there are accessible ways to incorporate these protocols into a wellness routine. While professional frequency therapies are available, certain approaches can also be applied at home.

PEMF Therapy for Daily Detox

PEMF devices, which emit low-frequency electromagnetic waves, are becoming popular for home use. A 20-minute PEMF session on the low setting can help stimulate cellular activity, supporting daily detoxification.

DIY Tips:
- Start with 5-10 minutes on a low-frequency setting.
- Place the PEMF device near the liver or abdomen to target primary detox organs.
- Use 2-3 times per week to maintain consistent

detoxification.

Binaural Beats for Lymphatic Drainage

Binaural beats, created by playing slightly different frequencies in each ear, can stimulate brainwave states that encourage relaxation and detoxification. While often used for mental health, binaural beats in the range of 1-4 Hz may also stimulate lymphatic drainage.

DIY Tips:
- Listen to binaural beats designed for detoxification using headphones for at least 15-20 minutes.
- Lie down with feet elevated to encourage lymphatic flow.
- Use daily during detox phases for optimal results.

Rife Frequencies for Heavy Metal Detox

Rife machines are used to generate frequencies believed to resonate with specific toxins or pathogens, disrupting their structure. Though typically administered by practitioners, some portable Rife machines are available for personal use.

DIY Tips:
- Consult a trained practitioner to determine the right frequencies for heavy metal detox.
- Start with 5-10 minute sessions, gradually increasing as recommended.
- Drink plenty of water before and after sessions to help flush out toxins.

A Vision for the Future: Personalized Frequency Detox Plans

Imagine a world where detox protocols are tailored to each individual's unique vibrational profile. Advances in AI and biofeedback technology could soon make this possible, creating a future where people have access to detox frequencies specifically calibrated for their bodies. We are only at the beginning of understanding the intricate relationship between our cells and frequencies, yet the potential is vast. Personalized frequency detox could support ongoing wellness, prevent toxin buildup, and even restore balance for individuals facing chronic toxicity.

Balancing Science and Belief

Though frequency-driven detox shows promise, it's essential to approach it with balanced scepticism. Many of these methods are still emerging, and research, though promising, is ongoing. Critics argue that some claims lack rigorous evidence. However, anecdotal experiences and early studies provide a compelling case for further exploration. As science progresses, we may see more studies supporting frequency-driven detox as a viable option alongside traditional detox methods.

Reflection: The Resonance of Cleanliness

There's a timeless truth in the notion that wellness begins with cleanliness—inside and out. Frequency detox aligns with this truth, offering a subtle, vibrational approach to cleanse and renew. It's a reminder that the body, much like an instrument, performs best when free of impurities, resonating at its natural frequency. Frequency detox can be viewed as a return to this natural resonance, enabling us to live healthier lives in harmony with our body's needs.

Incorporating frequency-based detox methods into one's lifestyle doesn't just mean cleansing the body; it involves tuning the body to its inherent state of harmony. As we explore these methods, we're not only discovering ways to remove toxins but also unveiling the deeper resonance of health itself—one that supports both body and mind.

Final Thoughts: A Detox Journey Beyond the Physical

Frequency-driven detox encourages a paradigm shift. Detoxification is no longer limited to the physical removal of toxins; it's also about harmonizing our bodies on a vibrational level. This journey is as much about releasing as it is about renewal. It teaches us that true health involves nurturing the frequencies that sustain us, fostering both physical vitality and inner clarity.

For those ready to embrace this approach, frequency detox may be the key to unlocking a new dimension of wellness—one where the body's natural rhythms are restored, and the burdens of modern living are gently lifted, one frequency at a time.

PART XIV: INTEGRATING FREQUENCY HEALING WITH OTHER MODALITIES

CHAPTER 62: COMBINING FREQUENCY THERAPY WITH HERBAL MEDICINE

In our exploration of healing, we find two ancient practices emerging as complementary forces: frequency therapy and herbal medicine. Like two interwoven threads, they harmonize, amplifying each other's potential to support the body, mind, and spirit. The marriage of frequency therapies with herbal remedies is not only a new frontier in alternative medicine but a return to something timeless—a synthesis of natural energy and the healing power of plants.

This chapter will explore the scientific and practical ways in which frequency therapy can enhance the effects of herbal medicine. By resonating with the body at the cellular level, frequency therapy may increase the bioavailability of herbal compounds, help the body better absorb and use them, and even target the unique needs of each individual for a more tailored approach to healing.

Understanding the Synergy of Frequencies and Plants

Herbs and plants have been used medicinally for thousands

of years, providing remedies for ailments ranging from digestive issues to stress. Each herb has a unique chemical makeup, and modern science has shown that the active compounds within plants—such as alkaloids, flavonoids, and terpenes—have tangible effects on human physiology.

Frequency therapy, on the other hand, works by using specific sound, light, or electromagnetic frequencies to enhance cellular function, communication, and repair. When these two approaches are combined, the potential for synergy is powerful. Think of it as two musicians playing in perfect harmony, each enhancing the other's melody. In practical terms, frequencies can optimize how herbs work within the body, making the effects more profound and sustained.

Case Study: Lavender and Frequency for Anxiety Relief

Let's consider Sarah, a woman in her mid-30s, who has been battling chronic anxiety. Conventional treatments left her feeling tired and unfocused, leading her to explore herbal remedies and frequency therapy as an alternative. Sarah started using lavender, a well-known herb for relaxation, while also incorporating binaural beats—a form of sound therapy—at a frequency of 8 Hz, which is associated with calming effects on the brain.

Over a few weeks, Sarah reported feeling an improvement that went beyond what lavender alone had achieved. The combination of lavender's active compounds with the specific calming frequency seemed to boost her relaxation, deepening the herb's impact and creating a lasting sense of calm. Sarah's experience illustrates how frequency therapy can act as an amplifier, enhancing the calming effects of herbal remedies.

How Frequency Therapy Enhances Herbal Healing

1. Increased Bioavailability: Frequencies can help increase the bioavailability of herbs, meaning that the body can more easily absorb and use their active compounds. PEMF (Pulsed Electromagnetic Field) therapy, for example, is thought to enhance cell permeability, allowing herbal compounds to penetrate cells more efficiently.

2. Targeted Cellular Resonance: Each cell type resonates at a particular frequency, and frequency therapy can help direct herbs to where they are needed most in the body. When specific frequencies are applied, they can create a resonance that draws the healing properties of herbs to particular tissues or organs.

3. Detoxification and Preparation: Certain frequencies can prepare the body for herbal therapy by stimulating detoxification pathways. For example, using detoxifying frequencies before introducing an herbal regimen can help the body clear out old toxins, creating a clean slate for the herbs to work more effectively.

4. Enhanced Blood Flow and Circulation: Frequencies can stimulate blood flow, which in turn aids the distribution of herbal compounds throughout the body. This is especially useful for herbs that support cardiovascular health, as the combination of frequency and herb can improve overall circulation and heart function.

Practical Tips for Combining Frequency and Herbal Therapies

For those interested in integrating these methods, here are practical steps and guidelines:

Choosing Your Frequency

Identify the goal of your herbal therapy and select a complementary frequency. Here are a few common combinations:

- Lavender + 8 Hz (Theta waves): Excellent for relaxation and stress relief, this combination helps deepen the calming effects of lavender.
- Ginger + PEMF at 10-15 Hz: For digestive health, ginger's anti-inflammatory properties can be boosted with PEMF therapy, which supports cellular health and circulation.
- Turmeric + 528 Hz: Known as the "love frequency," 528 Hz is believed to promote cellular regeneration. When used alongside turmeric, a powerful anti-inflammatory, this combination can support tissue repair and joint health.

Preparing the Body

Start by using detoxifying frequencies, such as those associated with lymphatic flow (1-3 Hz), to prepare the body before introducing herbal remedies. This can help clear the body of impurities and improve the absorption of the herbs.

DIY Sound Therapy and Herbal Blends

If professional frequency devices aren't accessible, you can incorporate sound therapy with the help of sound healing apps or playlists designed for specific frequencies. Pair your chosen herb with the frequency as follows:

- Brew a tea of your chosen herb, such as chamomile for sleep, and listen to a frequency playlist that aligns with your healing goal (e.g., delta waves for sleep).
- Create a relaxing environment by diffusing essential oils while using binaural beats. This can create an immersive experience that combines aromatherapy with frequency therapy.

A Glimpse into the Future of Frequency and Herbal Medicine

Imagine walking into a clinic where an herbalist and frequency therapist collaborate. After a biofeedback analysis, the practitioner prescribes not only a custom herbal blend but also a specific frequency session tailored to enhance the herbs' effects. This is the future we may be heading toward— a truly holistic approach that acknowledges the vibrational nature of both the human body and the plant world.

In the near future, AI technology could help develop personalized frequency and herbal protocols based on each person's unique resonance, making healing more precise and effective. This approach could make it possible to tune therapies in a way that not only addresses physical ailments but also harmonizes the body's subtle energy fields.

Reflection: Harmonizing with Nature's Frequencies

At the heart of both frequency therapy and herbal medicine lies a profound truth: healing is not solely a physical process but a vibrational one. Plants grow from the earth, absorbing the frequencies of sunlight, water, and minerals, and they carry that energy into our bodies when we consume them. Frequency therapy, in its essence, is a reminder that everything, including our cells, is energy vibrating at specific rates.

By combining herbal medicine with frequency therapy, we are aligning ourselves with nature's rhythms, creating a holistic approach that honours both the body's need for nourishment and its need for resonance. This symphony of herbs and frequencies offers a pathway to health that is both timeless and innovative, a union of ancient wisdom and modern science.

Closing Thoughts: Embracing the Symphony of Healing

As we conclude our exploration of combining frequency therapy with herbal medicine, let's take a moment to appreciate the elegance of this partnership. Each herb, each frequency, is like an instrument in a symphony, playing a unique note that contributes to the body's harmony. Through this chapter, we've unveiled not just a method of healing but a way of living in alignment with the energies around us.

In blending the resonance of frequencies with the life force of plants, we are invited into a deeper relationship with nature—a relationship that nurtures, heals, and restores. This approach to healing reminds us that wellness is not a solitary journey but one deeply connected to the world, vibrating with the same frequencies that give life to the earth and to us all.

CHAPTER 63:
ACUPUNCTURE
AND FREQUENCY
SYNERGY

Imagine a symphony where each instrument contributes a unique sound, yet together they create a melody that resonates far beyond the individual notes. This is the essence of combining acupuncture with frequency-based therapies. While acupuncture has ancient roots, dating back thousands of years in Traditional Chinese Medicine (TCM), modern frequency-based tools bring a new dimension to this art, amplifying its effects and targeting healing in deeper, more profound ways.

In this chapter, we'll explore how the convergence of acupuncture and frequency therapy is transforming the landscape of medicine. Through a blend of storytelling, scientific exploration, and practical tips, we'll dive into the potential of this synergy and the insights it reveals about the body's natural energy systems.

The Ancient Art of Acupuncture Meets Modern Frequencies

Acupuncture is based on the idea that the body's life force, or "Qi," flows along pathways known as meridians. When these pathways become blocked, according to TCM, illness or pain arises. By inserting fine needles into specific points along these meridians, acupuncture seeks to restore balance, allowing energy to flow freely and support healing.

On the other hand, frequency-based therapies—whether through sound, light, or electromagnetic pulses—target cellular processes at a fundamental level. They operate on the principle that every cell, tissue, and organ vibrates at specific frequencies. When these frequencies are disrupted, dysfunction and disease can occur. By applying the correct frequency, healing can be facilitated, and cellular functions can be restored to their optimal state.

Combining acupuncture with frequency therapy amplifies each method's effect, bridging ancient wisdom with modern science. Think of acupuncture needles as the "tuning forks" of the body, while frequencies provide the resonance needed to amplify the healing. This creates a powerful synergy that supports the body's natural healing mechanisms.

Case Study: Relief from Chronic Pain

Consider the story of James, a man in his late 40s who suffered from chronic back pain due to a herniated disc. Traditional pain medications offered him little relief, and he wanted to avoid surgery. James turned to an acupuncturist trained in frequency therapy, who used needles in combination with pulsed electromagnetic field (PEMF) therapy. The needles targeted specific meridians to relieve pain, while the PEMF device was set to a frequency known to support nerve and tissue repair.

After a few sessions, James noticed a substantial decrease in his pain. Unlike previous treatments, this approach didn't just mask his symptoms; it addressed the underlying causes, supporting his body's natural healing process. Within weeks, James experienced a renewed sense of vitality and ease that he hadn't felt in years. His case exemplifies how the combined use of acupuncture and frequencies can offer relief even when other treatments fall short.

How Frequencies Amplify Acupuncture's Effects

1. Enhanced Cellular Communication: Frequency therapy stimulates cells to communicate more effectively. Acupuncture points are known for their high conductivity, acting like "energy portals" in the body. When frequencies are applied, they help facilitate faster and more efficient cellular responses, enhancing the overall effect of acupuncture.

2. Increased Blood Flow and Oxygenation: Frequencies, such as those used in low-level laser therapy (LLLT), improve blood flow and oxygenation. This not only helps deliver nutrients to tissues but also enhances the removal of metabolic waste. Acupuncture already promotes circulation, but adding frequencies amplifies this effect, accelerating recovery and healing.

3. Targeted Pain Relief: Certain frequencies can directly impact the nervous system, blocking pain signals and promoting the release of endorphins. When combined with acupuncture, which itself has pain-relieving properties, the effects can be compounded. This dual approach is particularly beneficial for those with chronic pain conditions, as it provides relief without relying on medications.

4. Balancing the Nervous System: Both acupuncture and frequency therapy have been shown to balance the sympathetic and parasympathetic nervous systems. This balance is essential for reducing stress, promoting relaxation, and supporting immune function. Using frequencies in tandem with acupuncture creates a calming effect on the body, helping to lower cortisol levels and foster a state of deep relaxation.

Practical Tips for Combining Acupuncture and Frequency Therapy

For those interested in experiencing this powerful combination, here are some ways to incorporate frequencies into an acupuncture session:

- Laser Acupuncture: Low-level laser therapy, or "laser acupuncture," uses lasers at specific frequencies on acupuncture points instead of needles. This non-invasive approach is ideal for individuals sensitive to needles and has shown promising results in pain management and inflammation reduction.

- Sound Therapy with Acupuncture: Binaural beats or specific sound frequencies can be played during an acupuncture session. For example, frequencies in the range of 174 Hz, known for pain relief, can support an acupuncture treatment focused on musculoskeletal pain. A playlist with healing frequencies can enhance relaxation and deepen the therapeutic experience.

- PEMF and Acupuncture: Pulsed electromagnetic field therapy can be applied during acupuncture to boost its effects, especially for tissue repair and nerve function. Practitioners often apply PEMF devices to the area surrounding the acupuncture needles to enhance energy flow and cellular repair.

Reflective Insights: Harmonizing the Body's Energy

The synergy between acupuncture and frequency therapy reveals profound truths about the nature of healing. At a fundamental level, both modalities recognize that the body is not merely a physical entity but a complex energetic system. This perspective encourages us to see health not as the absence of disease but as a harmonious flow of energy within the body.

Combining these therapies is an invitation to deepen our understanding of health, aligning the ancient insights of Eastern medicine with modern discoveries in bioenergetics. It speaks to a universal truth that has resonated throughout history: healing is a process of returning to harmony, of bringing the body back into resonance with its natural frequencies.

Vision for the Future: Integrating Eastern and Western Wisdom

As interest grows in integrative medicine, the combination of acupuncture and frequency therapy has the potential to reshape how we approach health and wellness. Imagine a future where acupuncturists routinely use frequency devices alongside needles to tailor treatments to the unique vibrational needs of each patient. Such an approach could help bridge the gap between Eastern and Western medicine, offering a more comprehensive path to healing.

Technology may soon allow for highly personalized treatments based on real-time biofeedback, enabling practitioners to adjust frequencies in response to the body's immediate needs. We may witness the development of "smart" acupuncture devices that use AI to detect energetic imbalances and apply frequencies that address them with pinpoint accuracy.

Closing Reflections: Embracing a New Paradigm of Healing

The combination of acupuncture and frequency therapy invites us to embrace a new paradigm of healing—one that sees the body as a dynamic interplay of energy and matter. It's a reminder that true healing is not about suppressing symptoms but about restoring balance and resonance.

As you explore the possibilities of this synergy, consider the power of harmonizing ancient and modern practices. Just as acupuncture points serve as gateways for energy, frequencies can serve as the tones that tune our body's inner symphony. Together, they offer a holistic path to well-being that honours the body's inherent wisdom, helping us reconnect with the harmony that is our natural state.

In embracing this union, we open ourselves to a future where healing is not just about treating illness but about enhancing life's vibrational quality, helping us live in resonance with ourselves and the world around us.

CHAPTER 64:
FREQUENCY AND
CHIROPRACTIC CARE

Imagine a finely tuned guitar: each string perfectly adjusted, creating harmonious notes that resonate deeply. The human body is no different. Our bones, muscles, and nerves work in delicate balance, resonating with natural frequencies to keep us in a state of harmony. But when physical or energetic misalignments occur, this balance is disrupted, creating dissonance that can manifest as pain, stiffness, or even illness. Chiropractic care, with its hands-on adjustments, has long aimed to restore this harmony, particularly within the musculoskeletal and nervous systems. Now, as vibrational therapies enter the mainstream, they offer chiropractors powerful tools to complement traditional adjustments, supporting the body's natural alignment and enhancing overall health.

In this chapter, we'll explore the synergy between frequency therapy and chiropractic care, with a focus on how frequencies can amplify the benefits of physical adjustments. We'll dive into case studies, practical applications, and visionary perspectives to offer a full understanding of this integrative approach to musculoskeletal health.

The Fundamentals of Chiropractic Care: Aligning the Body

At its core, chiropractic care is built on the understanding that the body's alignment is integral to health. Misalignments, often called "subluxations," can interfere with nerve function and disrupt the body's natural ability to heal. Chiropractors use adjustments to correct these misalignments, helping to alleviate pain, restore range of motion, and improve bodily functions. The philosophy behind chiropractic care goes beyond pain relief; it embraces a holistic view of the body, where the nervous system is key to maintaining overall wellness.

Yet, even after adjustments, patients may still experience residual tension, inflammation, or muscle imbalances. This is where vibrational therapy can play a significant role, addressing these underlying factors by resonating with tissues at a cellular level.

Vibrational Therapy: Resonating with the Body's Natural Frequencies

Vibrational therapy, which includes sound, electromagnetic fields, and light frequencies, works on the premise that every part of our body has an ideal vibrational frequency. Muscles, bones, nerves, and organs all operate within specific ranges, and disturbances in these frequencies can lead to pain or disease. By applying frequencies that match the natural vibrations of these tissues, vibrational therapy can restore balance and promote healing at a cellular level.

The combination of chiropractic adjustments and vibrational therapy acts like a "double resonance"— addressing both structural and energetic imbalances, and providing patients with relief that lasts beyond their time on the chiropractor's table.

Case Study: Enhancing Spinal Health with Frequency Therapy

Consider Sarah, a young athlete with recurring lower

back pain. After numerous chiropractic adjustments, she found only temporary relief, with the pain often returning after intense training sessions. Her chiropractor decided to integrate vibrational therapy using a pulsed electromagnetic field (PEMF) device, applying low frequencies directly to her lower back muscles before each adjustment.

The results were astounding. The PEMF therapy helped her muscles relax and reduced inflammation, making the chiropractic adjustments more effective and longer-lasting. Within weeks, Sarah was experiencing not only reduced pain but improved mobility and a greater sense of relaxation after each session. For her, the integration of frequency therapy was the missing link that allowed her body to retain its alignment.

How Frequencies Support Chiropractic Adjustments

1. Reducing Inflammation: Frequencies, particularly those from PEMF and low-level laser therapy, have anti-inflammatory properties. Applying these frequencies to muscles and joints can reduce swelling and improve circulation, making chiropractic adjustments easier and more effective.

2. Relaxing Muscles: Often, muscle tightness around a misaligned joint can limit the effectiveness of an adjustment. Frequencies such as those from ultrasound therapy can penetrate deep into muscle tissue, promoting relaxation and preparing the body for realignment.

3. Improving Nerve Function: Chiropractic care focuses on freeing up nerve pathways by aligning the spine. Frequency therapies can complement this by targeting specific nerve frequencies, supporting nerve regeneration and conductivity. This dual approach can be particularly effective for patients with nerve compression or sciatica.

4. Promoting Healing Post-Adjustment: After an adjustment, the body needs time to settle into its new alignment. Frequency therapies can provide gentle stimulation to tissues, supporting healing and reducing post-treatment soreness. Techniques like low-level laser therapy (LLLT) and red light therapy have been shown to accelerate cellular repair, helping the body adapt to its new structural balance.

Practical Tips for Integrating Frequency Therapy into Chiropractic Sessions

For those interested in experiencing the combined benefits of chiropractic care and vibrational therapy, here are some options to consider:

- PEMF Therapy Before Adjustments: Applying pulsed electromagnetic field therapy before a chiropractic session can relax tense muscles and reduce inflammation, enhancing the effectiveness of adjustments.

- Laser Therapy on Targeted Areas: Low-level laser therapy (LLLT) or red light therapy can be used on specific muscles or joints after adjustments to promote healing. These therapies are especially effective for those with chronic inflammation or joint pain.

- Sound Therapy for Relaxation: Frequencies in the range of 528 Hz, known as the "miracle frequency," can create a calming effect and help patients release tension. Sound therapy can be played during or before adjustments to help prepare the mind and body for treatment.

Reflective Insights: The Body as a Symphony of Frequencies

The integration of chiropractic adjustments with frequency therapy invites us to consider the body not just as a mechanical structure but as a symphony of vibrations. Each part, from our bones to our cells, resonates in harmony when we're healthy, and disruptions to this harmony can lead to discomfort and disease. Chiropractic adjustments restore structural alignment, while frequency therapy tunes the body's energetic vibrations, creating a comprehensive approach to wellness.

This perspective aligns with the emerging field of bioenergetics, which views health as a matter of balance across physical, biochemical, and energetic dimensions. When we see the body in this holistic way, chiropractic care becomes more than a mechanical realignment—it becomes an energetic reset, supported and enhanced by frequencies.

Vision for the Future: Personalized, Frequency-Enhanced Chiropractic Care

As technology continues to evolve, we may soon see frequency therapy becoming a routine part of chiropractic care. Imagine a future where chiropractors use frequency analysers to identify the exact vibrational imbalances within a patient's body before adjustments. With this information, they could apply tailored frequency treatments alongside manual adjustments, targeting the precise frequencies that support structural health.

This vision isn't far-fetched; tools like biofeedback and frequency analysis are already on the rise, with the potential to bring chiropractic care into a new era. This integrative approach would allow practitioners to work not just with the physical alignment of their patients but with their energetic harmony, providing deeper and more lasting healing.

Final Reflections: Aligning with Health's Natural Frequency

The combination of chiropractic care and frequency therapy is a reminder that health is not merely the absence of pain but the presence of harmony. Each adjustment, each frequency pulse, is a note in the melody of well-being.

Together, they create a holistic approach that nurtures both the body and its energy field, helping individuals reconnect with their natural state of resonance.

As you explore this synergy, consider your body as an orchestra, with each part playing its role in harmony. Chiropractic care aligns the physical "instruments," while frequency therapy tunes the energetic resonance, creating a symphony of health that resonates at the deepest levels. In embracing this integrated approach, we open ourselves to a world where healing is not just about fixing what's broken but about restoring the body's natural song.

CHAPTER 65:
MASSAGE THERAPY
WITH FREQUENCY
TECHNIQUES

Massage therapy has long been a trusted method for relieving tension, enhancing circulation, and promoting relaxation. But what if we could elevate this ancient practice to work on a deeper, cellular level? By integrating frequency-based techniques such as sound and vibration, massage therapy is now taking on new dimensions. These methods enhance traditional massage benefits by working with the body's natural vibrations, helping to release tension held deep within muscles and even restoring balance to the nervous system.

In this chapter, we explore how frequency-based devices, such as sound baths, tuning forks, and vibration plates, are used alongside massage techniques to provide a holistic experience. By looking at the science of frequency-based massage, real-life case studies, and practical applications, we'll uncover how these methods allow therapists to go beyond the surface, transforming massage into a journey of deep relaxation and healing.

The Science of Frequency and Massage: A Symbiotic Relationship

Massage and frequencies share a common goal: to return the body to a state of equilibrium. While massage physically releases tension through kneading and pressure, frequencies work on the cellular and energetic levels, encouraging harmony in a way that hands alone may not achieve. Each muscle, organ, and tissue in our body vibrates at a unique frequency, and imbalances in these vibrations can manifest as stress, pain, or fatigue. Frequency-based therapies aim to restore these vibrations to their optimal state, helping the body recalibrate.

Imagine your body as a musical instrument that, over time, has gone slightly out of tune. Massage provides a hands-on approach to ease tension and encourage relaxation, while sound and vibration can be seen as the "tuning" process, helping each part of the body find its natural frequency. Together, these therapies create a holistic approach to health, addressing both the physical and energetic aspects of healing.

Real-Life Impact: A Case Study in Frequency Massage

Consider the case of James, a 40-year-old athlete who struggled with chronic muscle soreness and stiffness despite regular massage therapy. His massage therapist decided to incorporate sound therapy, specifically using a tuning fork set to 528 Hz—a frequency known for its calming and healing properties—directly on his shoulders and back. The vibrations from the tuning fork penetrated deep into his muscles, helping to release stubborn tension that traditional massage techniques couldn't reach.

After just a few sessions, James noticed that his muscle soreness was significantly reduced, and he felt a sense of calm and mental clarity that he hadn't experienced from massage alone. The frequency-based approach allowed him to experience deeper relaxation, showing how frequency techniques can enhance the traditional benefits of massage therapy.

Tools and Techniques: How Frequencies Amplify Massage Benefits

1. Tuning Forks: Tuning forks set to specific frequencies can be applied to muscle groups, joints, and pressure points. The vibrations generated by the forks stimulate cellular activity, improving circulation, and encouraging the release of deeply held tension.

2. Sound Baths: Using sounds from Tibetan singing bowls, crystal bowls, or gongs, sound baths create an immersive experience where clients are "bathed" in healing frequencies. These sounds work on both the mind and body, inducing a state of relaxation that enhances the massage experience. This is especially helpful for clients who carry tension from stress or anxiety.

3. Vibration Plates: Vibration plates create gentle oscillations that stimulate blood flow and muscle relaxation. These can be used as a pre-massage treatment to warm up the muscles, making them more receptive to deeper work.

4. Low-Frequency Sound Devices: Some massage therapists are incorporating low-frequency sound devices, which create gentle, inaudible vibrations designed to penetrate deeply into tissues. These devices are particularly beneficial for clients with chronic pain or muscle tightness.

Practical Tips for Integrating Frequency into Massage Therapy

For therapists and clients interested in blending massage with frequency techniques, here are some helpful tips:

- Start with Tuning Forks on Tension Points: Applying a tuning fork to tension-prone areas, like the neck, shoulders, or lower back, can help release tightness before beginning the massage. Frequencies between 432 Hz and 528 Hz are often used for relaxation and balance.

- Use Sound Baths for Mental Relaxation: Playing a sound bath in the background can create an ambient environment that puts the client in a state of deep relaxation. This allows the body to be more receptive to touch, helping to achieve a more effective massage.

- Apply Vibration Therapy Pre-Session: Using a vibration plate or device on the client's lower back, feet, or legs before the massage can stimulate blood flow and prepare muscles for a deeper massage.

The Power of Combining Frequencies and Touch

The combination of touch and frequency therapy creates an experience that goes beyond relaxation, touching deeper realms of healing. Imagine a massage session where, after warming up the muscles with light strokes, the therapist places a warm tuning fork along your spine. The gentle hum of the tuning fork reverberates through your muscles, encouraging tight spots to release. As the vibrations reach each cell, the body begins to recalibrate, moving closer to its natural, balanced state.

This synergy between touch and sound frequencies allows the body and mind to experience deep relaxation, creating a sense of harmony that stays with the client long after the session ends. Clients often report feeling lighter, more centred, and even emotionally rejuvenated after a frequency-enhanced massage, highlighting the holistic benefits of these techniques.

Reflection and Vision: The Future of Massage with Frequency Therapy

As frequency-based massage therapy continues to gain

recognition, we can expect to see even more advanced techniques and technologies enter this space. In the future, massage therapists might use frequency analysers to determine a client's specific vibrational needs, customizing frequencies to their unique body composition. This level of personalization could create a new era of therapeutic massage, one that not only addresses physical tension but also resonates with each client's specific energetic needs.

DIY Tips: Using Frequencies at Home for Relaxation

For those who wish to explore frequency-enhanced massage at home, here are some simple techniques:

- DIY Sound Bath: Create a sound bath by playing Tibetan singing bowl music or binaural beats. Lie down, close your eyes, and let the sound waves wash over you. This can be a powerful way to calm the mind and body after a long day.

- Home Massage with Tuning Forks: Invest in a 528 Hz tuning fork and use it on sore muscles by gently pressing it onto tight areas. The vibration can provide a gentle massage effect that releases muscle tension.

- Guided Frequency Meditation: Listen to a guided meditation with frequencies in the background. As you breathe deeply, allow the frequencies to resonate through your body, helping you achieve a state of deep relaxation.

Final Reflections: A New Path in Healing Touch

As we explore the possibilities of combining frequency and massage therapy, we're reminded of the interconnectedness of mind, body, and energy. Every touch, every frequency, is an opportunity to bring balance and harmony to the body. In this synergy, massage therapy evolves from a physical practice to a holistic healing experience.

The power of frequency-based massage therapy lies in

its ability to go beyond superficial relaxation, helping us connect with the deeper rhythms within ourselves. As we tune our bodies to these natural frequencies, we embrace a path of healing that acknowledges both our physical and energetic selves. It's a reminder that true healing is not only about alleviating pain but also about harmonizing our inner frequencies with the world around us.

PART XV: CASE STUDIES IN FREQUENCY HEALING

CHAPTER 66: CASE STUDY - FREQUENCY HEALING FOR TRAUMA RECOVERY

In the exploration of healing, few realms are as complex and elusive as trauma. Trauma, whether physical or emotional, leaves its mark on both the mind and body, often creating invisible wounds that affect a person's ability to feel safe, present, and connected. Conventional therapies focus on managing symptoms, yet an emerging approach is gaining recognition for its profound effects on trauma recovery: frequency healing. This chapter delves into a compelling case study that reveals the potential of frequency healing to unlock hidden depths of resilience and recovery.

Through a blend of scientific understanding, real-life application, and patient experience, we explore how sound waves and vibrational frequencies have transformed lives burdened by trauma, illuminating new paths for healing that may one day complement or even replace traditional methods.

Setting the Scene: The Science of Trauma and Frequency

When a person experiences trauma, their body undergoes a physiological response that's hardwired for survival.

However, this response often becomes ingrained, leaving individuals in a prolonged state of tension or hypervigilance. Trauma can disrupt the natural rhythms of the body—such as heartbeat, breathing, and even the subtle frequencies emitted by cells. These rhythms, which help maintain balance, are thrown off, resulting in a state of discord within the body.

Frequency-based therapies work by reintroducing harmonious vibrations, helping the body recalibrate to its natural state. Just as a musical instrument out of tune can be gently tuned back into harmony, the body can be coaxed back to balance through resonance. This process involves using specific frequencies to reach deep within cellular structures, promoting a sense of relaxation that enables the body to release tension and stored emotional energy.

A Journey of Transformation: Emma's Story

Emma, a 34-year-old woman with a history of complex post-traumatic stress disorder (PTSD), had tried nearly every form of traditional therapy. Though talk therapy and medication had provided some relief, she continued to experience frequent anxiety attacks, flashbacks, and an overwhelming sense of numbness. Her trauma had impacted every part of her life, leaving her with feelings of isolation and exhaustion.

Emma's journey into frequency healing began with a sound therapist who introduced her to binaural beats—a form of sound therapy that uses two slightly different frequencies in each ear to create a third tone within the brain. This third tone, or "beat," induces a state of brainwave entrainment, calming the mind and reducing anxiety. Emma found that her body responded to the low, pulsing vibrations, feeling a sense of calm and safety she hadn't felt in years.

Encouraged by this experience, Emma's therapist gradually

incorporated vibrational therapy tools, such as tuning forks set to frequencies between 528 Hz and 396 Hz, known for their healing and grounding properties. The tuning forks were applied to key points on her body, including her chest and back, allowing vibrations to travel through muscle tissue and connective fibres. Over several months, Emma reported a growing sense of ease and the ability to remain present without triggering anxiety. The trauma she had carried for years began to soften, allowing her to embrace moments of peace.

Tools of Frequency Healing: Binaural Beats, Tuning Forks, and Beyond

The tools used in Emma's healing journey illustrate the diversity and adaptability of frequency-based therapies in trauma treatment. Here's a closer look at some of these tools:

1. Binaural Beats: By delivering two slightly different frequencies to each ear, binaural beats create a third frequency that helps synchronize brainwave patterns. Frequencies within the theta range (4–7 Hz) are often used for relaxation, while alpha waves (8–12 Hz) encourage a meditative state. For trauma survivors, these beats can serve as a bridge to a calmer mental space, enabling them to access memories without reactivating emotional distress.

2. Tuning Forks: Tuning forks allow precise application of vibrational frequencies to specific areas of the body. For instance, 528 Hz, often called the "love frequency," is believed to reduce stress, while 396 Hz is associated with liberating feelings of guilt and fear. Placing these frequencies on the body gently vibrates tissues and organs, fostering a state of physical and mental relaxation.

3. Sound Baths and Gongs: Sound baths immerse individuals in a cocoon of soothing frequencies. Instruments like crystal singing bowls or gongs produce continuous waves that

envelop the body, inducing a state of relaxation. For trauma survivors, sound baths offer a refuge where they can let go of fear and simply experience safety within sound.

Practical Tips for Applying Frequency Healing in Trauma Recovery

For those considering frequency healing, here are some practical approaches:

- Create a Calming Environment: Trauma survivors are often sensitive to sensory inputs. It's crucial to create a quiet, comfortable space where the person can feel secure. Sound therapy works best when the individual is able to focus on the vibrations without distraction.

- Start with Low Frequencies: Some frequencies, especially higher ones, can be stimulating and might initially feel overwhelming. Begin with gentle, grounding tones like 396 Hz to encourage relaxation. Gradually introduce other frequencies based on individual comfort.

- Use Guided Meditations: For those new to sound therapy, listening to guided meditations with background frequencies can provide structure. Guided meditations with theta or alpha waves are particularly helpful for grounding and mindfulness.

Reflections and Future Possibilities

Emma's story is one among many that highlight how frequency healing has the potential to revolutionize trauma recovery. Her ability to reconnect with her body, release tension, and find a sense of peace that had eluded her for years points to the deep-seated potential of these therapies. Science is only beginning to scratch the surface of how frequencies impact the body on both a cellular and energetic level, but Emma's experience gives us a glimpse of what the future might hold.

In a world where trauma is an all-too-common experience, frequency-based therapies offer a non-invasive, compassionate approach to healing. Unlike conventional methods that often require revisiting painful memories, frequency healing focuses on creating an environment of harmony and balance, allowing trauma to dissolve rather than be relived.

The Vision Forward: Frequency Healing as Mainstream Care

Imagine a future where trauma recovery centres routinely include sound and vibration rooms designed to help clients recover at their own pace, free from the stigma and discomfort that can accompany conventional approaches. Therapists might be equipped not only with talk-based skills but also with an understanding of frequencies and how they influence the nervous system. This vision of trauma care holds the promise of a world where people can find relief without having to revisit their pain over and over.

Final Thoughts: Trauma as a Portal to Inner Harmony

Healing from trauma is a deeply personal journey, and frequency healing offers a unique approach by bypassing the mind's defences and speaking directly to the body. It reminds us that beneath the noise of our daily lives lies a profound symphony—a harmony that is our natural state. Trauma may disrupt this balance, but through the power of sound and vibration, we can remember and return to this place of inner peace.

Emma's transformation reveals that frequency healing doesn't simply mask trauma; it has the potential to rewrite the patterns left behind, allowing us to step into a version of ourselves that is both healed and whole. In this way, trauma becomes a doorway, inviting us to explore and resonate with the deeper harmonies within us. Frequency healing might just be the key to unlocking the hidden power we each carry to heal from within, resonating with the true harmonies of life.

CHAPTER 67: CASE STUDY – FREQUENCY-BASED TREATMENT FOR ARTHRITIS

Arthritis is a condition that afflicts millions worldwide, causing chronic pain and limiting mobility, often robbing people of their vitality and independence. Traditional approaches to arthritis management typically focus on pain relief through medication, physical therapy, or even surgery. But a new, non-invasive approach is emerging in the realm of frequency medicine, offering hope for those seeking relief without side effects. This chapter explores the story of Marie, a 65-year-old woman who found unexpected freedom from arthritis pain through sound therapy, revealing the profound potential of frequency-based treatments for this common condition.

A Harmonious Approach: How Frequency Interacts with the Body

To understand the science behind Marie's transformation, it's essential to grasp how frequency-based therapies work. In essence, the human body operates much like a symphony, with each cell, tissue, and organ vibrating at its own unique frequency. When disease or injury occurs, these frequencies fall out of harmony, creating pain, inflammation, and dysfunction. Sound therapy, specifically tailored to address arthritis, aims to reintroduce harmony at a cellular level, reducing inflammation, easing pain, and restoring function.

Low-frequency sound waves, like those used in Marie's therapy, are believed to stimulate cellular activity and promote blood flow. Just as a soothing melody can ease a troubled mind, these frequencies seem to ease the body's internal discord. Scientific studies have shown that certain frequencies, particularly in the range of 40 to 60 Hz, can decrease inflammation and promote cellular repair, a perfect match for conditions like arthritis that involve chronic inflammation and tissue degradation.

Marie's Journey: From Pain to Relief

Marie's battle with arthritis began in her early fifties, with stiffness in her hands and knees that gradually worsened. Daily tasks like holding a coffee cup or climbing stairs became challenging, and she found herself avoiding activities she once enjoyed. Despite trying various treatments—including anti-inflammatory drugs, physical therapy, and even dietary changes—Marie found only temporary relief, and the pain always returned.

In her quest for alternatives, Marie came across frequency-based sound therapy and was intrigued by its promise of pain reduction without medication. Her initial scepticism faded when she read case studies and testimonials from other arthritis patients who had experienced significant improvements. With an open mind, she began sessions with a sound therapist who specialized in arthritis treatment.

The Therapy Process: Tuning in to Healing

Marie's sound therapy sessions involved lying comfortably

while low-frequency sounds were applied directly to her joints through a device resembling a small speaker. These frequencies—set between 40 and 60 Hz—were designed to target inflammation and promote cellular repair. The device emitted vibrations that resonated deeply within her tissues, creating a gentle massage-like sensation. Marie described feeling a warmth and relaxation that she hadn't experienced in years.

Initially, she attended three sessions per week. After the first two weeks, Marie noticed subtle changes: her stiffness began to ease, and her pain levels were decreasing. By the sixth session, she could grip a pen comfortably and even managed a walk through her favourite park without the usual throbbing in her knees. Encouraged by her progress, Marie continued with weekly sessions for three months.

Scientific Insights: Why Frequency-Based Therapy Works for Arthritis

While Marie's story is inspiring, it's supported by emerging science that explains how frequencies impact arthritis. Chronic inflammation is at the root of most forms of arthritis, causing pain and joint degeneration over time. Studies indicate that sound frequencies, specifically those in the low-frequency range, reduce the production of inflammatory markers like cytokines, molecules that fuel inflammation and pain.

Additionally, frequency therapy appears to stimulate the production of collagen, a protein crucial for joint health. As cells are exposed to these frequencies, their energy levels increase, enabling them to repair damaged tissues more effectively. This regenerative effect, combined with the anti-inflammatory properties of low-frequency vibrations, makes frequency-based therapy an ideal candidate for managing arthritis pain and restoring joint function.

Practical Tips for Arthritis Patients Considering Frequency Therapy

If you're considering frequency-based therapy for arthritis, here are some practical tips:

1. Consult a Specialist: It's essential to work with a therapist who has experience in frequency-based treatments for arthritis. They can tailor the frequencies to your specific condition and adjust the treatment as you progress.

2. Consistency is Key: Like physical therapy, frequency therapy requires consistency to see lasting results. Aim for regular sessions—ideally once or twice a week—to maintain the benefits.

3. Combine with Gentle Movement: Sound therapy can improve joint flexibility, and gentle exercises like stretching or yoga can enhance these effects, helping you regain mobility more quickly.

4. Hydrate: Sound therapy promotes cellular activity and circulation, and staying hydrated supports these processes, aiding in the elimination of toxins released during treatment.

Reflections on Frequency-Based Therapy and Its Potential

Marie's journey is just one example of how frequency therapy can transform the lives of those living with arthritis. Her experience suggests a future where arthritis management doesn't have to rely solely on medication or surgery but can include therapies that restore balance within the body at a cellular level. Marie regained not just the use of her hands but also a sense of independence and joy she thought she had lost.

While this therapy may not be a cure, it offers a path to relief that feels more holistic, aiming to bring harmony to the body's innate frequencies rather than suppress symptoms. The idea that sound can heal might seem like a revelation, yet it's deeply rooted in centuries of healing traditions that used drums, singing bowls, and chanting to promote health and balance.

A Vision for the Future: Expanding Access to Frequency Therapy

Imagine a world where frequency-based therapies are as accessible as physical therapy. Clinics would offer frequency therapy as part of standard arthritis care, with insurance plans covering these non-invasive treatments. Hospitals could even incorporate sound therapy rooms, where patients receive tailored frequencies that support their recovery from various conditions. Such a vision isn't far-fetched; it's a logical progression as we deepen our understanding of how sound and vibration affect human biology.

Final Thoughts: Rediscovering Inner Harmony

Marie's story reminds us that healing doesn't always come from an external intervention; sometimes, it's about rediscovering harmony within ourselves. Just as a misaligned instrument produces discordant sounds, the human body out of balance expresses itself in pain and disease. Frequency therapy has the potential to gently tune the body back into alignment, offering relief in a way that feels natural and profound.

Arthritis may be a chronic condition, but Marie's experience reveals a path forward that's full of promise. Through frequency therapy, she learned that her body could resonate with healing frequencies, transforming pain into peace, rigidity into fluidity, and dependence into freedom. For arthritis patients everywhere, this approach offers a new perspective—one that not only addresses the symptoms of their condition but empowers them to explore the harmonious, healing potential within their own cells.

With frequency-based therapy, the journey to relief becomes a journey of self-discovery, where each session reveals the body's capacity to heal, resonate, and thrive.

CHAPTER 68: CASE STUDY - FREQUENCY AND IMMUNE SYSTEM SUPPORT

When we think of the immune system, we picture a highly efficient, complex network that defends our bodies from threats. But for those with compromised immune systems, this internal army doesn't always function as intended. For decades, treatments have centred around boosting immunity through medication, supplements, and lifestyle adjustments. However, a lesser-known path is emerging—one that uses sound and frequency to gently harmonize and empower the body's defence mechanisms. This chapter delves into the journey of David, an individual with a compromised immune system who experienced remarkable improvement through frequency-based therapies, showing us how the vibrational language of frequencies might hold the key to immunity.

The Language of Frequencies: Harmonizing with the Body's Immune Symphony

Imagine the body as a grand orchestra, with each system, cell, and organ playing a specific note that, when in harmony, creates the melody of health. When the immune system becomes weakened or compromised, it's like an instrument falling out of tune, disrupting the entire body's rhythm. Frequency therapy, by reintroducing balance at a cellular level, aims to "retune" the body, allowing each part to play its part in the symphony of health.

Scientific research has shown that certain frequencies can stimulate cellular activity and even encourage the production of immune cells. Studies reveal that frequencies between 432 Hz and 528 Hz are particularly beneficial for calming the nervous system and supporting immune function, as these frequencies resonate deeply within the body. By introducing frequencies that the immune system naturally aligns with, frequency-based therapies may help the body heal itself from within.

David's Journey: From Chronic Infections to Vibrant Health

David, a 45-year-old graphic designer, had struggled for years with a compromised immune system. He experienced frequent colds, recurrent infections, and chronic fatigue, all of which took a toll on his quality of life. Every winter brought a series of colds and flus that lingered for weeks, often leaving him bedridden. Despite following a healthy diet, exercising, and taking immune-boosting supplements, David saw little improvement. His doctors had no clear answers, suggesting only symptom management through medication.

David's introduction to frequency-based therapies came through a recommendation from a holistic health practitioner. Intrigued by the idea of an approach that didn't involve more medication, he decided to try a series of sessions focused on strengthening his immune system. At his first session, David's therapist explained the principles of frequency healing in a way that resonated deeply with him. She used the analogy of tuning a piano, describing how specific frequencies could "tune" his immune system, restoring its resilience.

The Therapy Process: Frequencies as Immune Boosters

David's therapy sessions involved lying in a relaxed state as low-frequency sound waves were directed toward his body, particularly focusing on areas associated with immune function, like the thymus gland and spleen. These frequencies, set between 528 Hz and 741 Hz, were chosen to stimulate immune cell production and reduce stress—two critical factors for immune support.

The therapy room was equipped with speakers and a frequency-generating device. As the sessions began, David could feel gentle vibrations resonating through his body, inducing a profound sense of relaxation. Over the course of his sessions, he described the experience as "being bathed in sound," feeling a warmth and lightness spreading throughout his body. After just a few sessions, David noticed subtle changes: his energy levels were increasing, and he no longer felt the lingering fatigue that had been his constant companion.

The frequency therapy regimen consisted of twice-weekly sessions for three months. After six weeks, David reported a remarkable shift—he no longer experienced the frequent sore throats and sinus infections that had plagued him. By the end of the treatment plan, he felt stronger, healthier, and more resilient than he had in years.

The Science Behind Immune System Support Through Frequency

Why did David's immune system respond so well to sound therapy? Emerging research provides some insights. Studies suggest that certain frequencies, particularly those within the range of 528 Hz, can reduce inflammation—a major factor in immune system dysfunction. Inflammation not only wears down the immune response but also diverts the body's resources from fighting off infections.

Additionally, frequency therapy has been shown to improve lymphatic flow, enhancing the removal of toxins and waste products that burden the immune system. By stimulating the lymphatic system, sound frequencies encourage the body's natural detoxification processes, making it easier for the immune system to focus on protecting the body.

Beyond the physical effects, frequencies in the 528 Hz range have been linked to reduced stress. High-stress levels weaken immunity, and calming frequencies can activate the body's relaxation response, allowing it to repair and regenerate. This dual effect of reducing stress and promoting detoxification creates an environment where the immune system can thrive.

Practical Tips for Immune System Support Through Frequency Therapy

For those interested in exploring frequency therapy as a means to support immune function, here are some practical tips to get started:

1. Find a Certified Frequency Therapist: Working with a trained therapist who understands immune-focused frequency treatments is crucial. They can customize the frequencies and duration to suit individual needs.

2. Consistency Matters: Frequency therapy isn't a one-time fix. To see meaningful results, commit to regular sessions, ideally once or twice a week over several months.

3. Combine with Mindfulness Practices: Frequencies resonate deeper when the body and mind are relaxed. Simple mindfulness techniques, such as deep breathing or visualization, can enhance the effects of frequency therapy.

4. Use At-Home Sound Tools: For ongoing support, consider using sound tools at home, such as tuning forks or frequency apps set to immune-supportive frequencies like 528 Hz. These tools can help maintain the benefits between professional sessions.

Reflections on the Potential of Frequency Therapy for Immune Health

David's story opens up a new perspective on immune health, revealing that sometimes the best solutions are the ones that help the body heal itself. Frequency therapy offers a path that doesn't seek to replace the immune system's natural functions but rather amplifies them, making it easier for the body to defend itself against infections.

Imagine a future where frequency-based therapies become a common immune support tool, integrated into preventive health care. Sound therapy clinics could work in tandem with traditional practices, offering non-invasive, side-effect-free options to those in need of immune system support.

Final Thoughts: Rediscovering Health Through Harmony

The idea that sound can influence immune function might seem radical, but David's experience and emerging research paint a compelling picture. Just as an orchestra must be finely tuned to play harmoniously, our bodies need internal balance to function at their best. Frequency therapy offers a way to restore that balance gently, tapping into the body's inherent wisdom.

For David, frequency therapy wasn't just a treatment; it was a journey toward a new understanding of health and resilience. It taught him that healing doesn't always come from an external intervention—it can arise from aligning with the natural frequencies that empower the body's own defences. His story serves as an inspiring reminder that our bodies, when given the right tools, have an extraordinary capacity to heal, thrive, and resonate in harmony with life.

CHAPTER 69: CASE STUDY - FREQUENCY HEALING FOR CHRONIC FATIGUE

Chronic Fatigue Syndrome (CFS) is a mysterious, often debilitating condition. Those who suffer from it know the frustration of endless exhaustion, where even the simplest tasks feel insurmountable. Conventional medicine has provided some relief, but its focus on symptoms often leaves patients searching for a deeper, more holistic path to recovery. Frequency-based therapy—an approach that treats the body as a system of vibrations and rhythms—has begun to offer hope for those in search of an alternative. This chapter introduces Laura's journey with chronic fatigue and her discovery of the revitalizing power of frequency healing.

Frequency Healing: Resonating with the Body's Natural Rhythm

Imagine the body as a finely tuned orchestra, each organ and cell playing its part in harmony. For individuals with chronic fatigue, this harmony is often disrupted, as if some instruments have fallen out of sync. Frequency healing aims to retune the body, restoring a sense of internal balance and enabling the body's systems to work together once more.

Frequencies, often delivered through sound waves, electromagnetic fields, or vibrations, tap into the body's own natural frequencies. Research shows that certain frequencies can enhance cellular activity, encourage energy production, and calm overactive nervous systems—all crucial for individuals dealing with chronic fatigue.

Laura's Story: The Quest for Energy

Laura, a 36-year-old teacher, had battled chronic fatigue for nearly five years. It began subtly, with occasional exhaustion and brain fog, but within a year, her symptoms had progressed to the point where she struggled to get through a single workday. Weekends were no longer times of relaxation, as she spent them mostly in bed, recovering from the week's demands. Despite countless visits to specialists, strict dietary regimens, and even leaving her teaching position temporarily, her energy levels remained low.

In her search for answers, Laura encountered a holistic practitioner who introduced her to frequency healing. Sceptical at first, she listened as the practitioner explained the science of cellular resonance. "Think of your body like a guitar," he said. "Every cell has its own frequency, and sometimes those frequencies go out of tune. Frequency therapy can help bring them back in harmony."

The Treatment Process: A Symphony of Healing Frequencies

Laura's journey with frequency therapy began with a thorough assessment to determine which frequencies would best support her recovery. Chronic fatigue often involves imbalances in the nervous and immune systems, so her therapist focused on frequencies between 7 Hz and 10 Hz—known for their calming effects on the body and their ability to encourage restful sleep. Additionally, frequencies around 528 Hz were applied, as studies suggest these can aid cellular repair and regeneration.

The sessions themselves were simple. Laura lay on a comfortable mat while gentle frequencies were delivered through speakers and tuning forks. As the sound waves vibrated through her body, she described a sensation of deep relaxation, as though her body was "letting go" of stress she hadn't realized she was holding. During her first few sessions, Laura noticed subtle shifts—a slight lift in her mood, deeper sleep, and moments of mental clarity.

After a month of twice-weekly sessions, Laura's improvements became more pronounced. Her energy levels were still fluctuating, but she experienced more days where she felt able to take on light physical activities or enjoy a social outing. By her fourth month, she had resumed part-time work and was once again enjoying simple pleasures like cooking and walking in nature.

How Frequency Therapy Affects Chronic Fatigue

Why did frequency therapy have such a profound effect on Laura's fatigue? Studies suggest that certain frequencies stimulate the body's parasympathetic nervous system, which governs our rest-and-digest response. Chronic fatigue often keeps the body in a state of "fight or flight," where stress hormones flood the system and inhibit recovery. By promoting relaxation and shifting the body's focus to repair, frequency therapy allowed Laura's body to redirect its energy toward healing.

Additionally, frequencies around 528 Hz have shown promise in enhancing mitochondrial function. Mitochondria are the powerhouses of our cells, producing energy through a process called ATP synthesis. For individuals with chronic fatigue, mitochondrial function is often impaired. Frequency therapy, by restoring cellular energy production, can address one of the core issues in chronic fatigue.

Practical Tips for Using Frequency Therapy for Fatigue

If you're considering frequency therapy as part of your journey to overcome fatigue, here are a few steps you can take:

1. Seek a Certified Frequency Therapist: Finding a knowledgeable practitioner who understands chronic fatigue is crucial. They can guide you in selecting the most beneficial frequencies.

2. Practice Patience: Frequency healing is cumulative. Results build over time, so give yourself at least three to six months to see significant changes.

3. Complement with Mindfulness: Deep breathing or meditation exercises enhance the body's receptiveness to frequencies, amplifying the therapeutic effects.

4. Incorporate Home Practices: Use frequency apps, tuning forks, or playlists set to 528 Hz and other restorative frequencies at home. This can help sustain the benefits between sessions.

Reflection: The Future of Frequency Healing for Fatigue

Laura's story hints at a future where individuals with chronic fatigue and other invisible illnesses might find a new path to wellness. Frequency healing provides a non-invasive, side-effect-free alternative to standard fatigue treatments, one that aligns with the body's natural rhythms.

Envision a world where every health clinic offers frequency rooms, inviting patients to sit, relax, and simply resonate with healing vibrations. As technology advances, we may even see wearable devices that deliver personalized frequencies, enabling us to recalibrate our systems at any time.

Final Thoughts: Restoring Harmony Within

For Laura, frequency healing was a revelation—not just because it gave her energy back, but because it deepened her understanding of wellness. She learned that health is not merely the absence of disease but a dynamic state of balance and resonance. Frequency healing reminded her that her body, much like an orchestra, performs best when each part is in tune.

Her journey offers hope for others dealing with chronic fatigue, revealing that sometimes healing isn't about forcing change but gently encouraging the body to remember its own harmony. Through frequency, we can invite the body to realign itself, to embrace its natural rhythm, and to resonate once more with the song of life.

Laura's experience shows us that when we listen to our bodies, in both silence and sound, we can uncover profound insights and perhaps, one day, awaken the vibrant energy within.

CHAPTER 70: CASE STUDY - ENHANCED RECOVERY IN ATHLETES

For high-performance athletes, the difference between victory and defeat can hinge on the ability to recover quickly and efficiently. While rigorous training regimes and strict diets form the backbone of athletic success, recent advances have pointed to a surprising ally in recovery: frequency-based therapy. Imagine a tool that could harmonize every muscle, joint, and cell after intense exertion, accelerating healing and restoring energy. This chapter will delve into how frequency therapies are reshaping the athletic world, using the story of an elite runner, Evan, to illustrate this remarkable integration of science and sound.

A New Frontier in Athletic Recovery: Frequencies and Performance

The body of an athlete is like a finely tuned machine, each component—muscle fibres, nervous system, respiratory function—working in perfect synchrony. When one part of this intricate system breaks down due to overexertion, it can disrupt the entire mechanism. Traditional recovery methods like ice baths, massage, and physical therapy are staples of athletic training, yet frequency-based therapies introduce an innovative approach by tapping into the body's own energy fields.

Frequency therapy employs targeted sound waves, electromagnetic fields, or vibration to stimulate cellular activity and encourage tissue repair. Research in this area has shown that specific frequencies can accelerate muscle recovery, reduce inflammation, and improve circulation. These physiological benefits are crucial for athletes whose bodies endure repeated stress and demand rapid recovery to sustain high performance.

Evan's Journey: From Injury to Recovery Through Frequency Therapy

Evan, a marathon runner and aspiring Olympian, had trained for years to reach his peak. But as he pushed himself harder, he began to experience strain injuries in his legs and lower back. He tried every recovery method available: cryotherapy, deep tissue massage, even acupuncture. While these therapies provided some relief, none of them were enough to allow him to train at his full capacity without pain. At the recommendation of a teammate, Evan decided to explore frequency therapy, despite being initially sceptical .

His first session introduced him to pulsed electromagnetic field (PEMF) therapy, a frequency-based technique that targets damaged tissue at a cellular level. "Think of your cells like batteries," his therapist explained. "When they're run down, they need a recharge. PEMF helps restore energy to your cells, so they can repair themselves more effectively."

After just a few sessions, Evan noticed a difference. The lingering aches in his legs began to subside, and he felt more energized after his workouts. Encouraged by these results, he incorporated frequency therapy into his regular recovery routine, coupling PEMF with sound therapy sessions designed to stimulate relaxation and enhance circulation.

Over the following months, his performance improved, and he found himself able to sustain intense training with fewer setbacks.

How Frequency Therapy Supports Athletic Performance

The benefits Evan experienced can be attributed to the unique physiological effects of frequency therapy:

1. Enhanced Cellular Repair: Frequencies used in PEMF therapy help energize cells, specifically targeting mitochondria—the powerhouse of the cell. By optimizing cellular energy production, PEMF accelerates the healing of muscle fibres and connective tissue, which are often stressed by repetitive athletic exertion.

2. Reduced Inflammation: Frequencies between 5 Hz and 15 Hz are known to calm the nervous system and reduce inflammation. This is crucial for athletes like Evan, as chronic inflammation is a leading factor in injury and prolonged recovery time. By minimizing inflammation, frequency therapy prevents minor injuries from escalating.

3. Improved Circulation and Oxygenation: Sound therapy, particularly using frequencies around 528 Hz, is shown to promote circulation and oxygen uptake, both of which are vital for endurance athletes. Enhanced oxygenation means that muscles are better supplied during exertion, allowing for longer periods of peak performance with less fatigue.

Practical Applications: Tips for Athletes Integrating Frequency Therapy

For athletes curious about using frequency therapy as part of their recovery and performance strategies, consider the following practical tips:

1. Create a Routine: Just as you have a set training schedule, consistency is key for frequency therapy. Aim for sessions

twice a week and adjust based on the intensity of your training.

2. Use Frequency Therapy After Workouts: Studies suggest that frequency therapies like PEMF and sound therapy are most effective when used immediately after exercise. This timing aids in rapid inflammation reduction and cellular repair.

3. Combine with Traditional Recovery: Frequency therapy works well as a complement to other recovery techniques. Ice baths, massages, and stretching are still beneficial, and frequency therapy can enhance their effectiveness by helping the body heal at a cellular level.

4. Invest in Portable Frequency Devices: Many athletes are finding benefits from portable frequency devices that deliver therapeutic frequencies directly to targeted muscle groups. Devices like mini-PEMF mats or vibration plates can be used from home, providing convenient access to recovery between sessions with a therapist.

Vision for the Future: Frequency Therapy in Professional Sports

Imagine a world where every major sports team has a dedicated frequency therapist, where locker rooms are equipped with sound healing chambers, and where PEMF devices are as common as foam rollers. This vision is no longer just speculative; several professional teams have already begun integrating frequency therapy into their recovery programs, recognizing the immense value it brings to both physical and mental resilience.

Beyond recovery, frequency therapy is also proving valuable in mental training for athletes. Sound therapies, particularly binaural beats, are being used to synchronize brain waves, reducing anxiety and sharpening focus. For athletes who rely on both mental acuity and physical endurance, these therapies represent a holistic approach to peak performance.

Reflection: A New Paradigm of Recovery

Evan's story reveals an emerging truth: that healing and performance can be achieved not through brute force but through harmony. Frequency therapy encourages the body to recalibrate itself, aligning its natural rhythms and energies. This approach contrasts sharply with the "no pain, no gain" mentality that has long pervaded athletic culture, introducing instead a gentler, more attuned way to health and peak performance.

As frequency therapies continue to evolve, they challenge us to reconsider what it means to push the limits of human potential. Could it be that the path to ultimate strength lies not in pushing harder, but in listening closer—to our own bodies and to the subtle frequencies that guide our health?

Final Thoughts: Tuning the Body's Orchestra

In frequency therapy, there is an underlying philosophy: that each part of the body, from cells to systems, is part of a greater whole, a harmonious orchestra that must be tuned carefully to perform at its best. For athletes like Evan, embracing this harmony meant finding strength not only in his muscles but also in the quiet power of resonance and alignment.

The story of Evan's recovery and enhanced performance is a testament to the transformative potential of frequency healing. It invites us to imagine a future where every individual, whether an elite athlete or an everyday health enthusiast, can unlock the body's natural rhythms, restoring balance and resilience from within. By embracing frequency therapy, we may find that the most profound gains in health and performance come not from adding force, but from fostering flow.

PART XVI: DIY FREQUENCY HEALING PRACTICES

CHAPTER 71: SIMPLE FREQUENCY TOOLS FOR HOME USE

Imagine being able to tap into the world of healing frequencies right from the comfort of your own home. What if, with just a few simple tools, you could harmonize your mind and body, relieving stress, promoting cellular repair, and nurturing your overall well-being? In this chapter, we will explore accessible frequency healing tools like tuning forks, singing bowls, and mobile apps, offering you a gateway into the profound world of vibrational healing. With these tools, frequency healing can become part of your daily routine, transforming your home into a sanctuary of healing harmonies.

The Science of Sound: How Everyday Tools Can Heal

Think of your body as an orchestra, with each cell vibrating at its unique frequency. Just as an orchestra relies on every instrument to be in tune, our bodies rely on each cell and organ to resonate in harmony. Frequency tools like tuning forks and singing bowls are designed to bring each "instrument" within us back into alignment. These tools create sound waves that, when directed toward specific parts of the body, can restore balance, reduce stress, and encourage natural healing.

This concept is based on the phenomenon of resonance.

When two objects vibrate at similar frequencies, they amplify each other's vibrations, creating a harmonious effect. For instance, if a part of the body is out of sync—due to stress or physical discomfort—introducing sound waves from a tuning fork or a singing bowl can help that part of the body "re-tune," encouraging it to return to its natural state of harmony.

Tools for Personal Healing

Let's dive into the specific tools and how you can integrate them into your everyday life.

1. Tuning Forks: Precision in Sound

Tuning forks are among the most accessible tools for frequency healing. Made of metal, they emit a specific pitch when struck, producing sound waves that resonate with particular frequencies. Different forks are calibrated to different frequencies, allowing you to target specific areas of the body or emotional states.

How to Use Tuning Forks at Home:
- Targeted Healing: Strike a tuning fork gently and hold it close to the area of your body you wish to target. The vibration from the fork creates waves that travel into your tissues, helping to alleviate pain, reduce tension, and even encourage faster healing.
- Calming the Mind: Tuning forks in the range of 136.1 Hz, often called the "OM" frequency, are excellent for calming the nervous system. Strike the fork and place it near your temples or ears to allow the sound waves to relax your mind and bring you into a meditative state.

2. Singing Bowls: A Symphony in a Bowl

Singing bowls, traditionally used in Tibetan and Himalayan cultures, produce a sustained harmonic sound when played. Made of metal or crystal, these bowls create a rich, resonant tone that can fill a space with calming energy.

How to Use Singing Bowls:
- Creating a Healing Atmosphere: Gently tap the side of the bowl with a mallet and then glide it around the rim. The bowl will emit a sustained note that can resonate throughout the room, filling it with calming, healing energy.
- Self-Healing Practices: Place a singing bowl near a specific area of discomfort in your body, or simply let it sit in front of you during meditation. As you focus on the sound, visualize the vibrations traveling through your body, promoting relaxation and harmony.
- Types of Bowls: Crystal bowls are often tuned to specific chakras (energy centres in the body), making them useful for targeted healing. For example, a bowl tuned to 432 Hz is said to promote heart health, while bowls in the 528 Hz range are associated with DNA repair and regeneration.

3. Mobile Apps: Digital Tools for Modern Healing

In the digital age, mobile apps have brought frequency healing to our fingertips. Many apps use binaural beats, isochronic tones, or ambient sounds to generate specific frequencies aimed at relaxation, focus, or sleep improvement.

How to Use Frequency Apps:
- Binaural Beats for Meditation and Sleep: Binaural beats work by playing two slightly different frequencies in each ear, creating a third "phantom" frequency in the brain that aligns with specific mental states. Use headphones and set the beat to a low frequency (e.g., 4 Hz) for deep relaxation or sleep, or a higher frequency (e.g., 14 Hz) for focus.
- Isochronic Tones for Stress Relief: Isochronic tones involve single tones that pulse at specific intervals, encouraging the brain to sync with a calming or energizing frequency. These tones don't require headphones and are ideal for background

listening during work or relaxation.

- Apps for Chakra Balancing: Some apps provide audio tracks tuned to the frequencies of different chakras, allowing users to focus on one chakra at a time to release blocked energy and promote emotional well-being.

Practical Tips for Getting Started with Frequency Tools

Incorporating frequency tools into your daily life doesn't require extensive training or a major investment. Here are a few practical tips for integrating these powerful healing tools:

- Start Simple: Begin with a basic tuning fork or a small singing bowl. Familiarize yourself with the sounds and observe how your body and mind respond.
- Create a Routine: Choose a time each day to use your frequency tool. For example, you might begin the day with a short tuning fork session for focus or end the evening with a singing bowl session to unwind.
- Listen to Your Body: Pay attention to how your body feels during and after each session. Over time, you may find that certain frequencies or tools resonate more deeply with your unique needs.

The Future of Frequency Healing at Home

The accessibility of frequency tools suggests a future where healing harmonies become a standard part of home wellness routines. With portable and user-friendly devices on the rise, it's becoming easier than ever for people to tap into the body's natural resonance for relief from stress, pain, and fatigue.

Envision a world where homes are equipped with frequency tools, and where individuals reach for a tuning fork or a sound app as readily as they might for a cup of tea to soothe a weary mind or aching body. As more people explore and adopt these practices, frequency-based healing may transition from the realm of alternative therapy into a foundational part of holistic health.

Reflection: A New Paradigm for Personal Empowerment

The rise of accessible frequency tools represents a profound shift in how we view health and healing. Instead of relying solely on external treatments, individuals are now empowered to become active participants in their well-being, harnessing the power of sound and frequency in everyday life.

In the quiet resonance of a tuning fork, the lingering hum of a singing bowl, or the rhythmic pulsing of a binaural beat, we discover that healing is not confined to hospitals or clinics. It is a personal journey, accessible to all, and rooted in the timeless vibrations that connect body, mind, and spirit.

By embracing frequency tools in our lives, we tune ourselves to a state of harmony, where balance is not something we seek outside but something we cultivate within. The journey begins with a single sound, a single intention, and a willingness to listen—to our bodies, our energies, and the healing harmonies within.

CHAPTER 72: HEALING THROUGH DAILY SOUND ROUTINES

Imagine beginning each day with the subtle hum of healing frequencies, letting vibrations flow through you like an invisible stream, gently aligning body, mind, and spirit. Just as our ancient ancestors turned to rhythm and melody to connect with their surroundings, we too can tap into the harmonious power of sound in our daily routines. In this chapter, we'll uncover practical ways to integrate sound frequencies into our lives, transforming ordinary moments into healing rituals that nurture wellness from within.

The Science of Sound in Everyday Wellness

Think of your body as a symphony, with each cell and organ vibrating at its own unique frequency, harmonizing to create a balanced state of health. When we experience stress or illness, this harmony is disrupted. Sound frequencies help restore our body's natural rhythm by resonating with our cells and tissues, gently guiding them back into balance. Scientific studies have shown that sound waves can influence everything from heart rate and brain waves to mood and immune function.

Integrating sound healing into our daily lives allows us to constantly recalibrate our inner symphony, making it easier to maintain mental clarity, emotional balance, and physical health. And the beauty of this approach lies in its simplicity —you don't need complex devices or special skills to start. With basic tools like tuning forks, singing bowls, and even digital apps, you can weave healing frequencies into your daily routine.

Morning Awakening: Sound to Start the Day

Starting your day with sound frequencies can set a positive tone, helping you feel grounded and energized.

Simple Morning Practices:
1. Tuning Fork Activation: Use a tuning fork tuned to 128 Hz or 432 Hz—frequencies known for grounding and harmony. Strike the fork gently and hold it near your temples or chest. Take a few deep breaths, letting the vibrations awaken your senses and bring mental clarity.
2. Chanting or Humming: Your own voice is a powerful tool. Start your day with a gentle hum or chant, focusing on the "OM" sound, which vibrates at a frequency that promotes relaxation and mental clarity.
3. Play a Sound Bath Track: Many streaming platforms offer sound bath recordings featuring crystal bowls, gongs, or chimes. Listen for just five minutes to feel the calming effects. This short practice can help you feel centred before stepping into the busyness of the day.

Midday Balance: Sound to Restore Focus

Incorporating sound into the middle of the day can help you regain energy and focus, especially if you feel stress or mental fatigue setting in.

Practices for a Balanced Day:
1. Binaural Beats for Focus: Listen to binaural beats set to

frequencies between 12 and 30 Hz, known as the beta range, which is associated with alertness and concentration. You can find many apps that offer guided sessions for focus—just plug in your headphones, and let the sound refocus your mind.

2. Desk-Friendly Tuning Forks: Keep a small tuning fork at your desk. When you feel tension building, take a moment to strike it and hold it near your body. The vibration can help release tension and restore calm.

3. Breathing with a Singing Bowl: If you work from home or have a quiet space, keep a small singing bowl nearby. A few minutes of playing the bowl, paired with deep breathing, can clear mental fog and reset your focus.

Evening Release: Sound to Let Go and Relax

Sound frequencies can help signal to your mind and body that it's time to unwind, making evening the perfect time for restorative sound practices.

Evening Relaxation Practices:

1. 432 Hz Sound Bath for Calm: Listening to music tuned to 432 Hz, often called the "miracle tone," can have a deeply calming effect on the body. Play a 10-15 minute track as you prepare for bed, letting the vibrations ease away the day's stress.

2. Guided Frequency Meditation: Many meditation apps offer sessions with isochronic tones or binaural beats in the theta range (4-8 Hz), which promote relaxation. A 20-minute guided meditation with these frequencies can help prepare your mind and body for restful sleep.

3. Soft Humming for Self-Soothing: Before bed, take a few moments to hum softly, feeling the vibration in your chest. This simple act can be incredibly grounding and is a natural way to signal to your body that it's time to rest.

Integrating Sound Into Other Daily Activities

Even everyday tasks can become opportunities to incorporate healing frequencies, transforming mundane moments into soothing rituals.

In the Shower
The shower offers an ideal setting to explore sound resonance. Try humming or singing, letting the sound echo against the tiles. Feel the vibrations in your chest and throat; this resonance can be very calming and helps release stress.

While Cooking
Play background music tuned to specific frequencies while you cook. Songs tuned to 528 Hz, for example, are said to promote healing and positivity. Cooking to a rhythm or humming along can also infuse your food preparation with mindful energy.

During Exercise
Incorporate frequency-based music in your workout routine. For example, listen to upbeat tracks tuned to 396 Hz or 417 Hz, which are associated with liberation from negative energy and personal empowerment. These frequencies can keep you motivated and aligned with your body's rhythm during physical activity.

DIY Sound Routine: Designing Your Own Frequency-Based Day

Creating a daily sound routine can be as unique as you are. Here's a sample schedule to inspire you:

- Morning: 5 minutes of tuning fork meditation followed by gentle humming.
- Midday: A quick 10-minute binaural beat session for focus after lunch.
- Evening: Play 432 Hz music as you wind down for bed, paired with a few deep, calming breaths.

Consistency is key. With time, these small practices will become second nature, forming an invisible web of healing harmonies that supports you day in and day out.

Vision for a Harmonious Future

Imagine a future where sound-based routines are woven into the fabric of our daily lives—where we instinctively turn to frequencies for balance and healing. What if hospitals used sound baths to calm patients, workplaces integrated binaural beats for focus, and schools encouraged children to begin their day with sound-based meditation?

This vision is not so far-fetched. As we understand more about the power of sound to influence well-being, the integration of frequency medicine into our lives becomes not just possible but natural. Healing will become less about reactive measures and more about daily nurturing practices that prevent illness and promote sustained harmony.

Reflection: Tuning Into the Rhythms of Life

As you explore sound frequencies in your daily life, consider it an invitation to tune into the larger rhythms that guide us all. Each hum, each beat, each note becomes a reminder of our connection to a greater harmony, an eternal pulse that flows through all living things.

In the end, healing through sound is not merely about the tools or the tones; it's about becoming aware of life's rhythms and aligning with them. As you bring sound into your life, you may find yourself feeling more present, more attuned to your own needs, and more connected to the world around you.

This is the promise of sound healing—a return to harmony, not only within yourself but within the symphony of life itself. Let each note guide you, and may you always find your way back to your own healing harmony.

CHAPTER 73: CREATING A PERSONAL SOUND BATH AT HOME

Imagine lying comfortably in your own space, feeling sound waves ripple through the air, gently washing over you like an invisible, healing tide. This is the essence of a sound bath: an immersive experience where vibrations of carefully chosen frequencies help you release tension, restore balance, and renew your energy. The beauty of a sound bath is that you don't need to be a professional or have specialized equipment to create this powerful healing experience at home. In this chapter, we'll explore the transformative effects of sound baths, provide step-by-step guidance on setting one up, and uncover the deeper potential of sound as a tool for healing and self-care.

The Science Behind Sound Baths

A sound bath is much more than a relaxing experience; it's a scientifically grounded method of healing that taps into the body's natural resonance. Think of each cell in your body as a tiny instrument, vibrating at specific frequencies to maintain harmony and health. When we encounter stress, illness, or injury, these frequencies can become "out of tune." By exposing our body to external frequencies, such as those generated by singing bowls or tuning forks, we can help realign our cells' vibrations, restoring balance and promoting healing.

Recent studies have shown that sound baths can reduce anxiety, lower blood pressure, and enhance sleep quality. This is largely due to the ability of sound waves to influence our brainwaves. Frequencies between 4 and 7 Hz, for instance, can stimulate theta brainwaves—those associated with deep relaxation and meditation. By immersing ourselves in these frequencies, we encourage our brain to enter states of calm, allowing our body's natural healing processes to take over.

Setting Up Your Home Sound Bath

Creating a sound bath at home is surprisingly simple. With just a few basic tools and a little preparation, you can set up a healing oasis in any room.

Step 1: Choose Your Space

Find a quiet, comfortable space where you won't be disturbed. This could be your bedroom, living room, or even a cozy corner. Dim the lights or light a few candles to create a calming atmosphere. Many people also find it helpful to add natural elements like plants, crystals, or a small water fountain to enhance the healing environment.

Step 2: Gather Your Instruments

You don't need a full collection of singing bowls to create a powerful sound bath; even a single instrument can be transformative. Here are some accessible tools:

- Tuning Forks: Easy to use and available in a variety of healing frequencies, tuning forks are a popular choice for beginners.
- Singing Bowls: Tibetan or crystal singing bowls produce rich, resonant tones. Each bowl typically corresponds to a specific chakra or frequency, making it easy to target areas

you wish to heal.

- Chimes or Bells: These can add gentle, soothing tones to your sound bath and are excellent for creating a soft, meditative environment.

- Digital Sound Apps: If you don't have physical instruments, there are many sound bath apps and tracks available online that provide high-quality recordings of singing bowls, gongs, and other instruments.

Step 3: Set an Intention

Sound baths are most powerful when guided by intention. Take a few deep breaths and reflect on what you hope to achieve—whether it's relaxation, emotional release, or physical healing. Holding an intention helps focus the mind and directs the healing energy toward your needs.

Step 4: Begin with a Grounding Sound

Start your sound bath with a grounding tone, like the deep hum of a low-pitched tuning fork or singing bowl. Strike the instrument gently and hold it close to your body. Allow yourself to feel the vibrations reverberate, anchoring you in the present moment. You can repeat this grounding sound several times to fully settle in.

The Experience: Immersing in Healing Frequencies

As you play each instrument, focus on how the sound feels in your body. Close your eyes and let the vibrations wash over you, noticing any areas of tension or discomfort. Imagine these vibrations penetrating deeply into each cell, releasing stored stress and restoring balance.

Moving Through Different Frequencies

For a full-body experience, use instruments tuned to different frequencies that target specific areas of the body. For example:

- Low Frequencies (e.g., 174 Hz): These are excellent for grounding and physical pain relief. Use them at the beginning and end of your sound bath to create a sense of rootedness.

- Mid-Range Frequencies (e.g., 432 Hz): Known as the "miracle frequency," 432 Hz is said to promote emotional healing and is ideal for relieving stress and encouraging relaxation.

- Higher Frequencies (e.g., 963 Hz): Known as the "frequency of the gods," this high frequency is often associated with spiritual connection and mental clarity. Use it toward the end of your sound bath for a sense of elevation and renewal.

Tips for Deepening the Experience

1. Practice Mindful Breathing: Sync your breath with the sound. Breathe in deeply as you strike the instrument, and exhale slowly as the sound fades. This practice helps anchor you in the rhythm of the sound and enhances relaxation.

2. Visualize Healing: Imagine the sound waves traveling through your body, reaching areas that need healing. Visualization amplifies the effects of the sound bath by engaging the mind in the healing process.

3. Stay Present: It's natural for the mind to wander, but gently bring your focus back to the sound whenever you notice your thoughts drifting. This process of returning to the sound is itself healing, helping you practice mindfulness and presence.

Ending the Sound Bath: Grounding and Reflecting

When you feel ready to end the sound bath, bring back the grounding sound you started with. This could be the same low-pitched tone or a gentle chime to help ease your body back to the present. Take a few moments to lie still, reflecting on any sensations, emotions, or insights that emerged during the experience.

Many people find it helpful to journal afterward, noting any thoughts or feelings that came up. This reflection is part of the integration process, allowing you to carry the healing energy of the sound bath forward into your daily life.

Vision: Sound as a Pillar of Future Wellness

As you explore sound baths and other frequency-based healing techniques, consider the possibility that sound could one day be a primary tool in medicine. Imagine clinics where sound baths are prescribed for stress relief, schools where children start their day with tuning forks, or workplaces that offer frequency-based relaxation sessions. The future holds limitless potential for integrating sound into our well-being practices, bringing us closer to natural, non-invasive methods of healing.

Sound baths are just one facet of this growing field, yet they provide a glimpse into the profound healing power of frequencies. By incorporating these practices into our lives, we are not only tapping into ancient wisdom but also embracing a forward-thinking approach to health and wellness.

Reflection: Embracing Sound as a Lifelong Ally

Sound healing is about more than momentary relaxation; it's a journey toward deeper alignment with ourselves and the world around us. Each sound bath is an opportunity to tune into the vibrations that connect us all, reminding us that healing is not just about the physical body but also the mind, spirit, and environment.

As you continue exploring sound baths at home, know that you're participating in an age-old tradition, one that celebrates the inherent wisdom of nature. Allow each vibration to guide you back to harmony, reminding you of the beauty in simplicity and the profound healing available to all through the power of sound. Let this chapter serve as your invitation to embrace sound as a lifelong ally on your path to wellness.

CHAPTER 74: USING FREQUENCIES FOR MEDITATION AND MINDFULNESS

In a world filled with constant noise and busyness, finding moments of peace and self-awareness can feel elusive. Yet, as ancient wisdom and modern science increasingly reveal, meditation and mindfulness are powerful tools for nurturing mental clarity, emotional balance, and overall well-being. Imagine, then, if we could amplify these benefits further with the simple addition of sound frequencies. By tuning into specific frequencies during meditation, we can deepen our connection to the present moment, heighten our mindfulness, and tap into a profound state of relaxation. This chapter explores the role of sound frequencies in meditation, shares techniques for incorporating them into your practice, and unveils the scientific magic behind this powerful combination.

The Science Behind Frequencies in Meditation

Meditation works by slowing down brainwave activity, shifting our brain from higher frequencies associated with daily alertness and stress (beta waves) to lower frequencies that promote relaxation and creativity (alpha and theta waves). Frequencies in the range of 4–8 Hz, for instance, help the brain enter a theta state, which is associated with deep relaxation, enhanced creativity, and profound insight. When sound frequencies are introduced, such as through singing bowls, tuning forks, or even digital binaural beats, they gently guide the brain into these lower frequencies, making it easier to achieve and sustain meditative states.

This concept of "brainwave entrainment" occurs when external frequencies synchronize with our brain's natural rhythms, essentially helping our minds "tune in" to the desired state more quickly. Studies have shown that using frequencies during meditation can help reduce anxiety, improve mood, and enhance focus. With this understanding, we can begin to see frequencies not merely as sound, but as a supportive tool that can unlock deeper meditation.

Crafting Your Frequency-Based Meditation Practice

Creating a meditation practice that incorporates frequencies is simple and accessible. Whether you're a beginner or an experienced meditator, you can use sound to amplify your journey inward.

Step 1: Choose Your Frequency Tool

The first step in creating a frequency-based meditation practice is selecting a tool that resonates with you. There's no one-size-fits-all here, so feel free to explore different options until you find what feels most effective and soothing. Some popular choices include:

- Singing Bowls: Tibetan and crystal singing bowls are tuned to specific frequencies, often corresponding to the body's chakras. The resonant tones help calm the mind and promote healing within specific energy centres.
- Tuning Forks: These are small, portable, and produce pure tones. They are commonly tuned to 528 Hz, a frequency associated with transformation and DNA repair, making them ideal for meditative healing.

- Binaural Beats: Using headphones, binaural beats create a phenomenon where two slightly different frequencies are played in each ear, leading the brain to perceive a third "phantom" frequency. This allows you to target specific brainwave states (theta for deep meditation, alpha for relaxation, etc.).
- Gongs or Chimes: The deep, reverberating sound of a gong or the delicate chime of a bell can signal your mind to enter a relaxed state. These instruments add a grounding element to the practice.

Step 2: Set an Intention

Begin your meditation by setting a clear intention. Why are you meditating today? Perhaps you're seeking relaxation, clarity, or emotional release. By focusing on an intention, you give your practice a purpose that the frequencies can work toward, guiding both your conscious and subconscious mind.

Step 3: Find Your Posture and Breathe

Sit or lie in a comfortable position. Close your eyes, and start by taking several deep breaths. Feel your breath as it flows in and out, anchoring you to the present moment. This simple act of focused breathing allows you to prepare for the sound frequencies, creating a receptive state for them to work.

Frequency Techniques for Deepening Meditation

Now that you're settled in, it's time to explore techniques to bring sound into your meditation practice. Here are a few methods that combine frequency with mindfulness, offering unique ways to connect with each sound and its effect on your mind and body.

The "Whole-Body Resonance" Technique

Imagine your body as a vessel, ready to be filled with sound. Gently strike a singing bowl or tuning fork, and place it close to your heart or other chakras. As the sound vibrates through the air, visualize it traveling through your body, dissolving tension, and aligning each cell to its natural state. You might sense the vibration lingering in certain areas longer than others—this is your body's way of telling you where healing or release is needed.

Binaural Beat Mindfulness

If using binaural beats, select a track that corresponds with your meditation goal. For instance, a 6 Hz beat can guide you into a theta state, which promotes creativity and emotional insight. As you listen, let the beat wash over you, tuning your brain into that specific frequency. When thoughts arise, gently bring your awareness back to the sound, allowing it to anchor you in the present moment. This practice deepens focus, encouraging a calm and centred state of mind.

The "Sound Sweep" Technique

Using chimes, tuning forks, or a gong, start by creating a gentle sound and imagine it sweeping through your entire body. Visualize the sound gathering any distractions, stress, or negativity, and carrying them away as the tone fades. With each repetition, feel your body becoming lighter, your mind quieter, and your awareness more refined. The sound sweep is excellent for beginners, as it offers a tangible way to "clear" mental clutter and ease into mindfulness.

Practical Tips for Enhancing Your Frequency Meditation Practice

1. Practice Consistency: Set aside a few minutes each day to work with sound frequencies. The cumulative effect of consistent practice can lead to profound shifts in mood, energy, and mental clarity.
2. Experiment with Different Times: Early morning or just before bedtime are often ideal, as our minds are naturally

closer to meditative brainwave states at these times.

3. Observe How Your Body Responds: Notice how different frequencies affect your energy. You may find that lower frequencies calm you, while higher ones energize. Tailor your practice based on what you need that day.

4. Create a Sacred Space: Dedicate a quiet corner or room for your sound meditation, enhancing the practice's impact by associating the space with peace and mindfulness.

A Vision for the Future: Sound as the Key to Inner Balance

Imagine a world where frequencies are as commonly prescribed as medications. Where schools and workplaces offer sound meditation spaces, and where stress relief doesn't come in a pill but through a tuning fork or a sound bath. Sound frequencies could become a cornerstone of preventive mental health care, addressing issues like anxiety, depression, and burnout in natural, accessible ways. As we learn to harness the potential of frequencies, we inch closer to a society where well-being is not just about fixing what's broken, but nurturing what is whole.

Reflection: The Journey of Sound and Silence

In the end, the most profound aspect of using frequencies in meditation is the simplicity of it all. It reminds us that healing doesn't have to be complex or invasive; sometimes, all it takes is a quiet moment, a gentle tone, and the willingness to listen. Frequencies give us a tool to tune ourselves, to find our inner rhythm, and to reconnect with the essence of who we are beneath the noise of daily life.

As you explore these techniques, remember that the goal is not to "achieve" something but to experience presence. Each sound, each silence that follows, is an invitation to return to yourself. May this journey into frequency-based meditation be a path that leads you not only to peace but to a deeper understanding of the power within.

CHAPTER 75:
FREQUENCY-BASED
YOGA PRACTICES

In yoga, the aim is often to create a harmony between the body, mind, and spirit, uniting them through breath, movement, and mindfulness. Yet, there is an ancient, often forgotten ally that can deepen this journey toward inner balance: sound. Imagine incorporating specific frequencies into your yoga practice, where each note and vibration works in tandem with each movement and breath. With frequency-based yoga practices, we not only deepen relaxation but also enhance the energetic impact of each posture, guiding the mind into deeper states of calm and the body into profound relaxation. This chapter will reveal how incorporating frequencies into yoga creates a more immersive experience that harmonizes the entire being.

Understanding Frequencies and Their Effect on the Body

The science of sound and vibration tells us that each frequency has a unique impact on the body. Just as different yoga poses activate various muscles and energy centres, specific frequencies can stimulate certain states of consciousness and even particular regions within the body. Lower frequencies, for instance, often have a grounding effect, helping one feel rooted and connected to the earth. Higher frequencies, on the other hand, may evoke clarity and a heightened sense of awareness.

This synergy between frequencies and yoga aligns perfectly with the practice of "Nada Yoga" or the "Yoga of Sound," which has been explored for thousands of years in Indian spiritual traditions. In this approach, sound becomes a powerful tool for aligning body and mind. Scientific studies are beginning to reveal how sound frequencies can entrain brainwaves, helping practitioners achieve deeper states of meditation and facilitating relaxation by activating the parasympathetic nervous system.

Integrating Frequencies into Your Yoga Practice

Integrating frequency-based elements into yoga doesn't require complex equipment. With tools like tuning forks, singing bowls, or even binaural beats, you can easily transform your practice. Below are some practical ways to weave these frequencies into each stage of your yoga routine.

1. Beginning Your Practice with Grounding Frequencies

Starting your practice with grounding frequencies can help centre the mind and calm any restlessness in the body. Frequencies like 40 Hz, known for grounding, or 136.1 Hz, which is said to resonate with the earth's vibration, are ideal choices. Begin your session seated in a comfortable cross-legged position. Strike a tuning fork or gently ring a Tibetan bowl, allowing the sound to permeate the space and guide your breath into a steady rhythm.

Practical Tip: Focus on the vibration as it resonates, feeling it in your lower body. Imagine the sound grounding you, rooting you to the earth beneath you. With each inhale, draw in stability, and with each exhale, release any distractions.

2. Flowing with Frequencies: Moving Through Asanas

As you flow through yoga postures, frequencies can amplify

the physical and energetic effects of each movement. For example, incorporating a frequency like 528 Hz, often associated with love and healing, can enhance heart-opening poses such as Cobra or Camel.

Playing a recording of a singing bowl tuned to 528 Hz as you move into a heart-opening posture invites an emotional release and a sense of compassion. When moving into a pose such as Child's Pose, lower frequencies like 174 Hz can aid relaxation, encouraging the muscles to let go of tension and allowing you to sink deeper.

Practical Tip: As you hold each pose, consciously focus on how the sound resonates within your body. Notice if the frequency enhances your awareness of any tightness or creates a sense of release in certain areas. Let the sound support your journey inward, as though it is gently guiding you to connect with your inner self.

3. Enhancing Breathwork (Pranayama) with Binaural Beats

Binaural beats—tones created by playing slightly different frequencies in each ear—are a great complement to pranayama, or breath control. By using headphones, you can use binaural beats to align your brainwaves with the meditative frequency of your choice.

For pranayama exercises, consider using a 6 Hz binaural beat to stimulate theta waves, which promote relaxation and introspection. Begin with Nadi Shodhana (alternate nostril breathing), focusing on the rhythm of your breath and allowing the binaural beat to guide you into a calm, centred state.

Practical Tip: Breathe slowly, inhaling to a count of four, holding for four, exhaling for six, and holding again for four. The beats can act as a metronome, pacing your breath and easing you into a meditative state where both body and mind feel balanced.

4. Deepening Savasana with Higher Frequencies

Savasana, or the final resting pose, is a perfect moment to introduce higher frequencies associated with clarity and transcendence. Frequencies like 963 Hz, known as the "frequency of divine consciousness," can help the mind feel lighter and more open, creating an expansive sensation as you lie still.

Place a crystal singing bowl near your head or listen to a recording of this frequency. As you lie in Savasana, let the sound envelop you, each tone feeling like a gentle wave that washes over your entire body. Visualize the sound clearing your mind and freeing your spirit from daily concerns.

Practical Tip: Imagine that each tone is filling your body with light, each cell vibrating in harmony with the sound. Allow the frequency to hold you in a state of peace, helping you transcend beyond the physical and into a state of deep stillness and unity.

A Reflective Practice: Harmonizing Body, Mind, and Spirit

Frequency-based yoga goes beyond physical poses and breathwork—it becomes a holistic practice that touches all aspects of being. Just as yoga teaches us the interconnectedness of our mind, body, and spirit, frequencies guide us toward balance, showing us that healing is a full-body experience.

Imagine a future where studios integrate frequencies into every class, offering yoga as both a physical and vibrational healing practice. In this vision, practitioners are no longer just moving through poses but are immersed in a sound bath that aligns them at every level. This integration holds immense potential, offering both mental and physical healing that echoes far beyond the yoga mat.

Reflecting on the Healing Power of Frequency in Yoga

When we blend frequency with yoga, we deepen our practice in ways we may not even consciously recognize. Each sound, each vibration, is a reminder of the natural rhythms within us—our heartbeats, our breaths, our thoughts. With each session, we step closer to understanding ourselves, feeling more whole, present, and grounded.

As you bring frequencies into your yoga practice, remember that each sound is an invitation to connect with yourself. Each vibration is an invitation to heal, to release, and to grow. In this space of harmony, we find that yoga is more than movement; it's a journey back to balance, a return to our own unique rhythm.

May your practice be enriched by this union of body, breath, and sound, leading you to discover the power of frequencies in harmonizing your life. With each tone, may you find yourself a step closer to the peace that lies within.

PART XVII: EXPLORING CONTROVERSIES AND ETHICAL CONSIDERATIONS

CHAPTER 76: THE DEBATE OVER FREQUENCY HEALING

As the world of medicine evolves, innovative therapies often spark spirited debates, and frequency-based healing is no exception. The concept of using sound waves, electromagnetic fields, and specific vibrational frequencies to influence health is an exciting frontier, yet it's one that's met with both enthusiastic support and cautious scepticism within the medical community. In this chapter, we'll explore the various perspectives, concerns, and ongoing debates that surround frequency healing. We'll dive into the scientific evidence, share real-world case studies, and examine the reasons why frequency healing has yet to gain universal acceptance in mainstream medicine.

Setting the Stage: Why Frequency Healing Faces Resistance

The scepticism surrounding frequency healing largely stems from its departure from conventional medical practices. Traditional Western medicine has long been grounded in principles of biochemistry and pharmaceuticals, with interventions designed to target specific diseases at a molecular level. Frequency healing, on the other hand, operates from the premise that the body is an intricate energy field and that by tuning into certain frequencies, we can influence cellular health, mental well-being, and even pain levels.

For some in the medical field, the idea of healing through vibration and sound seems nebulous or ungrounded, too intangible compared to the measurable effects of drugs and surgery. Additionally, some forms of frequency therapy still lack the robust scientific validation required to meet the rigorous standards of modern medicine.

However, there are also many practitioners and researchers who argue that our bodies are not just biochemical entities but also bioelectrical ones, and that frequencies can indeed impact health in ways that traditional medicine has only begun to explore. In fact, medical devices like ultrasound and MRI rely on similar principles, as do therapies such as transcranial magnetic stimulation (TMS) for depression. The question then becomes: why are some forms of frequency healing embraced by the medical community, while others remain on the fringes?

Exploring the Science: A Balancing Act of Evidence and Belief

One challenge frequency healing faces is the need for scientific studies that meet the "gold standard" of randomized, double-blind, placebo-controlled trials. While there is research showing that frequencies can influence biological processes—such as studies on electromagnetic fields promoting bone growth and sound therapy reducing stress—many argue that larger, more conclusive studies are needed.

For example, a study published in Pain Research and Management found that pulsed electromagnetic field (PEMF) therapy significantly reduced chronic pain in patients with osteoarthritis. Similarly, research on sound therapy suggests that frequencies between 4–8 Hz, which align with theta brainwaves, can promote relaxation and reduce anxiety. Yet, critics argue that many studies are still preliminary and that placebo effects cannot be entirely ruled out.

This ongoing need for high-quality research leaves frequency healing in an in-between space—a therapy that shows potential but requires more evidence to gain widespread trust. This gap in research leads many doctors to adopt a cautious approach, as they strive to balance innovation with patient safety.

Case Studies: Real-World Applications and Mixed Outcomes

Despite the scepticism, many individuals have found relief through frequency-based therapies. Consider the story of Sarah, a 45-year-old woman suffering from chronic fibromyalgia. Traditional painkillers offered limited relief, and the side effects were challenging. Frustrated, she turned to PEMF therapy, which she credits with reducing her pain by 50%. Sarah describes the therapy as "feeling like a gentle buzz through her muscles," and although her doctor initially dismissed the approach, her results were undeniable. "I'm finally able to go through my day without constantly fighting pain," she says.

Similarly, sound therapy has been transformative for individuals struggling with anxiety. One patient, a former combat veteran, found that listening to binaural beats at specific frequencies helped ease his symptoms of PTSD. He describes the experience as "finding an inner calm" that traditional talk therapy couldn't reach. For him, sound therapy became a powerful supplement to his conventional treatment plan.

However, not all cases are successful. There are patients who have tried frequency therapies with little or no effect, leaving them disappointed and sometimes even more sceptical . These mixed outcomes are a significant factor in the debate, with proponents arguing that individual differences play a role, while sceptics maintain that the lack of consistent results raises questions about the therapy's validity.

Addressing Scepticism: Bridging the Gap

For frequency healing to gain broader acceptance, advocates argue that it's essential to engage sceptics with open, science-based conversations. This includes discussing both the limitations and the potential of frequency-based therapies. Many practitioners believe that a balanced approach—acknowledging areas where frequency healing is still in its infancy, while also highlighting credible research and positive patient outcomes—could help shift the narrative.

For example, explaining the science behind PEMF therapy and showing parallels between frequency-based treatments and other accepted medical technologies, like MRI or TMS, can help make frequency healing more relatable and credible. Additionally, transparent conversations about the limitations of current studies can foster trust, showing that the frequency healing community is committed to scientific rigor and improvement.

The Path Forward: Integration or Divergence?

As we look toward the future, the question remains whether frequency healing will integrate into mainstream medicine or continue to exist as a complementary approach. One potential path is through integrative medicine, which combines conventional treatments with alternative therapies to offer patients a holistic care model. By bringing frequency healing into integrative medicine practices, doctors could monitor and document outcomes, contributing valuable data to the scientific community.

Moreover, advancements in technology could pave the way for more precise, personalized frequency-based therapies.

Imagine a future where wearable devices can scan your body's frequencies and deliver targeted sound waves or electromagnetic pulses to promote healing. Such a future could bring together sceptics and advocates alike, as frequency healing becomes more precise, measurable, and grounded in real-time data.

A Reflection on Health, Healing, and Harmony

At its core, the debate over frequency healing touches on a deeper question: What is healing? Is it only the measurable effects on our bodies, or does it also encompass the more subtle, intangible aspects of well-being? Sound, vibration, and energy have been used by ancient cultures for centuries to promote healing, and perhaps there is wisdom in these traditions that science has yet to fully uncover.

The truth may lie in a balanced perspective—one that honours both the rigor of science and the mysteries of human experience. Frequency healing invites us to expand our view of medicine, to consider that the vibrations of sound, like the beating of our hearts, may hold more significance than we currently understand. As we continue to explore this field, it's crucial to maintain a sense of wonder and openness, even as we apply critical thinking and scientific inquiry.

May we find ourselves moving toward a future where healing is not confined to one modality or approach but is as harmonious and diverse as the frequencies that resonate within us. The debate over frequency healing is, in many ways, a testament to the evolving landscape of medicine—a reminder that true healing often requires us to look beyond the known and venture into the mysteries that lie just beyond our understanding.

CHAPTER 77: ADDRESSING SAFETY AND RISKS IN FREQUENCY MEDICINE

In our exploration of frequency-based healing, it's easy to get swept up in the possibilities and potential benefits, but it's equally essential to consider safety. Frequency medicine, like any therapeutic intervention, carries risks that must be carefully examined, respected, and managed. This chapter delves into the considerations for safely applying frequencies for healing, from understanding the physiological limits of the body to the ethical responsibility of practitioners. Here, we'll lay out guidelines, share insights from medical professionals, and offer tips for those considering frequency-based therapies, creating a balanced approach to this promising yet delicate field.

The Power of Frequencies—and Their Risks

Imagine the body as an intricate orchestra where each cell vibrates to a specific frequency. Just as a musician's mistake can disrupt a symphony, frequencies that are too intense, sustained, or poorly applied can potentially disrupt the body's natural harmony. Therapeutic frequencies can be powerful tools, but that power comes with a responsibility to use them thoughtfully.

Much of the scientific basis behind frequency medicine relies on resonance—the idea that specific frequencies can harmonize with cells, organs, or systems in the body. However, if these frequencies are too intense or used incorrectly, they can lead to overstimulation, cellular damage, or other unintended effects. In the case of high-intensity ultrasound, for example, while it can help in medical diagnostics and treatment, excessive exposure can damage tissue due to its heating effect.

Real-World Examples: Cautionary Tales and Successes

Consider the story of Alan, a 55-year-old dealing with chronic joint pain. He turned to low-intensity PEMF (Pulsed Electromagnetic Field) therapy, hoping it might ease his discomfort without medication. For several weeks, he experienced significant relief, but as he increased the frequency of his sessions without professional guidance, he began to feel soreness and fatigue. It was a stark reminder that even seemingly benign therapies require moderation and guidance.

Another patient, Maria, used sound frequency therapy to alleviate her anxiety. She played binaural beats daily, focusing on frequencies aligned with calm brainwave states. However, when she switched to a higher frequency out of curiosity, she experienced headaches and heightened anxiety. Through trial and error, she learned that, while frequencies could aid her, they required personalization and careful attention to her body's responses.

These stories underscore the need for education, guidance, and respect for the limits of frequency-based therapies.

Understanding the Biological Limits

Just as the body has biochemical thresholds, it also has bioelectrical and vibrational limits. Frequencies that work for one person may not be suitable for another, as each body has a unique resonance and tolerance. Research on safe frequency ranges for therapeutic use is still developing, and much remains unknown about how individual differences— such as age, health status, and environmental factors—affect responses to frequency-based treatments.

For instance, studies have shown that PEMF can stimulate bone repair at specific frequencies, but at higher intensities, it may interfere with cellular function. The FDA-approved devices for bone healing and post-surgical recovery use carefully regulated intensities, underscoring that such therapies, while powerful, need to be applied within controlled parameters.

Guidelines for Safe Frequency-Based Therapy

To safely explore frequency medicine, whether in a clinical setting or at home, here are some practical guidelines:

1. Start with Low Intensities: Especially for beginners, it's best to start with low-intensity frequencies and gradually increase as you observe your body's response. Just as you wouldn't start a new exercise routine at maximum effort, frequencies should be approached with caution.

2. Limit Session Durations: Begin with short sessions—5 to 10 minutes for sound therapy or PEMF, for example— and increase only if you notice positive effects without any discomfort. Frequency healing is more effective as a gentle process than an aggressive one.

3. Seek Professional Guidance: Always consult with a trained practitioner or healthcare provider before beginning any frequency-based therapy, particularly if you have underlying health conditions. Professionals can provide personalized

recommendations and monitor for any adverse effects.

4. Monitor for Side Effects: Be vigilant for any adverse reactions, which can range from mild discomfort to more significant symptoms, like headaches or fatigue. If you experience these, reduce the frequency, intensity, or duration of your sessions, or take a break until your body adjusts.

5. Stay Informed on Device Regulations: Many frequency devices, such as PEMF mats and sound therapy apps, are available to the public. However, not all devices meet safety standards. Research the efficacy and safety records of any device before use, and choose FDA-cleared or clinically validated options whenever possible.

Ethical Considerations in Frequency-Based Therapies

As frequency medicine grows in popularity, ethical considerations become paramount. Practitioners have a responsibility to educate themselves on the nuances and risks of these therapies. Unlike prescription drugs, frequency devices are often marketed directly to consumers, which raises ethical questions about the transparency of risks and the role of professionals in guiding their use.

Moreover, practitioners must exercise caution in treating vulnerable populations, such as the elderly or those with compromised immune systems. For instance, while studies suggest that PEMF may aid in tissue repair, it might not be suitable for everyone, and treatments should always be tailored to the individual's health profile.

A Future of Informed Frequency Medicine

The future of frequency medicine holds promise as our understanding of safety thresholds, individual responsiveness, and optimal usage grows. Innovations in technology, like AI-powered biofeedback devices, could allow real-time monitoring of the body's response to frequency treatments, reducing risks and enhancing benefits. Such advancements could enable practitioners and patients to achieve a precise balance between therapeutic intensity and safety, creating a future where frequency healing is not just effective but inherently safe.

Reflecting on the Path Forward

In closing, the field of frequency medicine calls for a balanced approach, combining curiosity with caution. Like a tuning fork that aligns with the natural resonance of an instrument, effective frequency healing resonates with the body's inherent rhythms without disrupting them. When we explore frequencies with respect and understanding, we tap into a profound potential that supports the body's innate ability to heal.

As this field continues to expand, let us remember that with every frequency applied comes a responsibility— to ourselves, to our patients, and to the ethical pursuit of healing. Frequency medicine challenges us to redefine healing in new dimensions, adding layers of sound, energy, and vibration to our understanding of health. But at its core, it also reminds us that true healing is a harmonious act, requiring not only the right tools but the wisdom to use them wisely.

Let this chapter be a foundation for thoughtful exploration in frequency healing—a practice guided by knowledge, moderation, and the timeless principle of "do no harm." As we journey deeper into the world of vibrational medicine, may we always remember that healing is both a science and an art, one that demands our highest care, respect, and integrity.

CHAPTER 78: ETHICAL IMPLICATIONS OF FREQUENCY-BASED INTERVENTIONS

As we step further into the world of frequency medicine, the journey is not simply about unlocking new methods for healing but about the ethical responsibilities that come with such power. Frequency-based therapies, from sound healing to electromagnetic treatments, bring up questions that are essential to the field's growth. How can these therapies be made accessible? How do we balance cost with fair access? What level of training should practitioners receive before they apply these techniques? In this chapter, we'll navigate these complex ethical considerations, inviting readers to ponder the broader implications of frequency medicine in society.

Healing for All: The Challenge of Accessibility

In an ideal world, health technologies would be accessible to everyone. But frequency-based interventions—especially those involving advanced technology, like PEMF (Pulsed Electromagnetic Field) or high-frequency ultrasound—come with high costs. This issue brings us to an ethical crossroads:

how can we ensure that those who could benefit from frequency healing actually have access to it?

Consider Sarah's story. A teacher from a small town, she suffered from chronic pain for years. Standard medical treatments proved ineffective, and her limited income meant she had no access to alternative, frequency-based therapies that might alleviate her suffering. This is not just Sarah's story but the story of countless individuals whose access to advanced healing is limited by geographic and financial barriers.

One of the major ethical questions in frequency medicine is how to address this divide. Organizations, healthcare providers, and even policy-makers will play pivotal roles in ensuring that the technology's benefits do not remain the privilege of a select few. Ideas such as subsidized treatments, mobile clinics, and partnerships between tech companies and community health organizations could pave the way for more equitable access.

The Cost of Frequency Medicine: Balancing Innovation with Affordability

The cost of frequency-based therapies often reflects the expenses of developing, manufacturing, and maintaining advanced devices. For example, PEMF devices, which can aid in pain relief and tissue regeneration, range from hundreds to thousands of dollars. These prices make it challenging for everyday individuals to experiment with frequency healing without significant financial commitment.

Yet, there are hopeful signs. As technology evolves, the cost of developing and producing frequency devices is likely to decrease. Take cell phones as an analogy: once prohibitively expensive, they are now accessible to a majority of people. The same trajectory might hold true for frequency medicine as economies of scale make devices more affordable. In addition, there is room for local healthcare systems to advocate for insurance coverage and funding, particularly for patients who have exhausted conventional treatments without success.

Practitioner Training: Ensuring Safety and Ethical Use

In frequency-based healing, proper training and understanding are critical to safety and effectiveness. However, as this field is still evolving, there is no universally accepted standard for training practitioners in the use of frequency therapies. This raises concerns about ensuring that those administering these treatments are both competent and ethical.

Consider David, a well-meaning alternative healer who recently integrated frequency therapy into his practice. Without formal training, he started using a frequency device he bought online. However, his first few sessions left clients uncomfortable, as he didn't fully understand how to calibrate the device. Stories like David's highlight the need for standardized training, both for safety and to establish trust with patients.

Establishing minimum qualifications for practitioners, similar to those in acupuncture or massage therapy, would benefit both patients and providers. Certifications or even formal education programs could guide practitioners in best practices, ethical considerations, and potential risks associated with frequency-based interventions. As more people adopt frequency medicine, these guidelines would ensure that treatment is administered safely and effectively.

The Importance of Informed Consent

As with any medical intervention, informed consent is paramount in frequency-based therapies. Patients should be made aware of how frequency treatments work, potential risks, and any uncertainties surrounding the practice. While frequency healing is often gentle and non-invasive, it's essential that patients understand both its potentials and its limits. This is particularly important because many individuals turning to alternative medicine are seeking solutions for chronic or unresolved health issues and may be more vulnerable to unproven claims.

For instance, a patient considering PEMF for chronic pain should know that while studies show promising results, frequency medicine isn't a miracle cure and may not work for everyone. Ethical practitioners will take the time to explain these nuances, empowering patients to make informed choices about their health.

The Future of Frequency Medicine: Looking Toward Ethical Innovation

The future of frequency medicine could see a world where everyone has access to safe, effective, and affordable frequency-based treatments. Imagine a healthcare system where a patient experiencing anxiety can access sound frequency therapy, while someone with chronic pain can turn to electromagnetic therapies as easily as they would any conventional treatment.

Achieving this vision requires innovation that keeps ethical concerns at the forefront. Emerging technologies like AI and personalized medicine could help tailor frequency-based interventions, offering customized treatments based on individual physiology. At the same time, advancements should not overshadow the ethical imperatives of accessibility and safety.

It's essential that frequency medicine doesn't become another divide in healthcare but a bridge to more holistic, inclusive health solutions. This will involve ongoing dialogue between practitioners, regulatory bodies, and communities to adapt standards and practices as the field progresses.

Reflecting on Responsibility in Healing

Ultimately, frequency-based medicine asks us to reconsider what it means to heal responsibly. At the heart of this practice lies the idea that healing is not simply the elimination of disease but the restoration of harmony within the body. This concept encourages both practitioners and patients to look beyond quick fixes and invest in treatments that consider the body's holistic needs.

As frequency medicine grows, practitioners will need to hold themselves to high ethical standards, guided by the principles of respect, empathy, and humility. Frequency-based interventions offer a profound gift, but one that demands responsibility. Those who provide these therapies should continually strive to learn, to educate, and to keep patient well-being as their highest priority.

Concluding Thoughts

In exploring the ethics of frequency-based medicine, we come to realize that healing is a shared journey. It involves not only the technology and the practitioner but also the broader society and its commitment to making these advancements accessible, affordable, and safe.

Frequency medicine holds transformative potential, and with that comes the responsibility to wield it wisely. As we advance in this field, may we remember that true healing honours not only the physical body but the dignity, rights, and trust of each individual it seeks to serve. In this way, frequency medicine can indeed become a universal harmony, resonating across barriers and boundaries, uniting science with the deeper calling to care.

PART XVIII: FREQUENCY HEALING AND MODERN TECHNOLOGY

CHAPTER 79: WEARABLE FREQUENCY DEVICES FOR HEALTH MONITORING

Imagine slipping on a wristband in the morning that not only tracks your physical activity but also emits subtle frequencies tailored to enhance your body's natural rhythms. As technology advances, this vision is not far from reality. Wearable devices that monitor and emit frequencies are opening a new chapter in personal health management, blending science, accessibility, and futuristic thinking. In this chapter, we'll explore the science, stories, and possibilities behind these wearable frequency devices, revealing how they can bridge the gap between traditional medicine and the promise of non-invasive, daily health optimization.

A Symphony in Motion: How Wearable Frequency Devices Work

To understand how these devices operate, consider your

body as an orchestra, each cell vibrating at its unique frequency. When you are healthy, these cells harmonize like a well-tuned symphony. But, under stress or illness, certain parts of the orchestra fall out of sync. Wearable frequency devices act as a conductor, guiding your body back to its natural rhythm. By emitting specific frequencies, they help synchronize cell vibrations, promoting balance and resilience.

These devices, often in the form of wristbands, patches, or even wearable necklaces, rely on sensors and small transducers to both monitor and emit frequencies. Many of them use electromagnetic frequencies (EMFs) to stimulate cellular repair, support pain relief, and even alleviate anxiety. By monitoring biomarkers like heart rate variability, sleep quality, and physical activity, these wearables create a feedback loop, tailoring the emitted frequencies to the body's current needs.

The Evolution of Wearable Healing: A Look at Real-World Applications

Wearable frequency devices are transforming health for individuals across a variety of needs, from chronic pain sufferers to high-performance athletes. Take the story of James, a marathon runner who had struggled with recurring injuries. Traditional treatments only offered temporary relief. Frustrated, he turned to a wearable frequency device designed to promote tissue repair and pain relief. By wearing the device consistently, he reported reduced inflammation and accelerated recovery times, allowing him to get back on the track faster than he ever had.

Or consider Marie, a middle-aged woman managing chronic stress and anxiety. She started using a wearable device that emits calming frequencies designed to interact with her nervous system. Using the device daily, she found herself better able to manage stress, noticing an improvement in her sleep patterns and emotional resilience. For her, this small device became a quiet but constant reminder of wellness, a gentle push toward calm amidst the noise of daily life.

These stories illustrate that while wearable frequency devices may seem futuristic, they are already reshaping how individuals approach health. And they represent a new avenue for people to take charge of their wellness, transforming healthcare from something reactive to something proactive.

Practical Tips for Using Wearable Frequency Devices

If you're considering adding a wearable frequency device to your wellness routine, here are some practical tips to maximize its effectiveness:

- Start Small: Choose a device with basic functionalities, such as tracking heart rate variability or sleep patterns. These can provide valuable insights without overwhelming you with data.
- Consistency is Key: For optimal results, wear the device consistently. Many users report that benefits, such as improved sleep or reduced pain, become more pronounced with daily use.
- Stay Mindful of Feedback: Pay attention to how your body responds to the frequencies. If you notice an increase in energy, calm, or focus, take note of what settings or times work best for you.
- Use as a Supplement, Not a Substitute: While wearable devices can support health, they should complement, not replace, traditional medical treatments and healthy lifestyle habits.

Vision of the Future: The Convergence of AI and Frequency Medicine

The potential for wearable frequency devices expands dramatically with the integration of artificial intelligence. Imagine a wearable that not only monitors your biometrics but also learns from them, using AI to adapt the emitted frequencies in real time to support your specific needs. This vision represents a step toward precision wellness, where the device becomes a highly personalized companion in the journey to optimal health.

For instance, AI-driven wearables could monitor subtle changes in heart rate or temperature and recognize early signs of an illness before you even notice symptoms. By emitting immune-supporting frequencies, the device could work preventatively, potentially reducing the severity or duration of an illness. This convergence of AI and frequency medicine hints at a future where wearable devices are as attuned to our needs as any healthcare professional, only with continuous access and intervention.

In a broader sense, such devices could also contribute to public health efforts. Imagine a world where health data from millions of wearable devices could reveal trends and support proactive measures in healthcare on a societal level. A spike in respiratory issues in a specific region, for instance, could alert health officials to take preventive measures, while individuals in the area receive supportive frequencies.

Reflecting on the Ethical Dimension

As with any groundbreaking technology, wearable frequency devices raise important ethical questions. Who will have access to these devices? Will they be affordable and available to all, or remain a luxury only a few can enjoy? The responsibility lies with developers, regulators, and healthcare providers to ensure that wearable frequency devices do not widen healthcare disparities but instead serve as a universal tool for well-being.

Moreover, privacy is a concern. These devices gather and transmit personal health data, raising the question of how this data is used and stored. Striking a balance between personalized healthcare and data security will be crucial in making wearable frequency devices a trusted part of our lives.

Concluding Thoughts: A Revolution on the Horizon

Wearable frequency devices offer us a glimpse into a future where healing becomes as much a part of our daily lives as eating or breathing. As they evolve, these devices have the potential to make frequency-based healing an integral, accessible part of wellness, inviting each person to become an active participant in their health journey.

More than just technology, these devices reflect a deeper vision of health. Frequency-based wearables challenge us to see health not as a fixed state but as a dynamic harmony that can be nurtured. They remind us that wellness is not merely the absence of illness but the continuous, conscious cultivation of balance, resilience, and vitality.

In this age of healing harmonies, frequency-based wearables are not simply devices—they are the modern echoes of ancient wisdom, carrying forward the belief that harmony, once restored, can elevate every part of our lives.

CHAPTER 80: FREQUENCY APPS AND DIGITAL HEALTH PLATFORMS

In today's world, technology has transformed almost every aspect of our lives. It was only a matter of time before digital platforms and mobile applications extended into the realm of health and wellness, especially frequency-based healing. Imagine a scenario where relief from pain or support for relaxation is just a tap away, with apps capable of emitting specific frequencies designed to balance and restore. This chapter dives into the growing landscape of frequency-based digital health platforms, explaining how they work, sharing real-world stories of users, and offering practical advice on incorporating these tools into everyday life.

The Science Behind Frequency-Based Health Apps

At their core, frequency-based health apps leverage the principle of resonance, a fundamental concept in physics and biology. Resonance refers to the way certain frequencies can influence the vibrations of other objects or organisms, much like how a tuning fork can make a nearby instrument vibrate. In the human body, organs, cells, and even emotions have their own natural frequencies. By exposing the body to targeted frequencies, these apps aim to create an environment where cells and tissues can align and "re-tune," promoting healing and balance.

Many of these apps offer sound therapy through carefully crafted sound waves, ranging from binaural beats to isochronic tones. For instance, binaural beats involve playing two slightly different frequencies in each ear, which the brain interprets as a single, steady beat. This technique has been shown to alter brainwave patterns, promoting states of deep relaxation, focus, and even improved mood. Isochronic tones, on the other hand, use single beats that pulse on and off, creating a rhythm that may help listeners reach meditative states quickly.

Stories from Users: How Frequency Apps Are Changing Lives

As we explore the potential of frequency apps, real stories of users bring the science to life. Consider Sarah, a corporate executive who often struggled with stress and sleep issues. After reading about frequency-based therapies, she downloaded an app offering nightly sound frequencies designed for relaxation and sleep. At first, she was sceptical , but within days of consistent use, Sarah found herself falling asleep faster and experiencing fewer midnight awakenings. "It's like my body was craving this balance," she explained, "and these frequencies somehow tapped into something I didn't even know I needed."

For others, frequency apps have offered an alternative to pain management. Jacob, a former athlete dealing with chronic knee pain, turned to an app featuring frequencies intended for pain relief. The app's session for joint pain relief played frequencies that he listened to for 15 minutes each day. After a week, Jacob noticed he was reaching for painkillers less frequently. While the app didn't "cure" his pain, it became a supportive tool in his pain management routine.

These stories, while anecdotal, illustrate how frequency apps are becoming more than just wellness gadgets—they are companions on the path to improved quality of life.

Practical Tips: Choosing and Using Frequency Apps

For anyone interested in trying frequency-based health apps, here are some guidelines to get the most out of these digital tools:

- Choose Your Purpose: Not all frequency apps serve the same function. Some are designed for stress relief, while others target sleep, focus, or even physical pain. Choose an app that aligns with your primary goal and read reviews to learn about other users' experiences.

- Experiment with Session Length: Frequency exposure can vary in effectiveness depending on individual needs. Start with shorter sessions (5-10 minutes) to see how your body responds, then gradually increase the duration based on comfort and effectiveness.

- Create a Routine: The effects of frequency-based therapies can be cumulative. Consider setting aside the same time each day, like morning meditation or pre-sleep relaxation, to allow the frequencies to work in harmony with your body's natural rhythms.

- Headphones Matter: Some frequencies, especially binaural beats, require headphones to work correctly. Invest in a comfortable pair that enhances your experience without distracting you.

- Mind the Volume: Frequencies don't need to be loud to be effective. Often, a gentle, barely audible level is enough to deliver benefits. Higher volumes may distract from the therapeutic effects and could cause discomfort over time.

A Look into Popular Frequency Apps

As this field grows, several apps have gained popularity for their user-friendly design and frequency options. Here are some examples:

1. Brain.fm - Known for its unique blend of AI and neuroscience, Brain.fm offers sounds tailored to focus, relaxation, and sleep. Each track is crafted to guide the listener into a state conducive to their chosen goal.

2. Insight Timer - This app is a vast platform of guided meditations and sound therapy. It offers specific frequency sessions for relaxation, healing, and mental clarity, allowing users to experiment with a range of options.

3. Solfeggio Frequencies App - Based on ancient Solfeggio frequencies, this app provides a variety of soundscapes designed to reduce anxiety, promote healing, and balance the chakras. It's an excellent choice for those interested in more spiritual aspects of frequency therapy.

4. Binaural Beats Therapy - Specifically dedicated to binaural beats, this app allows users to select different beats for sleep, meditation, or relaxation, enabling them to personalize their experience based on preference and mood.

5. ThinkRight.me - While focused on mindfulness, this app incorporates frequency-based audio sessions that support relaxation and mental clarity, providing an integrated approach to wellness.

Reflecting on the Future of Frequency Health Apps

As technology and wellness continue to converge, the potential for frequency apps is vast. Imagine a future where these apps integrate biometrics, learning from each user's body responses to personalize frequency sessions further. Or, a scenario where apps sync with other health data—like heart rate, stress levels, and even diet—to deliver frequencies tailored to an individual's daily health needs.

These advancements could turn frequency apps into sophisticated digital health platforms, not just for personal wellness but also for supporting healthcare systems. Hospitals and clinics might prescribe specific frequency therapies for post-surgery recovery, mental health support, and pain management, making frequency medicine as accessible as a mobile device.

Final Reflections

Frequency-based health apps symbolize a new era in healthcare, where healing harmonies are woven into our digital lives. While some may still question the effectiveness of these tools, the potential they hold is undeniable. They invite us to rethink the boundaries between technology and well-being, offering a more personalized, accessible, and harmonious approach to health.

In a world often overwhelmed by noise—digital, emotional, and environmental—these apps serve as a reminder that healing doesn't always require complex solutions. Sometimes, it is the subtle, rhythmic frequencies, quietly pulsing through a device, that hold the power to restore balance, calm, and vitality.

CHAPTER 81: SMART HOME DEVICES FOR FREQUENCY THERAPY

Imagine a home where every room pulses with the gentle hum of frequencies designed to uplift mood, relieve pain, or even promote cellular repair. Today, this vision isn't science fiction. Smart home devices are already incorporating frequency-based therapies, offering convenient ways to integrate healing into daily life. These devices are at the forefront of a movement that combines ancient practices with modern technology, making the benefits of frequency medicine more accessible and adaptable than ever.

Understanding Frequency Therapy in the Home

To appreciate how smart devices are being used for frequency-based wellness, it's essential to understand the science behind it. Our bodies are deeply responsive to frequency and vibration. Cells, tissues, and organs each resonate at specific frequencies that correspond to their optimal states of function. When these frequencies are disrupted—by stress, injury, or illness—frequency-based therapies can help restore balance by "reminding" cells of their natural vibrations.

In the past, accessing such therapy required specialized equipment in clinics or spas. But advances in smart home technology now bring frequency healing into our living rooms. Devices that emit frequencies through sound, light, and electromagnetic fields are designed to provide benefits as subtle as mood enhancement or as targeted as pain relief.

The New Wave of Smart Frequency Devices

The market for smart frequency devices has expanded rapidly, with an array of gadgets tailored to various wellness goals. Here are some of the most promising devices transforming frequency therapy at home:

1. Frequency-Emitting Speakers: Equipped with the technology to play specific sound frequencies, these speakers go beyond typical music playback. They emit healing sounds like binaural beats or Solfeggio frequencies, often used for relaxation, meditation, and focus. When played consistently, these sounds can create a therapeutic environment in any room.

2. Smart LED Lights with Frequency Options: Light, like sound, has its own frequency. Certain LED smart bulbs now offer colour-changing options synchronized with specific frequencies. For example, red light frequencies are believed to stimulate cellular repair and collagen production, while blue light is known to improve mood and alertness. With these smart bulbs, users can set specific light frequencies to enhance their environment for relaxation, focus, or even physical healing.

3. Infrared and PEMF Mats: These mats emit pulsed electromagnetic fields (PEMF) and infrared light, which have been studied for their effects on pain management and inflammation reduction. Users can lie on these mats for targeted therapy, allowing the body to absorb healing

frequencies directly. This technology has been used for decades in clinical settings and is now available in consumer-friendly models for home use.

4. Smart Diffusers with Sound Frequency Options: Some essential oil diffusers now come with built-in frequency emitters that play low-frequency sounds, often too subtle to hear consciously but impactful on a cellular level. This combination of aromatherapy with frequency therapy provides a multi-sensory healing experience that can be both calming and therapeutic.

5. Wearable Frequency Devices: While technically wearable rather than home-based, these devices sync with smart home platforms, allowing users to monitor and adjust frequency settings from their phones. Devices like the Apollo Neuro, which emits low-frequency vibrations to reduce stress, can be integrated into daily routines and used alongside other smart home frequency technologies.

Real-Life Stories of Transformation

The best way to understand the impact of these devices is to look at real stories from users. For instance, Rachel, a mother of three, struggled with anxiety and insomnia for years. She decided to try a sound frequency-emitting speaker, setting it to play low theta waves—frequencies associated with relaxation and meditation—before bed. Within a week, Rachel noticed she was falling asleep faster and staying asleep longer. "It felt like my whole environment was finally supporting my need for peace and rest," she shared.

For others, the journey into frequency healing began with physical pain. Mike, an avid runner, experienced chronic knee pain that often interrupted his training schedule. After using a PEMF mat at a wellness centre, he decided to purchase a smaller version for home use. By dedicating 15 minutes each morning and evening to sit on the mat, Mike found he was able to manage his pain with less reliance on medication.

These experiences highlight how frequency therapy, once a luxury, is now becoming a daily, affordable practice for many.

Tips for Integrating Frequency Devices into Your Home

Creating a healing space with smart frequency devices doesn't have to be complicated. Here are some tips to help you make the most of these tools:

- Start Small and Build: If you're new to frequency therapy, start with a single device, like a frequency-emitting speaker or smart LED bulb. Observe how it affects your mood, energy, or sleep, and gradually introduce additional devices as you become more comfortable.

- Establish a Routine: Consistency is key. Set specific times to use these devices, whether it's listening to binaural beats during meditation, using infrared light therapy before bed, or lying on a PEMF mat in the morning. Consistent exposure helps the body adapt and respond more effectively.

- Customize for Each Room: Each space in your home can serve a different purpose for wellness. Consider setting up frequency therapy speakers in the living room for relaxation, using LED lights in your office to boost focus, and placing a PEMF mat in the bedroom for pain relief and recovery.

- Pair with Intentions: Healing frequencies work best when paired with mindfulness or intention-setting practices. As you turn on these devices, take a moment to breathe deeply, clear your mind, and set a specific intention for the session—whether it's relaxation, focus, or healing.

The Future of Smart Frequency Therapy

The potential of frequency-based smart home devices is vast and still largely untapped. We are only beginning to understand how customizable and integrative frequency therapy can become. Imagine a future where smart homes detect stress levels and automatically adjust room frequencies to promote calm, or where wearables sync with home devices to create an all-encompassing wellness network tailored to the needs of each resident.

We may one day see a world where frequency therapy is woven seamlessly into our daily lives, shifting our reliance from invasive interventions to preventive, restorative frequencies. As our understanding of resonance and cellular vibration deepens, we may find ourselves moving closer to a world where frequency-based health practices are not just an option but a fundamental part of wellness.

Reflective Insights

As we adopt these technologies, it's essential to remember that they represent a union between ancient wisdom and modern innovation. The use of frequencies for healing is as old as human civilization itself, but it has gained new depth with the advent of technology. By bringing these tools into our homes, we create spaces that are not only functional but also therapeutic.

In a world that often pulls us in a thousand directions, frequency-based smart devices remind us to slow down and listen—not only to the sounds they emit but also to the silent harmony they inspire within us. Healing, after all, isn't something external; it's a return to the body's natural resonance, a rediscovery of the symphony that is already playing within.

CHAPTER 82: ARTIFICIAL INTELLIGENCE IN FREQUENCY MEDICINE

Imagine a world where healing frequencies are so precisely tailored to our bodies that they adapt in real time. A world where artificial intelligence (AI) understands the unique language of our cells, using frequencies to enhance health and address imbalances before they even manifest as symptoms. This vision is no longer a distant dream but a tantalizing reality taking shape at the intersection of frequency medicine and AI.

As we journey through the potential of AI in frequency medicine, we uncover both the scientific marvels and philosophical implications of technology merging with ancient wisdom. In this chapter, we'll explore how AI can personalize and optimize frequency-based treatments, using biofeedback to create a truly individualized approach to health.

Bridging Ancient Wisdom with Modern Technology

Frequency medicine, rooted in the understanding that every

cell in our body resonates at specific frequencies, has been practiced in various forms for centuries. Ancient cultures recognized the healing power of sound and vibration, using techniques like chanting, Tibetan singing bowls, and rhythmic drumming to promote physical and emotional balance. Today, we have the scientific tools to deepen this understanding. We know, for instance, that certain frequencies can encourage cell repair, promote relaxation, or alleviate pain.

AI brings a level of precision and adaptability that was previously unimaginable. Through real-time data from our bodies—such as heart rate variability, brain wave activity, and even cellular response—AI can determine what frequencies our bodies need at any given moment, making treatments more effective and uniquely tailored to the individual.

The Role of AI in Frequency Medicine

1. Personalized Frequency Protocols: Each person's body resonates differently, and even within an individual, frequencies can shift due to factors like stress, illness, and mood. AI algorithms can analyse patterns in biofeedback data to identify the most beneficial frequencies for each individual. For example, an AI-based system might detect elevated stress levels through a wearable device and adjust frequencies accordingly, selecting calming theta waves for relaxation or alpha waves for focus and clarity.

2. Real-Time Biofeedback Integration: Biofeedback devices like heart rate monitors, EEG headsets, and skin sensors provide continuous data about our physiological state. AI systems can process this data instantly, creating a feedback loop where frequencies adjust in real-time. Imagine lying on a PEMF (pulsed electromagnetic field) mat designed to alleviate muscle tension. As your muscle fibres respond, the

AI adapts the frequency based on muscle relaxation levels, maximizing the therapeutic effect.

3. Predictive Healing: By analysing historical data, AI can also predict shifts in our physiological state and apply preventative frequency treatments. For instance, if your biofeedback data shows patterns that often precede migraines, the AI system could apply soothing delta waves to pre-emptively reduce the onset of pain. This approach shifts frequency medicine from treatment to proactive wellness.

Case Study: Emma's Journey with Personalized Frequency Healing

Emma, a 45-year-old architect, had battled chronic migraines for years. Traditional treatments offered only temporary relief, and she was desperate for a long-term solution. When she learned about a clinic integrating AI-driven frequency medicine, she decided to try it. The AI system analysed her biofeedback, identifying specific triggers in her brainwave patterns that typically led to migraines.

Through predictive analysis, the AI adjusted frequencies based on her body's signals, activating a soothing frequency tailored to ease tension and blood flow in her brain. Over time, Emma's migraines reduced in frequency and intensity. "It was like my body had finally found a rhythm that didn't fight against itself," she shared.

Emma's story exemplifies how AI-driven frequency medicine can offer solutions tailored to the individual—solutions that evolve in response to her body's changing needs.

Practical Applications of AI in Frequency Medicine

To bring the benefits of AI-integrated frequency medicine into everyday life, here are some practical applications:

- Wearable Devices: Wearables with integrated biofeedback sensors (such as heart rate variability and skin temperature) can now interface with frequency-emitting devices. A wristband might detect stress levels and communicate with an app that plays binaural beats or soothing sound frequencies directly through headphones, adjusting as the wearer's stress level changes.

- Home-Based Frequency Systems: Advanced home-based frequency devices can integrate with AI-driven apps, creating a complete wellness environment. For example, smart PEMF mats or sound frequency chairs can work in sync with personal data collected by wearables to deliver customized healing sessions right at home.

- AI-Powered Health Monitoring Platforms: As the Internet of Things (IoT) expands, many health monitoring platforms can sync with multiple frequency devices in a household. These platforms could serve as central hubs, analysing comprehensive health data from various sensors to deliver holistic, personalized frequency treatment plans.

Ethical Considerations and Future Reflections

The fusion of AI and frequency medicine holds vast potential, but it also raises essential ethical questions. When we allow AI to play such an intimate role in our health, it's crucial to consider who controls and has access to our data. Transparency in how data is collected, used, and stored is paramount, and future regulation may need to address these issues to protect users' privacy and autonomy.

There's also the question of accessibility. Will AI-integrated frequency therapies be available to all, or will they become an exclusive privilege for those who can afford high-end technology? For frequency medicine to reach its full potential, it must be accessible and affordable to everyone.

Looking Ahead: The Vision for AI in Frequency Medicine

Imagine a world where our homes are sanctuaries of personalized healing. A world where frequency devices, tuned by AI, detect subtle shifts in our bodies, providing healing frequencies precisely when needed. Picture a future where wearable devices communicate with our environment, creating an ecosystem of wellness that adapts to our every need.

The vision for AI in frequency medicine invites us to reconsider what health truly means. Instead of responding to illness reactively, we can cultivate balance and resilience. As our understanding of frequencies deepens and AI continues to evolve, we may find that true health lies in harmony—not only within ourselves but also with the technology designed to support us.

Practical Tips for Engaging with AI-Driven Frequency Healing

To start exploring the world of AI in frequency medicine, consider these tips:

- Research Reliable Devices: Look for trusted manufacturers of biofeedback and frequency devices. Quality and accuracy are vital for effective treatment.

- Start with Simple Tools: If you're new to frequency healing, start with devices that offer preset programs. Many AI-based systems have "beginner" options, making it easy to try frequency therapy without feeling overwhelmed.

- Monitor Your Own Responses: Pay attention to how your body responds to different frequencies. While AI can provide tailored suggestions, your awareness will help you make the most of your sessions.

- Combine AI with Mindfulness: Use AI-driven devices with

mindful intention. As you engage with frequency therapy, take a few moments to breathe deeply and visualize the healing frequencies permeating every cell in your body.

Final Reflections

AI in frequency medicine offers us a glimpse into a future where healing aligns effortlessly with our natural rhythms. This journey, however, is not solely technological. It is a call to reconnect with the essence of life itself—the vibrations that animate us and the harmonies that bring us into alignment. As we step into this brave new world, we are reminded that the true power of frequency healing lies not just in devices or data but in our innate capacity for balance and wholeness. In blending AI's precision with ancient wisdom, we create a healing symphony that has the potential to transform health for generations to come.

PART XIX: THE FUTURE OF FREQUENCY MEDICINE

CHAPTER 83: QUANTUM HEALING AND FREQUENCY TECHNOLOGY

Imagine a world where medicine is not merely a process of healing after the body breaks down, but rather a continuous dance of resonance, energy, and harmony that sustains health. This is the emerging vision of quantum healing—a field where frequency technology meets the principles of quantum mechanics to heal on an atomic level, addressing the energetic foundation of our very existence. In this chapter, we'll explore how quantum healing, powered by frequency technology, is opening doors to a deeper understanding of health, longevity, and balance.

A New Paradigm: Health as Resonance

To understand quantum healing, we must first recognize that we are more than just physical bodies. On a quantum level, every cell, atom, and particle in our body vibrates at unique frequencies. When we are healthy, these vibrations are harmonious, like instruments in a well-tuned orchestra. Illness, in contrast, may arise when cells vibrate out of sync, creating discord within the body's symphony.

Consider how a tuning fork, when struck, can make a nearby

object resonate at the same frequency. Similarly, quantum healing aims to use specific frequencies to bring dissonant cells back into harmony, restoring health at the most fundamental level. Unlike traditional medicine, which often focuses on symptoms, quantum healing seeks to address the energetic roots of dis-ease.

The Science of Quantum Healing

Quantum healing is rooted in principles of quantum mechanics, which reveal that particles exist not as fixed objects but as possibilities that can be influenced by observation and energy. This field offers a new way of thinking about the body and mind, seeing them as interconnected through a web of energy and frequency. While mainstream science is only beginning to accept these ideas, growing research suggests that the interaction of mind, body, and frequency could fundamentally transform our approach to healing.

Quantum healing devices typically use frequency waves to alter the state of atoms and molecules. Some devices emit specific electromagnetic frequencies to activate the body's self-healing mechanisms, potentially repairing DNA, enhancing cellular regeneration, and even altering the body's biochemistry.

Case Study: Sarah's Journey with Quantum Healing

Sarah, a 52-year-old woman battling chronic fatigue, had tried everything from medication to diet changes, with little relief. Frustrated, she turned to a quantum healing clinic. There, she was introduced to a device designed to deliver targeted frequencies aimed at restoring cellular energy and balance. After several sessions, Sarah reported not only increased energy but a profound sense of mental clarity and emotional resilience.

Sarah's case illustrates how quantum healing addresses not just the physical symptoms but the energetic misalignments that can underlie chronic conditions. By tuning into the quantum frequencies of her cells, she experienced a kind of healing that felt as though her entire being—body, mind, and spirit—had been realigned.

Quantum Frequency Devices: Healing at Home

As quantum healing technology advances, an increasing number of devices are becoming available for personal use. These devices, ranging from handheld frequency generators to wearable tech, allow individuals to experience the benefits of quantum frequency healing from the comfort of their homes. Here are some notable examples:

- Biofeedback Devices: These devices measure subtle changes in the body's energy and suggest specific frequencies to restore harmony. They can track everything from stress levels to metabolic changes, adjusting frequencies in real time.

- PEMF (Pulsed Electromagnetic Field) Therapy Mats: PEMF mats emit low-frequency electromagnetic waves, which have been shown to promote cellular repair, enhance circulation, and reduce inflammation.

- Rife Frequency Machines: Named after Royal Raymond Rife, a pioneer in frequency medicine, these machines deliver a range of frequencies believed to target pathogens, viruses, and unhealthy cells, restoring balance and vitality.

Practical Tips for Integrating Quantum Healing

Quantum healing offers an exciting avenue for those open to exploring the frontier of energy medicine. For those interested in trying quantum healing, here are a few practical tips to keep in mind:

1. Start Slowly: Like any new wellness practice, begin with shorter sessions and lower intensity to see how your body responds.

2. Stay Mindful of Your Reactions: Pay attention to how you feel after each session. Quantum healing is subtle and cumulative, meaning its effects may become more pronounced over time.

3. Consult Practitioners for Guidance: While many quantum healing devices are available for personal use, working with a knowledgeable practitioner can help you tailor treatments for your specific needs.

4. Practice Patience and Openness: Quantum healing, being rooted in energy, doesn't always yield immediate, visible results. Approach it with curiosity and patience, giving your body time to respond to the shifts.

Ethical and Philosophical Reflections

Quantum healing technology raises profound ethical and philosophical questions. If we can heal at the quantum level, are we moving toward a world where surgery and invasive treatments are no longer necessary? What does it mean to heal on an energetic level, and how do we measure success when the effects may be subtle or difficult to quantify?

There are also questions of accessibility. Quantum healing devices can be expensive, making them less accessible to lower-income individuals. If we are to embrace quantum healing as a new paradigm, it's essential to consider how these technologies can be made available to all, ensuring that health advancements do not become a luxury reserved for a privileged few.

Looking Ahead: The Future of Quantum Healing

Imagine a future where each home has a quantum healing

device, able to detect and correct energetic imbalances before they manifest as physical symptoms. Imagine hospitals equipped with quantum diagnostic tools that provide holistic assessments of patients' energetic states, allowing for interventions that prevent illness altogether. This future is not as far off as it may seem. With advancements in AI and bioengineering, we are inching closer to a world where quantum healing complements or even replaces certain forms of traditional medicine.

In this vision, health is not merely the absence of illness but a state of resonance where body, mind, and spirit align. Quantum healing encourages us to rethink health, moving beyond the physical and into the realms of energy, frequency, and consciousness.

Final Reflections: Embracing the Mystery

Quantum healing and frequency technology open doors to a level of health and awareness that feels both revolutionary and ancient. By connecting with the frequencies that underlie our existence, we can tap into a source of healing that transcends the physical. Quantum healing doesn't just treat; it tunes, aligning us with the vibrational essence of life itself.

As we embrace this journey, we are reminded that health is a mystery to be experienced rather than a problem to be solved. Quantum healing invites us to be co-creators in our wellness journey, engaging with the symphony of frequencies that sustain us. In this new paradigm, healing becomes a process of remembering our wholeness, as we resonate once more with the harmonies that have always been within us.

Resources and Guides

For those who wish to delve deeper, consider exploring

quantum healing resources that offer scientific studies, testimonials, and guided practices. Look for practitioners who specialize in quantum medicine and frequency technology, as well as online platforms that provide instructional content. Quantum healing is still emerging, but with openness, patience, and curiosity, we can each explore the potential of this profound field and bring its harmonies into our lives.

This journey into quantum healing and frequency technology may just be the beginning, but it is a beginning that promises to reshape our understanding of health, harmony, and the boundless power within each of us.

CHAPTER 84:
FREQUENCY-BASED
REGENERATION
OF TISSUES

Imagine if, instead of relying solely on traditional medicine, we could harness specific frequencies to repair and regenerate our bodies—bone by bone, cell by cell. This vision, while sounding like something out of science fiction, is coming closer to reality. Frequency-based regeneration taps into the innate power of sound and electromagnetic waves, reawakening the body's potential for healing in ways we are only beginning to understand.

In this chapter, we explore the science and potential behind using frequencies to regenerate tissues, from repairing damaged bones to restoring organs. We'll also look at practical case studies, inspiring stories, and a glimpse into the future where frequency medicine could become a mainstay in healthcare.

The Foundation: Frequency and Cellular Communication

To appreciate how frequency-based regeneration might work, let's consider how cells communicate. Think of each cell as a musician in an orchestra, vibrating at its own unique frequency. In a healthy body, these frequencies are harmonious, creating a "symphony" that represents our well-being. However, when cells are damaged, they lose this harmony, much like an out-of-tune instrument disrupts a musical ensemble.

Frequency medicine seeks to restore this harmony by sending specific signals to damaged cells, prompting them to return to their natural state. Studies have shown that certain frequencies can stimulate cellular activity, encouraging cells to divide, repair, and, in some cases, regenerate. This phenomenon, known as cellular resonance, forms the backbone of frequency-based tissue regeneration.

Pioneering Technologies in Frequency-Based Regeneration

One of the most promising applications of frequency-based medicine is in tissue and bone regeneration. Let's explore some cutting-edge techniques:

1. Pulsed Electromagnetic Field (PEMF) Therapy: PEMF therapy uses low-frequency electromagnetic waves to stimulate bone and tissue healing. Initially used to treat bone fractures, PEMF has expanded into treatments for soft tissues, like muscles and tendons, with impressive results. Clinical studies have shown PEMF can increase cellular repair activity, promoting faster healing and reducing inflammation.

2. Low-Level Laser Therapy (LLLT): Known also as "cold laser" therapy, LLLT uses specific wavelengths of light to stimulate cells. Although it may not involve sound in the traditional sense, light waves are forms of energy that interact with cells at a frequency level. LLLT has been effective in accelerating wound healing and reducing scar tissue, showing promise for broader regenerative applications.

3. Ultrasound Therapy: High-frequency sound waves can

be used to accelerate healing, especially in soft tissues. Ultrasound has long been a staple in physical therapy, but researchers are exploring its potential in promoting cell regrowth, especially for muscles and ligaments. With specific calibration, ultrasound waves could even play a role in the regeneration of more complex structures.

Case Study: Healing Bones with Frequency

Consider the case of James, a 45-year-old athlete who suffered a severe leg fracture. Traditional treatments were ineffective, and he faced the possibility of losing full use of his leg. As a last effort, James's doctors introduced him to PEMF therapy. Using frequencies designed to stimulate bone growth, James underwent a six-week course of PEMF. Remarkably, his bone began to heal at an accelerated rate, far beyond what doctors expected. Today, he credits his recovery to the power of frequency-based healing, a therapy that gave him back his mobility.

James's story highlights the potential of frequency-based treatments for bone regeneration. His case is one of many where frequency medicine has achieved results that would have seemed impossible a few decades ago.

DIY Frequency Therapy for Healing

For those intrigued by frequency-based healing, there are accessible methods for supporting the body's natural healing processes. Here are a few DIY practices you can try at home:

- Sound Healing with Tuning Forks: Specific tuning forks calibrated to healing frequencies, such as 528 Hz (known as the "miracle frequency" for DNA repair), can be used to stimulate areas of the body needing extra attention. Hold the vibrating fork close to the injured area, allowing the frequency to "bathe" the cells in healing energy.

- Binaural Beats: These are audio tracks that use two slightly different frequencies in each ear, creating a perception of a third frequency. Certain binaural beats, like those at 100 Hz, are believed to promote cellular repair. Listen to these tracks with headphones to experience the full effect.

- PEMF Mats: Portable PEMF mats for home use are becoming more accessible, allowing individuals to apply electromagnetic fields to sore or injured areas. Though the effects may vary, users often report a reduction in pain and quicker recovery times.

The Future of Regenerative Medicine

As we look toward the future, the possibilities of frequency-based regeneration are staggering. Imagine a world where, instead of invasive surgeries, we can use frequency technology to heal organs, repair spinal injuries, or even regrow limbs. Researchers are currently studying how frequencies might stimulate stem cells, the body's raw materials for regenerating damaged tissues. If successful, this could open doors to treatments for conditions that are currently untreatable.

One promising area is organ regeneration. Preliminary studies suggest that frequencies can "awaken" dormant cells in organs like the liver, prompting them to multiply and repair damaged tissue. While still in early stages, these findings hint at a future where conditions like liver disease or kidney failure might be managed with frequency-based therapies instead of relying on organ transplants.

Reflections on the Power and Potential of Frequency Medicine

Frequency-based regeneration challenges us to rethink our understanding of the human body. What if, instead of viewing it as a mechanical structure, we saw the body as an energetic ecosystem, where every cell vibrates, resonates, and communicates? Frequency medicine aligns with this vision, offering a holistic perspective on health and healing.

But as with all emerging fields, frequency-based healing faces scepticism and ethical considerations. Can everyone afford these technologies? Will they be available globally, or only to those with financial means? It is essential that as this field progresses, efforts are made to ensure accessibility, affordability, and inclusivity.

Embracing the Harmonic Future of Healing

In frequency-based regeneration, we glimpse a future that bridges science and mystery. This approach combines rigorous study with a reverence for the unseen harmonies that sustain us. Just as the cells in our bodies resonate to frequencies, so too do we resonate with the world around us, constantly influenced by the sounds, energies, and environments we inhabit.

This new era of medicine invites us to become active participants in our healing journey. By harnessing the power of frequency, we can learn to work in harmony with our bodies, aligning our energies, and embracing the potential within us all. Frequency-based regeneration is not just about restoring tissue; it's about reconnecting with the fundamental rhythms of life, where healing becomes a natural, almost musical process.

This journey is just beginning, but as we deepen our understanding, the power of frequency-based regeneration offers a hopeful glimpse into a future where healing is less invasive, more intuitive, and profoundly resonant with the very essence of life.

CHAPTER 85: APPLICATIONS IN SPACE MEDICINE – HARNESSING FREQUENCIES BEYOND EARTH

Imagine drifting through space, millions of miles from Earth, where traditional medicine's reach is limited. In the weightless expanse of space, the human body undergoes profound changes—bones weaken, muscles atrophy, immune systems falter. The physical toll on astronauts is immense, and traditional medicine can only do so much. What if frequencies could fill this gap, offering a non-invasive, versatile solution to these unique challenges?

As humanity ventures deeper into space, we need innovative ways to protect and restore health in environments that defy terrestrial limitations. Frequency-based medicine holds unique promise, providing safe, effective treatments for astronauts that can work without the need for complex surgical tools or heavy medical equipment. This chapter explores how frequency therapies are uniquely suited to the challenges of space medicine and how they could transform the health and resilience of future space travellers.

A New Frontier: Why Frequency Medicine Works in Space

In space, the human body experiences conditions unlike anything on Earth. The lack of gravity causes muscles and bones to degrade, blood circulation changes, and cellular activity diminishes. These challenges call for a medical approach that can work within these constraints, restoring cellular function, maintaining bone density, and promoting healing without requiring gravity or traditional interventions.

Frequency-based medicine is ideal for space for several reasons:
- Non-Invasive: Frequency therapies like ultrasound and electromagnetic waves can penetrate tissues without surgical incisions, reducing the risk of infection.
- Lightweight and Compact: Devices for frequency-based therapies are often small and portable, which is essential for space travel where cargo weight is at a premium.
- Self-Sustaining: Many frequency devices can be operated by astronauts without needing constant supplies, making them practical for long-duration missions.

The Science of Frequencies in Microgravity

In the absence of gravity, astronauts face rapid muscle and bone degeneration. Research has shown that specific frequencies can stimulate cellular processes that counteract these effects, offering a way to protect astronauts from the debilitating consequences of long-term space travel.

1. Bone Health: PEMF (Pulsed Electromagnetic Field) therapy, which uses low-frequency electromagnetic fields, has shown promise in promoting bone density by stimulating osteoblast activity, the cells responsible for bone formation. Studies on Earth have demonstrated its effectiveness in

osteoporosis patients, and the same principle could protect astronauts' bones in space.

2. Muscle Maintenance: Ultrasonic therapy can target muscle tissue, stimulating cellular metabolism and protein synthesis even in a zero-gravity environment. In microgravity, muscles deteriorate rapidly, but ultrasound at specific frequencies can encourage cells to maintain their structure and function.

3. Immune System Support: Frequency-based therapies, especially those using sound waves, have been found to modulate the immune system. Sound waves in the low-frequency range can stimulate white blood cells, potentially offering a way to boost immune function in space, where astronauts are at higher risk of infections due to changes in immune response.

Case Study: Frequency Devices on the International Space Station (ISS)

A few years ago, NASA introduced a small, portable PEMF device on the ISS to study its impact on bone density. Astronauts used this device for 15 minutes daily, targeting areas prone to weakening, like the hips and spine. The results were promising—after six months, astronauts who used the device showed a significantly lower rate of bone density loss compared to those who didn't.

This case represents just the beginning of what frequency-based therapies could achieve. The success of the PEMF study has led scientists to explore other frequencies and devices that could be adapted for space.

Practical Guide: Implementing Frequency Therapy for Space Travelers

While we're not all planning to head to space, the concept of frequency medicine in these extreme conditions can inspire ways to use frequencies for wellness here on Earth. The following techniques showcase how frequency-based therapies are being designed to support astronauts' health:

- PEMF for Bone Density: This therapy is increasingly available on Earth through portable devices. Those at risk of osteoporosis or with limited mobility may benefit from daily PEMF sessions, helping maintain bone health much like astronauts in space.
- Sound Therapy for Immune Health: Sound waves in the range of 40 Hz to 100 Hz have shown promise for immune system activation. Listening to specific binaural beats or frequency-based tracks could support immune health, particularly in environments where infections are more likely.
- Ultrasound for Muscle Health: High-frequency ultrasound devices are accessible for physical therapy on Earth and can be used to maintain muscle tone, especially for individuals who cannot exercise regularly due to injury or age.

The Visionary Future of Frequency Medicine in Space Exploration

Looking to the future, frequency medicine may do more than preserve astronauts' health; it could open new doors in space exploration. Imagine colonizing Mars or living on the Moon —places where human physiology would be tested to its limits. The frequency devices we are developing today could become the backbone of healthcare in these environments, reducing our reliance on Earth-bound medical facilities.

Consider the potential of using frequencies to enhance mental health during long missions. Frequencies that influence brainwave activity—such as alpha waves for relaxation or beta waves for alertness—could help maintain cognitive function and emotional well-being. This application of frequency medicine could be invaluable for astronauts who may spend months or years isolated from Earth.

In addition, advances in quantum frequency technologies, which operate on subatomic levels, may one day allow us to regenerate tissues or even control cellular growth remotely. Imagine a future where an astronaut with a broken bone or muscle injury could receive a dose of frequencies tailored to accelerate healing—all without any physical intervention.

Reflecting on the Journey Ahead

As we envision humanity's future in space, it's humbling to consider that ancient principles of frequency and vibration could become our most advanced tools for survival. What early civilizations understood intuitively about sound's impact on the body, modern science is now beginning to explore at a quantum level.

Yet, even as we contemplate these futuristic applications, frequency medicine's true potential lies in its ability to connect us to the fundamental harmonies of life. Whether on Earth or in the far reaches of space, we are all resonant beings, vibrating in harmony with the world around us. The possibilities unlocked by frequency-based medicine remind us of this profound truth.

Practical Takeaway: Using Frequencies to Support Well-Being on Earth

Inspired by space medicine, here are a few simple frequency-based practices you can integrate into daily life:

- Bone Health: Consider investing in a PEMF mat if you have concerns about bone density. Even a few minutes a day could help promote cellular health.
- Mental Clarity: Use binaural beats or other sound frequencies designed to match brainwave patterns for focus or relaxation. This is a simple yet powerful way to boost

mood and mental performance.

- Immune Support: Low-frequency sound therapies, such as Tibetan singing bowls or tuning forks, may help support immune health. These vibrations encourage your body to relax, indirectly strengthening immunity.

In conclusion, frequency-based medicine offers us an awe-inspiring glimpse of the future, a future where healing harmonizes with the very vibrations that underlie all existence. In space, where each heartbeat is precious, these therapies may one day be the difference between life and death. And here on Earth, they invite us to reconnect with an ancient wisdom—one that views health not merely as the absence of disease but as a symphony of well-being, resonant with life's natural rhythms.

CHAPTER 86: FREQUENCY THERAPY FOR ANTI-AGING – THE SCIENCE OF CELLULAR REJUVENATION

In a world obsessed with youth, the quest for anti-aging solutions has spanned centuries. But what if the secret to aging gracefully was less about external creams or surgeries and more about restoring the body's own cellular harmony? Frequency-based therapies offer a unique approach to anti-aging by targeting the body at a cellular level, rejuvenating cells, repairing tissues, and promoting holistic well-being. This chapter explores the science, techniques, and transformative potential of using frequencies to support cellular rejuvenation and slow the visible effects of aging.

The Symphony of Aging: Understanding the Body's Frequencies

Imagine the human body as a complex orchestra where each

cell, organ, and tissue vibrates at its own unique frequency, contributing to the harmony of our health. As we age, some instruments fall out of tune. Cellular processes slow down, DNA repair mechanisms become less efficient, and oxidative stress damages cells. These are the natural processes behind what we recognize as aging: wrinkles, reduced energy, and a gradual decline in physical and mental vitality.

At the heart of frequency therapy is the belief that, by "tuning" these biological processes with specific frequencies, we can encourage cells to return to optimal functioning. Unlike other anti-aging treatments that often focus on surface-level effects, frequency therapy addresses aging from within, at the cellular level.

How Frequency Therapy Affects Cellular Rejuvenation

1. DNA Repair and Cellular Regeneration: Frequencies between 528 Hz and 1,000 Hz have shown potential in supporting cellular repair mechanisms. Studies indicate that the 528 Hz frequency, often called the "miracle tone," can stimulate DNA repair, helping cells recover from oxidative stress and reduce damage. Over time, this could potentially slow down the aging process at the most fundamental level.

2. Collagen Production and Skin Elasticity: Collagen is the protein responsible for keeping our skin firm and youthful. As we age, collagen production declines, leading to wrinkles and sagging skin. Low-Level Laser Therapy (LLLT), which operates within specific frequency ranges, can stimulate collagen production. By using red and near-infrared light, LLLT encourages cells to generate more collagen, resulting in smoother and more resilient skin over time.

3. Improved Circulation and Oxygenation: Frequencies can also promote blood flow, delivering more oxygen and nutrients to cells and tissues. Pulsed Electromagnetic Field (PEMF) therapy, for instance, is known to improve

circulation, which helps skin cells regenerate and maintain a healthy, youthful glow. Enhanced circulation means that cells are better nourished and toxins are more efficiently removed, both of which are critical in the fight against aging.

4. Hormone Balance: Aging is often accompanied by a shift in hormonal balance, which affects energy levels, mood, and even skin health. Sound frequencies, especially through therapies like binaural beats, can influence the brain's endocrine functions, encouraging a balance in hormone production. By optimizing hormones like cortisol, dopamine, and melatonin, frequency therapy may help improve sleep quality, reduce stress, and enhance overall vitality—all of which contribute to a youthful appearance.

Case Study: Reversing Visible Aging with Frequency Therapy

Emma, a 60-year-old holistic wellness advocate, had spent years searching for natural ways to maintain her youth. She was introduced to a frequency therapy device that combined PEMF and LED light therapy, specifically targeting collagen production and circulation. After six months of consistent use, Emma reported significant improvements in skin elasticity, fewer visible wrinkles, and a remarkable increase in her energy levels. She described the experience as "feeling like I'm restoring my body's natural rhythm."

Emma's story reflects an emerging trend among those seeking age-defying solutions through non-invasive methods. While more research is needed to understand the full impact of frequency therapy on aging, the preliminary results are promising, inspiring a new wave of anti-aging enthusiasts to explore this approach.

Practical Tips for Integrating Frequency Therapy into an Anti-Aging Routine

If you're intrigued by the idea of using frequencies to support aging, here are some practical ways to start integrating these therapies into your life:

- Explore PEMF Devices: PEMF mats and wearable devices are now available for home use and are designed to enhance circulation, reduce inflammation, and promote cellular rejuvenation. Use these for 10-15 minutes daily, targeting areas where you may notice muscle or skin aging.

- Incorporate Sound Therapy: Tuning forks, binaural beats, and specific sound frequencies like 528 Hz can be part of a daily routine. These can be accessed through apps or meditation practices. Try listening to these frequencies for 20 minutes daily to promote relaxation and cell repair.

- LED Light Therapy for Skin: Red and near-infrared light devices are widely available and are specifically designed to promote collagen production in the skin. Use an LED light therapy mask or wand, applying it to the face for 5-10 minutes a few times a week.

- Mindfulness and Hormone Balance: Since stress accelerates aging, practicing mindfulness while using frequency therapy can be doubly beneficial. Listen to alpha and theta binaural beats while meditating, which can help balance cortisol and promote a more relaxed state of being.

Looking to the Future: The Visionary Promise of Anti-Aging Frequency Therapies

As technology advances, scientists are exploring how quantum frequencies and artificial intelligence could personalize frequency-based anti-aging treatments. Imagine a device that scans your unique cellular needs and delivers precisely the frequency combinations required to promote cell health, repair DNA, and restore youthful resilience.

Future frequency therapies may allow for:
- Personalized Anti-Aging Treatments: AI-driven frequency

devices that tailor treatments based on your specific cellular health metrics.

- Advanced Tissue Regeneration: By combining frequencies with regenerative medicine, we could potentially accelerate the regeneration of skin, muscle, and other tissues, offering more robust anti-aging benefits.

- Holistic Aging Prevention: Frequency medicine could extend beyond the physical, supporting emotional and mental well-being as part of an integrated anti-aging approach.

Reflective Insights: Embracing Aging with Grace and Resonance

While frequency therapy offers fascinating possibilities for delaying the signs of aging, it also invites us to redefine what it means to age gracefully. Aging is a natural process, a beautiful journey that reflects our experiences and growth. Frequency medicine aligns with this philosophy, offering not an escape from aging but a way to harmonize with it.

As you explore frequency therapies, remember that beauty and vitality come from within. This approach isn't about denying age but about supporting your body's natural rhythms, giving it the frequencies it needs to thrive and shine. Embrace each phase of life as part of your personal symphony, tuning in to the harmonies that best support your health and spirit.

Summary and Final Thoughts

Frequency-based therapies offer a groundbreaking approach to anti-aging, one that speaks to the body's innate ability to rejuvenate and heal. By targeting aging at the cellular level, frequencies can promote collagen production, improve circulation, support hormone balance, and enhance overall vitality.

Whether through PEMF devices, sound therapy, or LED light, frequency-based anti-aging is more accessible than ever, empowering you to take control of your well-being in a holistic, resonant way.

As you experiment with these tools, remember that aging gracefully is not about resisting time but about moving with it in harmony. Just as each instrument in an orchestra plays its part, every cell in your body resonates with a unique note. By nurturing this internal melody, you're not only supporting your physical appearance but cultivating a deeper, more resilient sense of health and harmony.

Let frequency therapy inspire you to see beauty in every phase of life—because true radiance, after all, is timeless.

CHAPTER 87: BIOTECHNOLOGY AND FREQUENCY-BASED IMPLANTS – THE NEXT FRONTIER IN CHRONIC HEALING

Imagine a world where a small implant inside your body emits healing frequencies to manage chronic pain, regulate inflammation, or even help regenerate damaged tissue. Such innovations may sound like science fiction, but the intersection of biotechnology and frequency medicine is opening new possibilities in chronic condition management. Frequency-based implants could become a non-invasive solution, continuously supporting the body's healing processes without the risks of medication dependency or repeated surgeries.

This chapter delves into the exciting potential of frequency-based biotechnology, blending scientific research with visionary exploration to offer readers a glimpse of what may be possible in the coming decades.

The Body as a Bio-Electrical System: A Foundation for Frequency Implants

To understand how frequency-based implants could work, it's essential to recognize the body as a bio-electrical system. Every cell communicates through electrical signals, creating a network of frequencies that govern everything from heartbeat rhythms to brain activity. When these frequencies are in harmony, the body functions optimally. However, chronic illnesses often disrupt these natural rhythms, leading to persistent inflammation, pain, or immune dysfunction.

Frequency-based implants would act as "tuners," restoring balance by emitting specific frequencies directly where they're needed. Think of them as miniature orchestrators within the body, guiding each cellular process back into harmony. The concept rests on the principle of resonance, where the frequency emitted by the implant aligns with the body's natural healing frequencies to promote repair and rejuvenation.

How Frequency-Based Implants Could Revolutionize Chronic Condition Management

1. Chronic Pain Relief: One of the most promising applications of frequency-based implants lies in chronic pain management. Pulsed Electromagnetic Field (PEMF) therapy has already shown success in reducing pain and inflammation, often used externally through wearable devices. Implants could take this a step further, delivering continuous PEMF therapy directly to pain-affected areas, providing relief without the side effects of long-term medication.

2. Anti-Inflammatory Support for Autoimmune Disorders:

Chronic inflammation is at the core of many autoimmune disorders, from rheumatoid arthritis to Crohn's disease. Frequency-based implants could emit anti-inflammatory frequencies, targeting specific areas like joints or the gastrointestinal tract to reduce inflammation at the cellular level. By balancing immune response and supporting cell repair, such implants could help manage autoimmune symptoms without immunosuppressive drugs.

3. Cardiovascular Health and Arrhythmia Management: Frequency-based implants could potentially support cardiovascular health, especially in managing conditions like arrhythmias. For example, implants could emit frequencies that help regulate heartbeat, promoting cardiovascular stability without relying solely on medication or surgical interventions. This application could reduce the risks associated with arrhythmia and improve overall heart health.

4. Neurological Support for Conditions like Parkinson's and Epilepsy: Neurological conditions such as Parkinson's disease and epilepsy could also benefit from frequency-based implants. By emitting brain-specific frequencies, implants could help regulate dopamine production, reduce seizure risk, or even support neuroplasticity. Imagine an implant that continuously supports brain health, improving quality of life for individuals managing chronic neurological conditions.

Case Study: Frequency Implants for Chronic Back Pain

Consider the journey of John, a 45-year-old who has struggled with debilitating back pain for over a decade. John tried physical therapy, pain medication, and even surgery, but his relief was temporary at best. He was introduced to a clinical trial for a frequency-based implant specifically designed for chronic pain.

The implant, surgically placed near the site of his pain, emitted a low-frequency pulsed electromagnetic field designed to reduce inflammation and encourage tissue repair. Within three months, John reported significant pain reduction, and his mobility improved dramatically. "It was like my body started healing from the inside," he shared, noting that he hadn't felt this pain-free in years. His story reflects the potential of frequency implants to provide a sustainable solution for chronic pain sufferers.

Practical Considerations: What Would Frequency Implants Look Like?

For those intrigued by the possibilities, it's worth considering the design and application of these implants:

- Size and Material: These implants would need to be small, biocompatible, and energy-efficient. They might resemble pacemakers in size but would need far less energy due to the low frequencies involved.
- Energy Source: Many frequency implants could utilize the body's own energy (bioelectricity) or draw from kinetic energy generated by movement. Advances in nanotechnology might allow for tiny batteries that recharge from body heat or subtle movements.
- Targeting Specific Conditions: Different conditions would require specific frequency ranges. For instance, lower frequencies might target muscle and bone repair, while higher frequencies could be tuned for neural support.
- Safety and Regulatory Challenges: For frequency implants to become widely accessible, rigorous safety testing is essential. Understanding long-term effects and ensuring implant reliability will be paramount, as will navigating regulatory approvals from health organizations worldwide.

Reflecting on the Future: Ethical and Societal Implications

As with any breakthrough technology, frequency-based implants raise ethical considerations. Who will have access to this technology? Could it widen the health inequality gap if costs are prohibitive? And how do we balance personal autonomy with medical intervention when implants continuously alter internal frequencies?

Furthermore, there are questions about privacy and data security. Since these implants would likely connect to external devices for monitoring, they would collect sensitive health data. Ensuring this data remains secure and respecting patient privacy will be critical challenges in implementing frequency-based biotechnology.

Inspiring a Vision: A Future Where Healing Is Seamless

Picture a future where chronic conditions no longer dominate lives, where a frequency implant works quietly within, allowing individuals to live without pain or fatigue. Visualize a world where aging bones and aching joints are rejuvenated by tiny emitters that restore cellular harmony. With each advancement, we move closer to a reality where healing is not a burden but an effortless, internal process, seamlessly integrated into daily life.

Practical Guide: Exploring Frequency-Based Devices Today

While implants may be a few years away, you can begin experiencing the benefits of frequency medicine through non-invasive devices already available:

- PEMF Mats and Wearables: Portable PEMF devices target pain and inflammation in specific areas. Many PEMF mats are available for home use, allowing you to try frequency therapy for chronic pain management.
- Sound Therapy: Sound frequencies are also accessible through tuning forks, sound bowls, and binaural beats apps, which can support mental and physical well-being.

- Red and Infrared Light Therapy: LED devices that emit specific frequencies of light are widely used for skin health, collagen production, and pain relief.

These tools offer a glimpse into the benefits of frequency-based healing and can support those with chronic pain or inflammation until frequency implants become available.

Final Reflections: Embracing the Frequency Frontier

Biotechnology and frequency medicine may seem like strange partners, but together they promise a radical transformation in how we manage health, especially chronic conditions. As we stand on the edge of this new frontier, we find ourselves not merely fixing symptoms but attuning the body to its own natural rhythms of healing.

Through frequency-based implants, we can envision a life where medicine no longer feels invasive but rather works in tandem with our bodies, seamlessly aligning our biology with the harmony that sustains life. Each heartbeat, each cellular repair, could be a note in the symphony of healing.

Let this be a reminder that healing isn't always about intervention; sometimes, it's about creating the conditions for the body to find its own harmony. As we explore frequency-based implants and biotechnology, we're not just pursuing a new medical solution; we're stepping into a future where we honour the body's inherent wisdom, allowing it to resonate, rejuvenate, and thrive in harmony.

CHAPTER 88: THE ROLE OF VIRTUAL REALITY IN FREQUENCY THERAPY – AN IMMERSIVE HEALING FRONTIER

Imagine stepping into a world where every sight, sound, and vibration is tailored to your body's healing needs—a world where virtual reality (VR) and frequency therapy unite to create immersive environments that guide your mind and body toward restoration. This is not just a vision for the future; it is a burgeoning field with enormous potential to change how we understand healing.

Virtual reality, once limited to gaming and entertainment, has expanded its reach into medicine. When combined with frequency therapy, VR can deliver a sensory experience that enhances physical and mental healing, particularly for those struggling with chronic pain, anxiety, and depression. In this chapter, we explore the science, applications, and possibilities of VR-based frequency therapy, weaving together real-world case studies, practical tips, and a forward-looking vision of a more holistic healthcare approach.

A New Dimension in Healing: VR and Frequency Therapy

VR has a powerful effect on the mind. By immersing individuals in alternative environments, VR can alter perception, reduce stress, and foster a sense of escape from pain or illness. But what if these virtual spaces could also be infused with healing frequencies that resonate with the body on a cellular level?

To understand this concept, consider the body as an intricate orchestra, with each cell vibrating at a specific frequency. Like a conductor guiding musicians to play in harmony, VR frequency therapy can "tune" the body's natural rhythms, helping each cell function at its best. This harmonization can support healing, reduce pain, and bring about a state of calm and relaxation that aids the healing process.

The Science Behind VR-Enhanced Frequency Therapy

Research has shown that both VR and frequency therapy can influence brain waves, impacting mental states and physical health. VR affects the brain's alpha, beta, and theta waves, which are tied to relaxation, focus, and creativity. Frequency therapy, meanwhile, uses sound waves or electromagnetic pulses to stimulate healing at a cellular level. When combined, VR and frequency therapy amplify each other's effects, engaging the mind and body simultaneously to promote deeper healing.

Imagine a VR headset transporting a person to a tranquil forest, with subtle frequency waves playing in the background. The individual may hear the rustling of leaves while feeling gentle vibrations calibrated to ease muscle tension or promote cellular regeneration. This multisensory experience connects the emotional and physical aspects of healing, allowing the person to experience relief on multiple levels.

Real-World Applications and Case Studies

1. Chronic Pain Management: For years, chronic pain has been treated with medication and physical therapy, but VR-based frequency therapy is proving to be a non-invasive alternative. Take, for example, Sarah, a 50-year-old woman with fibromyalgia who had experienced constant pain for years. Traditional treatments offered limited relief, but after trying VR frequency therapy, Sarah reported a significant reduction in pain. In her sessions, she entered a virtual beach setting with sounds of gentle waves layered over frequency pulses. The combination of visual immersion and targeted frequencies allowed her to relax deeply, relieving the tension that often exacerbates her pain.

2. Anxiety and PTSD Relief: Individuals with PTSD or chronic anxiety can benefit immensely from VR frequency therapy. VR can transport them to calming environments that might otherwise be inaccessible in their physical reality. Combined with low-frequency sound waves, this therapy helps regulate the nervous system, inducing calm and reducing cortisol levels. In clinical trials, veterans with PTSD have used VR scenarios that mimic safe, serene settings, paired with frequencies tailored to support heart rate and breathing regulation. Many report feeling "grounded" and "safe" during these sessions, allowing their minds to rest.

3. Improving Sleep Quality: Insomnia is a growing health issue, but VR-based frequency therapy may offer a novel solution. Some programs create a virtual environment resembling dusk or twilight, accompanied by theta waves, which are known to encourage relaxation and sleep.

Participants find themselves winding down as if they were watching a sunset in a quiet place, gradually transitioning to deep rest. With repeated use, many users experience improved sleep quality, proving that frequency therapy combined with VR can help restore natural sleep cycles.

How to Use VR Frequency Therapy in Daily Life

For those interested in exploring this therapy, here are some practical tips and DIY approaches to integrate VR frequency therapy into everyday life:

- Choose the Right VR Environment: Many apps and VR platforms provide settings ranging from forests and beaches to outer space. Select one that aligns with your healing goals. For example, a lush forest scene may work well for relaxation, while a beach scene might be more effective for pain management.

- Incorporate Binaural Beats or Frequency Playlists: For those without access to VR-based frequency therapy programs, try pairing a VR app with a frequency playlist that matches your needs. Frequencies around 528 Hz, often called the "love frequency," are linked with relaxation and healing, while 40 Hz has been shown to improve cognitive function.

- Use Consistent Practice: Like meditation or exercise, VR frequency therapy can offer cumulative benefits. Regular sessions—around 20 minutes a day, several times a week—help reinforce the relaxation and pain relief associated with these therapies.

Looking to the Future: A Vision of Personalized, Immersive Healing

The future of VR frequency therapy lies in personalization. Imagine a system where VR headsets are calibrated to each user's biometric data, such as heart rate, brain activity, and stress levels. As you immerse yourself in a

VR setting, sensors could adjust the frequency waves in real time to suit your body's specific needs. This dynamic, responsive approach would allow for a level of individualized therapy that could revolutionize both physical and mental healthcare.

Such technology could even integrate artificial intelligence, with VR avatars serving as virtual "guides" who offer real-time feedback and adjustment to frequencies based on a user's response. Imagine a scenario where, as your stress levels decrease, the VR environment subtly shifts to reinforce that state, prolonging the therapeutic benefits.

Reflections on VR Frequency Therapy: Bridging the Mind-Body Divide

VR frequency therapy offers a unique bridge between mind and body. By engaging all the senses, it harmonizes the emotional and physical layers of healing in a way that is unprecedented. In a sense, this therapy is not about escaping reality but about creating an environment where true healing can occur. It allows us to step into a space where the mind is calm, and the body responds accordingly.

There is also a profound philosophical element to VR frequency therapy. In many ways, it mirrors ancient practices that emphasized creating harmonious inner states for healing. The only difference is that today, we use modern technology to recreate these inner landscapes, reminding us that the wisdom of healing has always been within us.

Final Thoughts: Embracing VR Frequency Therapy as a Path Forward

As we continue to explore the union of VR and frequency medicine, we are reminded of our limitless potential for innovation in healing. VR frequency therapy opens doors not only to pain relief and relaxation but also to a future where healing is immersive, personalized, and accessible to all.

This chapter invites readers to consider VR frequency therapy not merely as a treatment but as an experience— a journey into realms of peace and wellness that may be closer than we think. By embracing this new paradigm, we can look forward to a future where healing becomes an all-encompassing experience, transforming how we perceive health and, ultimately, ourselves.

PART XX: PRACTICAL GUIDES AND TOOLS FOR READERS

CHAPTER 89: CHOOSING THE RIGHT FREQUENCY FOR YOUR NEEDS

Imagine standing in front of a symphony orchestra, where each musician plays a specific note, each instrument vibrating at a particular frequency. Now, envision that your body is the orchestra, with each cell, organ, and system resonating to its own tune, creating a harmonious symphony of health and well-being. When that harmony is disrupted—whether by stress, illness, or environmental factors—disease and discomfort emerge. The right frequencies can restore that lost balance, returning the body to its natural, harmonious state. This chapter explores how you can select the right frequencies for specific healing needs, offering a practical, science-backed guide to frequency therapy.

Understanding Frequency Therapy: The Basics

Frequency therapy works on the principle that each part of the body has an ideal vibrational frequency. When cells or tissues deviate from their natural frequency, disease and dysfunction can arise. By introducing sound, electromagnetic, or vibrational frequencies that resonate with specific bodily structures, it's possible to "tune" those areas back to their natural state, promoting healing and wellness.

Researchers have identified certain frequencies with therapeutic benefits:
- 40 Hz: Linked to improved cognitive function and memory.
- 174 Hz: Known for its analgesic properties, providing pain relief and reducing inflammation.
- 528 Hz: Often called the "Love Frequency," it is believed to promote DNA repair and cellular regeneration.

These are just a few examples, but with countless frequencies available, the question arises: how do you choose the right one for your unique needs?

Personalizing Frequency Therapy: Finding Your Sound

1. Begin with Intention
What is your primary goal for using frequency therapy? Are you seeking relaxation, pain relief, mental clarity, or perhaps a boost in energy? Defining your purpose will guide your choice, as different frequencies serve different functions. For example:
- Pain Relief: Frequencies between 174 Hz and 285 Hz are commonly used for pain relief and physical healing.
- Mental Clarity and Focus: Frequencies like 40 Hz stimulate brain activity, making them ideal for cognitive enhancement.
- Emotional and Spiritual Healing: Frequencies around 528 Hz and 639 Hz are used for emotional balance, love, and harmony.

2. Listen to Your Body's Response
Frequency therapy is a deeply personal experience; everyone's body responds differently. A frequency that energizes one person may feel overstimulating to another. Start by experimenting with lower frequencies—these are generally calming and safe for beginners. Notice how you

feel during and after each session. Are you more relaxed, energized, or perhaps even uncomfortable? Trust your body's feedback to find what works best.

Stories of Healing: Case Studies in Frequency Therapy

Throughout history, frequency-based healing has delivered remarkable results for people with a wide range of needs. Here are a few cases that highlight how selecting the right frequency can profoundly impact one's health.

Case 1: Relief from Chronic Pain with 174 Hz
Tom, a 55-year-old construction worker, had suffered from chronic back pain for years. Traditional treatments provided only short-lived relief, and he was eager for an alternative. After researching frequency therapy, Tom began using 174 Hz, a frequency known for its pain-relieving properties. Within weeks of daily 15-minute sessions, he noticed a substantial reduction in pain, allowing him to regain mobility and enjoy activities he had long avoided.

Case 2: Cognitive Enhancement with 40 Hz
Maria, a university student, struggled with focus and memory retention, especially during exam seasons. She discovered that 40 Hz, associated with cognitive improvement, could help. Using a frequency playlist that included 40 Hz, Maria would listen for 20 minutes while studying. Not only did she feel more focused, but she also retained information more effectively, ultimately improving her academic performance.

Case 3: Emotional Healing and Balance with 528 Hz
After a difficult divorce, Emily, a 45-year-old mother of two, felt emotionally drained and disconnected. A friend recommended she try 528 Hz, a frequency thought to aid in emotional and cellular healing. She found that listening to this frequency during meditation sessions helped her process her emotions and restore a sense of peace. Over

time, Emily's resilience and emotional balance returned, and she credited much of this transformation to the frequency therapy that helped her realign emotionally.

Practical Tips for Using Frequency Therapy at Home

For those eager to incorporate frequency therapy into daily life, here are practical tips to get started:

- Use a High-Quality Speaker or Headphones: The effectiveness of frequency therapy depends on the purity of sound. Invest in quality speakers or headphones to experience the full impact of each frequency.

- Start with Short Sessions: For beginners, start with short sessions (5-10 minutes) and gradually increase as your body adjusts. Pay attention to how you feel and avoid overuse, which can lead to overstimulation.

- Combine with Meditation or Breathwork: Pairing frequencies with deep breathing or meditation can enhance their effects. As you inhale deeply, visualize each cell in your body resonating with the frequency, releasing stress and restoring harmony.

- Create a Playlist Based on Your Needs: Different times of day may call for different frequencies. Create a morning playlist with uplifting frequencies like 639 Hz for energy and positivity, and a nighttime playlist with calming frequencies like 174 Hz to promote relaxation and sleep.

A Future Where Frequency Medicine is Mainstream

In imagining a future where frequency-based healing is mainstream, we envision a world where non-invasive therapies replace many of today's medical interventions. Hospitals and clinics might offer "frequency rooms" with specific settings for pain management, anxiety relief, and other needs, allowing patients to select environments that match their healing goals.

Imagine a hospital where patients recovering from surgery could visit a 174 Hz therapy room to accelerate tissue healing or where individuals with PTSD could relax in a 528 Hz chamber designed to foster emotional healing and reduce stress. This vision may seem futuristic, but it reflects the growing recognition of non-invasive healing methods that work with the body's natural rhythms.

Reflections on Frequency as a Path to Inner Harmony

There is something deeply philosophical in tuning our bodies to frequencies that restore natural balance. Frequency therapy reminds us that healing is more than physical; it connects us to deeper layers of self-awareness, prompting us to explore what it means to be in harmony—not just physically, but emotionally and spiritually as well.

As you experiment with frequencies, remember that this journey is not merely about targeting symptoms. It's about reawakening the natural rhythms that are the essence of health and vitality. The human body, much like the universe, vibrates with energy. By aligning with the right frequencies, we tap into a form of medicine that doesn't mask symptoms but harmonizes the core of who we are.

Empowering You to Explore Frequency Healing

Choosing the right frequency is an empowering act of self-discovery. As you delve into the world of frequency medicine, you are not merely a passive patient but an active participant in your own healing journey. Trust your intuition, listen to your body, and remain curious. The right frequency is out there, waiting to resonate with you and bring forth a new level of wellness.

In frequency healing, we find a reminder of the innate wisdom within our bodies—a wisdom that science is only beginning to understand. Through this exploration, may you uncover not only physical relief but also a profound connection to the harmony that exists within and all around us.

CHAPTER 90: FREQUENCY HEALING FOR BEGINNERS

Imagine yourself stepping into a world where each cell in your body hums with a unique frequency, contributing to the symphony that is your overall health. Just as an orchestra performs best when every instrument is in tune, so too does the body thrive when its internal frequencies are balanced and harmonious. This is the foundational principle of frequency healing—a natural, non-invasive way to restore wellness by aligning the body with its optimal vibrational state. For beginners, frequency healing may seem abstract, but with a gentle introduction, it becomes an accessible and powerful tool.

In this chapter, we will walk through the basics of frequency healing, offering a guide to understanding its benefits, exploring practical applications, and providing simple tips to incorporate these techniques into your everyday life.

What is Frequency Healing?

Frequency healing, sometimes referred to as vibrational or energy healing, is the practice of using specific sound waves, electromagnetic fields, or vibrations to restore balance in the body. Just as each instrument in an orchestra resonates with a particular note, each part of the body has a unique frequency. When these frequencies are disrupted—by stress, illness, or environmental factors—symptoms can appear. By reintroducing the correct frequency, we can encourage the body to return to its natural, balanced state.

An Analogy to Understand Frequency Healing

Think of a tuning fork. When struck, it produces a clear, steady sound that causes nearby objects to vibrate in harmony with it. Frequency healing works in much the same way. When we expose the body to a healing frequency, it "entrains," or aligns, with that frequency, which can stimulate healing, relaxation, or focus depending on the chosen sound.

Getting Started: Exploring Different Frequency Modalities

Frequency healing is diverse, with various techniques available for different goals. Here are some of the most accessible methods for beginners:

1. Sound Therapy
   - Overview: Sound therapy involves using specific sound frequencies, such as those produced by tuning forks, singing bowls, or even recorded audio, to affect the body's vibrational state.
   - Popular Frequencies for Beginners:
   - 174 Hz: Known for its grounding properties, this frequency is often used to alleviate physical pain and encourage relaxation.
   - 528 Hz: Called the "Love Frequency," it's believed to promote healing on a cellular level, supporting DNA repair and emotional balance.
   - 963 Hz: Linked to spiritual awakening and connection, this frequency can support mental clarity and peace.

2. Pulsed Electromagnetic Field (PEMF) Therapy

- Overview: PEMF therapy uses low-frequency electromagnetic waves to stimulate cellular repair and reduce inflammation.
- Benefits: PEMF is often used for muscle recovery, pain relief, and improving circulation. Many people find PEMF mats or devices helpful for their convenience and ease of use.

3. Binaural Beats

- Overview: Binaural beats are created by playing two slightly different frequencies in each ear, creating the perception of a third "beat" frequency. For instance, if you hear 200 Hz in one ear and 210 Hz in the other, your brain will perceive a 10 Hz beat.
- Benefits for Beginners: Binaural beats can improve focus, relaxation, and sleep. They're easy to try—many apps and online platforms offer free binaural beat tracks specifically designed for beginners.

Stories of Transformation: Real-Life Applications

Case 1: Relief from Anxiety with 528 Hz
Emma, a 30-year-old teacher, had been struggling with anxiety, especially during the pandemic. Searching for a non-medicinal approach, she came across sound therapy. She began listening to 528 Hz frequencies during her daily meditation sessions. Over time, Emma felt a noticeable reduction in her anxiety levels, describing the experience as "like a calm wave washing over me." Her story illustrates how even small, daily sessions with the right frequency can create profound mental and emotional shifts.

Case 2: Chronic Pain Relief through PEMF
James, a retired athlete, had endured years of knee pain following an injury. Traditional physical therapy and medications provided limited relief. A friend recommended PEMF therapy, and he decided to try a portable PEMF device.

After a few weeks of consistent use, James found that his pain had significantly diminished, allowing him to enjoy walks again. This experience demonstrated the tangible physical benefits of frequency healing for chronic pain.

Tips for Safe and Effective Frequency Healing

For beginners, frequency healing should be a gentle and exploratory journey. Here are some tips to get started safely:

- Start Slow: Begin with short, 5-10 minute sessions to gauge your body's response to each frequency. Over time, you can gradually increase session lengths.
- Use Quality Sound Sources: If you're using sound frequencies, choose high-quality recordings or devices. Good headphones or speakers can make a big difference in the effectiveness of the experience.
- Consistency is Key: Just like exercise, frequency healing is most effective when practiced consistently. Try setting aside time each day, even if only for a few minutes, to integrate these frequencies into your routine.
- Listen to Your Body: Frequency healing should feel comfortable. If a frequency feels unsettling or if you experience discomfort, it's okay to switch to a different one or take a break.

DIY Guide: A Beginner's Frequency Healing Session

1. Set Your Intention
   Start by setting a clear intention for your session. Are you seeking relaxation, relief from pain, or mental clarity? A clear intention helps to direct the focus of the session.

2. Choose Your Frequency
   Based on your intention, select a frequency:
   - For pain relief: Try 174 Hz.
   - For mental clarity: Opt for a 40 Hz binaural beat.
   - For emotional healing: Experiment with 528 Hz.

3. Find a Quiet Space

Select a peaceful environment where you won't be disturbed. Close your eyes, take a few deep breaths, and allow yourself to relax.

4. Play the Frequency

Begin playing your chosen frequency through headphones or speakers. Allow the sound to wash over you. Imagine each cell in your body resonating with the frequency, releasing tension and restoring harmony.

5. Reflect on the Experience

After your session, take a moment to reflect. How did you feel? Did any areas of your body respond particularly strongly? Noting your observations can help you fine-tune future sessions.

The Future of Frequency Healing: Imagine the Possibilities

As we look to the future, frequency healing may well become a standard part of health and wellness. Imagine visiting a healthcare facility where patients can select from different frequency-based therapies for a range of needs, from pain management to emotional support. As more research is conducted, the world of frequency healing will likely expand, with new frequencies identified and new technologies developed to harness the body's natural ability to heal itself.

Reflections on the Power of Frequency

Frequency healing offers a reminder that we are inherently vibrational beings, interconnected with the rhythms of the universe. In embracing frequency healing, we tap into an ancient wisdom that understands health as a dynamic, harmonious state. By learning to work with these natural rhythms, we not only support our bodies but also nourish our minds and souls.

For beginners, the journey into frequency healing is one of discovery, empowerment, and connection. As you explore this path, may you find the frequencies that resonate with your deepest needs, bringing you a step closer to balance, wellness, and a harmonious life.

CHAPTER 91: CREATING A SOUND HEALING JOURNAL

Imagine embarking on a journey of self-discovery and healing, where each session of sound therapy becomes a chapter in your personal health narrative. Tracking your responses to different frequencies through journaling can provide powerful insights, helping you identify patterns, recognize improvements, and deepen your understanding of how frequency-based healing works for your unique needs. A sound healing journal serves as both a reflection tool and a scientific log, capturing the subtle ways sound influences your mind, body, and spirit over time.

In this chapter, we'll explore how to set up and maintain a sound healing journal, offering prompts, examples, and tips to help you document your experience. Whether you're new to sound healing or have already dabbled in different frequencies, this journal can empower you to become an active participant in your healing journey, turning passive listening into a rich, reflective practice.

Why Keep a Sound Healing Journal?

A sound healing journal allows you to track your progress and explore the effects of different frequencies on your well-being. Every individual resonates uniquely with various frequencies, so what works well for one person may feel different for another. By documenting your experiences, you gain a clearer understanding of which frequencies align with your goals, be it stress relief, pain management, or spiritual insight.

Benefits of a Sound Healing Journal:
- Identify Patterns: Recognize which frequencies consistently enhance your mood, relieve stress, or help you sleep better.
- Enhance Awareness: Develop a deeper understanding of your physical and emotional responses to each session.
- Set and Achieve Goals: Use your journal to set intentions and measure progress toward healing objectives.
- Promote Consistency: Recording sessions encourages regular practice, which is essential for frequency healing to work effectively.

Setting Up Your Sound Healing Journal

Your sound healing journal doesn't need to be complex. A simple notebook will do, or you can use a digital journal if you prefer. You may also want to create sections or templates to organize your entries, especially if you're experimenting with different frequencies and techniques.

Suggested Sections for Each Entry:
1. Date and Time: Note when you begin each session, as time of day can affect your experience.
2. Frequency Used: Record the specific frequency or sound therapy method (e.g., 528 Hz for healing, 174 Hz for pain relief, binaural beats for relaxation).
3. Session Duration: Document the length of time you engaged with the frequency.
4. Environment: Describe your surroundings and any tools used (e.g., headphones, speakers, quiet room).
5. Physical and Emotional State: Jot down how you feel before the session—both physically and emotionally.

6. Intentions for the Session: State your goal (e.g., to relieve anxiety, promote relaxation, or gain clarity).

7. Observations During the Session: Describe sensations, emotions, or thoughts that arise as you listen to the frequency.

8. Post-Session Reflections: Note any changes in mood, energy, or physical sensations following the session.

Sample Entry for a Sound Healing Session

To get started, here's an example of a sound healing journal entry. Use this as inspiration to create your own unique entries:

Date: April 3, 2023
Time: 7:00 PM
Frequency Used: 528 Hz (Love Frequency)
Session Duration: 15 minutes
Environment: Dim lighting, comfortable chair, listening through noise-cancelling headphones
Physical and Emotional State (Pre-Session): Feeling a bit anxious and physically tense from a long workday
Intentions for the Session: To calm my mind and release built-up tension in my body
Observations During the Session: As the frequency started, I felt a slight tingling sensation in my chest area. My breathing naturally deepened, and I noticed my shoulders relaxing. After about 10 minutes, my thoughts became quieter, and I felt more present.
Post-Session Reflections: I feel noticeably calmer and less tense. My anxiety has decreased, and I feel more open and receptive. I'll try using this frequency again tomorrow to see if I experience similar effects.

Journaling Prompts for Deeper Reflection

To help you reflect on your sound healing journey, consider using these prompts:

- How did this session differ from previous ones?

Explore any new sensations, emotions, or insights that came up during this session and how they relate to past experiences.

- What intentions did I set, and were they achieved?

Assess if the frequency helped you reach your goal for the session. If not, think about what could be adjusted next time.

- Did any memories or emotions surface?

Sound frequencies often bring subconscious thoughts or feelings to the surface. Write down anything that emerged and reflect on what it might mean.

- How does my body feel now compared to before the session?

Note any changes in physical sensations, such as relaxation, pain relief, or increased energy.

- What would I like to explore in my next session?

Use your reflections to set intentions for future sessions, building on your experiences.

Tips for Maintaining a Consistent Practice

1. Set Aside a Regular Time

Establish a daily or weekly routine to help you stay consistent with your sound healing practice. This consistency is key to noticing the benefits over time.

2. Create a Comfortable Environment

Choose a quiet, comfortable space for each session to minimize distractions and help you focus on the experience.

3. Experiment with Different Frequencies

Don't be afraid to try various frequencies, from 174 Hz for grounding to 963 Hz for spiritual clarity. Your journal will help you identify which ones resonate best with you.

4. Review Your Journal Regularly

At the end of each week or month, go back through your entries. This practice can reveal patterns and insights that might not be obvious after a single session.

Reflecting on Your Sound Healing Journey

After several weeks or months of journaling, take time to reflect on your progress. Has your mood improved? Are there frequencies that consistently benefit you? Have any unexpected insights emerged? Revisiting your entries may reveal profound insights about yourself, your health, and how sound influences your state of being.

Looking Forward: Envisioning Your Growth

As you become familiar with the practice of frequency healing, you may find your intentions and goals evolving. Some people start with physical healing and then move into spiritual exploration, while others might focus on emotional resilience or creative inspiration. Whatever your path, a sound healing journal acts as a companion, capturing your growth and reminding you of your progress.

In time, you may discover that your sound healing practice has become a central part of your wellness routine—a ritual that not only soothes the mind and body but also connects you to a deeper understanding of yourself. Let each entry in your journal be a step toward harmony, self-discovery, and the healing that frequencies can offer.

With dedication and an open mind, your sound healing journal will become more than a record; it will be a testament to the transformative power of sound, a map of your healing journey, and a personal reminder of the strength and beauty within you.

CHAPTER 92: GUIDED MEDITATION SCRIPTS FOR FREQUENCY HEALING

Imagine settling into a peaceful space, your mind calm, your breathing steady, and a resonant frequency filling the room. This is the foundation of guided meditation using healing frequencies. Guided meditation can enhance the effects of frequency healing, combining visualization with the subtle power of sound to create a profound experience of inner harmony and healing. This chapter offers meditation scripts designed to work with specific healing frequencies, guiding you on a journey that integrates sound with intention and awareness.

Through these scripts, you'll learn to connect deeply with the frequencies, allowing their healing vibrations to support relaxation, emotional release, and even physical rejuvenation. With each meditation, you'll be gently led to focus on different aspects of well-being, tapping into the science of sound healing while embracing a transformative journey within.

Meditation 1: Grounding and Centring with 174 Hz

The 174 Hz frequency is known for its grounding, pain-relieving qualities, and helps to bring a sense of physical and emotional security. This meditation is ideal when you're feeling stressed, disconnected, or need to root yourself in the present moment.

Preparation: Find a comfortable seated position or lie down. If possible, use headphones to fully immerse yourself in the frequency. Close your eyes, breathe deeply, and let the sound envelop you.

Script:
"Begin by taking a slow, deep breath in... and out. Feel the surface beneath you supporting your body, holding you with a sense of safety and stability. Inhale deeply, allowing your body to soften, and exhale, releasing any tension.

As the 174 Hz frequency hums softly, imagine this sound like a gentle wave, slowly flowing through your body, grounding you with every note. Envision the sound traveling from the crown of your head, down through your neck, your shoulders, and into your chest. With each exhale, let the sound move down further—into your abdomen, through your hips, and all the way to your toes.

Feel the gentle, grounding energy of this frequency rooting you, connecting you with the earth. With each breath, sink deeper into this connection, feeling calm, steady, and present. Let go of any worries, allowing them to dissolve into the earth beneath you. Rest here for as long as you need, anchored and at peace."

Meditation 2: Emotional Release with 396 Hz

The 396 Hz frequency is associated with releasing fear, guilt, and negative emotions. This meditation helps let go of old emotional baggage and encourages a lightness in both mind and body.

Preparation: Sit or lie down in a quiet space. Take a few deep breaths, then focus on any heaviness or emotional tension you may be holding.

Script:
"Close your eyes and take a deep, steady breath, filling your lungs completely. Hold it briefly, and then exhale, releasing tension. Allow the 396 Hz frequency to fill the space around you, enveloping you like a warm, comforting blanket.

As you listen, imagine this sound as a soft light moving gently into your heart. Picture it illuminating the shadows, the places where you hold fear, guilt, or other burdens. With each breath, let the light expand, loosening and releasing these emotions, as if they were clouds dissolving in the sky.

Stay with this feeling, breathing deeply, letting the sound and light clear away any weight that no longer serves you. Feel your heart becoming lighter, open, and free. Rest in this feeling, allowing the frequency to continue its gentle cleansing."

Meditation 3: Physical Healing with 528 Hz

Known as the "Love Frequency" or "Miracle Tone," 528 Hz is said to promote healing, DNA repair, and overall rejuvenation. This meditation is especially effective when you're feeling run down or in need of physical healing.

Preparation: Find a comfortable position, ensuring your body is fully supported. Allow yourself to relax completely and breathe deeply.

Script:
"Take a long, slow breath in... and out. Feel your body settling, relaxing into the space around you. Let the sound of the 528 Hz frequency wash over you like a gentle wave, calming and soothing every cell in your body.

Imagine a golden light, infused with this loving frequency, beginning at the top of your head and cascading down. As it moves, it fills you with warmth and healing energy, moving through each muscle, each bone, each cell. Envision this light as it flows into areas that need healing, gently nurturing and revitalizing them.

With each breath, feel this light dissolving any tension or discomfort, replacing it with a profound sense of peace and love. Stay here, basking in the warm glow of this frequency, feeling renewed and whole."

Meditation 4: Creativity and Focus with 741 Hz

The 741 Hz frequency is thought to enhance creativity, problem-solving abilities, and mental clarity. Use this meditation when you're looking to unlock new ideas, focus on a project, or stimulate creative thinking.

Preparation: Sit upright, with a sense of alert relaxation. Keep a notebook nearby to jot down any insights that may arise after the meditation.

Script:
"Take a deep, grounding breath, feeling the air as it fills your lungs and releases. Let the sound of 741 Hz begin to open and clear your mind. Imagine this sound as a spark of light, gently illuminating the corners of your consciousness, dissolving any mental fog or distraction.

As you breathe, let the light expand, moving through your

mind like a clear breeze, refreshing and energizing your thoughts. Picture your mind opening like a flower, each petal revealing new insights, ideas, and perspectives.

With each exhale, release any blocks or doubts that might hinder your creativity. Embrace this frequency, allowing it to inspire and energize you. When you feel ready, gently return to the present, carrying this sense of clarity and focus with you."

Meditation 5: Connection and Compassion with 963 Hz

Often referred to as the "Frequency of the Gods," 963 Hz is associated with spiritual awakening, connection to higher self, and universal consciousness. This meditation helps you tap into compassion and a sense of interconnectedness.

Preparation: Sit or lie down in a relaxed position. Close your eyes and let your breathing become soft and steady.

Script:
"Begin by taking a deep breath, feeling peace as it fills your body. Allow the sound of the 963 Hz frequency to flow into your being like a gentle stream, bringing with it a feeling of serenity and unity.

Imagine this frequency as a radiant, soft light at the centre of your chest, expanding outward with each breath. As the light grows, sense a connection forming—not just with yourself, but with everything around you. Feel your heart open, embracing compassion, understanding, and a deep connection to all beings.

Breathe in this sense of unity, letting it fill you with a peaceful strength. Let this frequency remind you of the interconnectedness of all life, carrying this wisdom with you as you move through the world."

Integrating Frequency Meditation into Daily Life

Incorporating these meditations into your routine can enhance your physical, emotional, and spiritual health over time. You may start with a specific frequency for a few weeks, observing its effects, before exploring others. You might also find it beneficial to journal your experiences, as each meditation can reveal unique insights and shifts.

Reflection

As you continue to practice, remember that frequency healing is both a journey and a discovery. Each session allows you to tune into the subtleties of your inner world, creating space for healing and growth. Allow yourself to be open, receptive, and inspired by the transformative power of these frequencies, bringing harmony to every level of your being.

CHAPTER 93: DEVELOPING A FREQUENCY HEALING ROUTINE

Imagine a life where tuning into a specific frequency becomes as natural as taking a deep breath—where healing isn't a last resort, but a daily practice that brings your mind and body into harmony. Creating a personalized frequency-based wellness routine is not only empowering but taps into the ancient art and science of vibrational medicine, blending it with modern understanding to elevate our daily well-being. This chapter will guide you through designing a frequency healing routine that is intuitive, customizable, and, most importantly, attuned to your unique needs.

The Power of Consistency in Frequency Healing

In the world of frequency medicine, consistency is key. Just as physical exercise gradually builds strength, regular exposure to specific frequencies can harmonize the body's energy fields over time. Scientific studies reveal that cumulative exposure to certain frequencies —such as those emitted through pulsed electromagnetic fields (PEMF) or sound waves—can promote cellular regeneration, reduce stress, and enhance mental clarity. By incorporating frequencies into your routine, you invite these transformative benefits to become a constant presence in your life.

Step 1: Assessing Your Wellness Goals

Before diving into the specifics, take a moment to identify what you want to achieve with frequency healing. Common wellness goals that frequency-based routines can support include:

- Reducing stress and promoting relaxation
- Enhancing focus and mental clarity
- Encouraging physical healing and pain relief
- Balancing emotional well-being
- Promoting deep, restorative sleep

Each goal aligns with specific frequencies that target different physiological and psychological states. Identifying your priorities will help you choose the frequencies and modalities that best support your wellness journey.

Step 2: Selecting the Right Frequencies

Once you've outlined your goals, the next step is selecting frequencies known for their healing properties. Below are some popular frequencies and their benefits, along with recommendations for incorporating them into a daily routine.

- 174 Hz – Known for its grounding and pain-relieving properties. This frequency is beneficial for individuals experiencing physical discomfort or those who seek a sense of calm in their routine.

- 396 Hz – Often used for releasing fear and anxiety, the 396 Hz frequency helps create a safe space for emotional processing. Consider this frequency if your goals include emotional balance or overcoming trauma.

- 528 Hz – Often called the "miracle tone," 528 Hz is

associated with DNA repair and physical rejuvenation. This frequency supports cellular health and is a good addition if your goals include physical healing or revitalization.

- 741 Hz – Known to enhance mental clarity, intuition, and creativity, 741 Hz is ideal for those seeking focus and mental stimulation, making it a perfect addition to morning routines.

- 963 Hz – Often referred to as the frequency of enlightenment, this tone encourages spiritual connection and unity. If your wellness journey includes a focus on mindfulness or spiritual growth, this frequency can add depth to your practice.
Step 3: Crafting Your Routine

A successful frequency healing routine should feel as natural as brushing your teeth. Here's a sample routine, divided into morning, midday, and evening segments, to provide a balanced approach to frequency-based wellness.

Morning Routine: Energize and Focus

1. Choose a Frequency (528 Hz or 741 Hz): Start your day by playing the chosen frequency as you prepare for the day. Many people find the 528 Hz frequency rejuvenating, while 741 Hz sharpens focus.

2. Mindful Breathing: Sit in a comfortable position, close your eyes, and focus on your breath. Let the frequency wash over you as you breathe deeply, setting an intention for the day. Imagine each breath filling your cells with energy.

3. Visualize Success: As the sound continues, picture your day ahead—infused with clarity, creativity, and calm. Let the frequency help you visualize success and prepare mentally for the tasks ahead.

Midday Routine: Reset and Release

1. Take a 5-Minute Frequency Break (396 Hz or 174 Hz): Around midday, take a break to release any accumulated tension. Play a grounding frequency like 174 Hz to reconnect with a sense of calm, or 396 Hz to let go of stress or fear.

2. Simple Stretching or Movement: While the frequency plays, stretch your body gently, releasing any physical tension. This practice combines the benefits of sound with movement, helping you reset physically and mentally.

3. Practice Gratitude: Use the sound as a prompt to focus on something positive in your day. This simple gratitude exercise, paired with frequency healing, helps shift your mindset toward positivity.

Evening Routine: Relax and Restore

1. Prepare for Sleep with a Relaxing Frequency (528 Hz or 963 Hz): In the evening, play a soothing frequency that encourages relaxation. Many find the 528 Hz frequency helps them unwind, while 963 Hz enhances a sense of peace and connectedness before sleep.

2. Guided Visualization: Imagine the frequency as a gentle light spreading warmth and healing throughout your body. Visualize any physical discomfort or tension dissolving as you prepare for rest.

3. Journaling or Reflection: Spend a few minutes reflecting on your day, recording any insights or moments of gratitude. Over time, this practice will help you notice the positive changes that frequency healing brings to your mental and physical well-being.

Tips for Building Consistency

- Start Small: If a full routine feels overwhelming, start with one session per day. Gradually add frequencies and practices as they become part of your rhythm.

- Use Technology: Many apps and devices now offer access to healing frequencies, allowing you to play them at specific times or set reminders for your sessions.

- Track Your Progress: Consider keeping a frequency journal to record how each session makes you feel. Over time, you may notice shifts in energy levels, mood, and physical sensations.

- Incorporate Frequency Tools: Tools like tuning forks, singing bowls, or frequency-emitting devices can enhance your practice, bringing variety and helping you attune more deeply to each frequency.

A Vision of Frequency Medicine's Future

As we imagine the future of medicine, the potential for frequency-based healing is profound. What if one day, frequency therapy is as common as a visit to the doctor? Picture a world where we tune into specific frequencies for targeted healing, where doctors prescribe sound baths as readily as they prescribe medication.

Emerging research on biofeedback, personalized frequency therapy, and AI-driven health devices suggest that this isn't far-fetched. Soon, we may be able to map the precise frequencies that our bodies need for optimal health, creating a symphony of sound tailored to each individual.

Reflection

Your journey with frequency medicine begins with these small, daily acts of tuning in. Whether your goals are physical, emotional, or spiritual, a personalized frequency healing routine can transform your life in subtle but powerful ways. Frequency medicine invites us to remember the natural harmony within our bodies and minds, guiding us back to balance with every tone, every vibration.

May your practice grow with intention, each session bringing you closer to the realization of your fullest, most harmonious self. Embrace the rhythms, trust the process, and let the healing harmonies guide you toward well-being that resonates at the deepest level.

PART XXI: FINAL THOUGHTS AND MOVING FORWARD

CHAPTER 94: THE GROWING GLOBAL MOVEMENT OF FREQUENCY MEDICINE

In recent years, a quiet yet powerful movement has been sweeping across the world, one that harnesses an age-old concept—frequency—for healing. What was once considered an esoteric practice is now edging into the mainstream, capturing the interest of scientists, doctors, and everyday individuals. Across continents, from cutting-edge labs to ancient wellness centres, frequency medicine is gaining traction as an integrative, non-invasive approach that challenges our very understanding of health and healing.

A Modern Revival of Ancient Wisdom

Imagine standing inside a temple in ancient Greece, where patients with ailments gathered not for surgery, but for sound baths created with gongs, flutes, and rhythmic chants. In cultures worldwide, from Tibetan singing bowls to Indigenous drumming rituals, sound has always played a role in healing. These practices recognized something profound: the human body is highly responsive to frequencies, capable of resonating with sounds that calm the mind and invigorate the spirit.

Today, frequency medicine is a blend of this ancient wisdom and modern technology. With tools like ultrasound therapy, pulsed electromagnetic field (PEMF) devices, and sound wave treatments, practitioners are rediscovering the immense therapeutic power of frequency. In the process, they are creating a new paradigm in medicine, one that redefines what it means to heal.

The Science Behind Frequency Medicine: A Global Awakening

To grasp the global interest in frequency medicine, it helps to understand the science that supports it. Every cell in the body has a natural vibration or "resonance," akin to a unique musical note. This frequency shifts in response to various stimuli, including stress, illness, and environmental factors. The theory behind frequency medicine is that, by exposing the body to certain frequencies, we can "retune" cells to their optimal state, much like tuning a musical instrument.

In Japan, a team of researchers is exploring the effects of low-frequency sound waves on tissue regeneration, with remarkable results in promoting faster healing. Across the globe in Germany, scientists are delving into electromagnetic frequency (EMF) therapies for pain relief and inflammation, showing measurable improvements in patients with chronic conditions. In the United States, frequency-based therapies like neurofeedback and transcranial magnetic stimulation (TMS) are used for mental health, helping individuals with depression and anxiety find relief without the need for medication.

These studies are building a body of evidence that confirms what ancient healers intuited: that frequency can be a powerful catalyst for healing.

Case Studies: Real-World Applications of Frequency Medicine

Case Study 1: Sarah's Journey with Chronic Pain

Sarah, a 52-year-old woman from Sydney, Australia, had been struggling with fibromyalgia for years. Traditional treatments offered limited relief, leaving her in near-constant pain. After a recommendation from a friend, she tried PEMF therapy. Within weeks, Sarah noticed a reduction in her pain levels. The frequency emitted by the PEMF device seemed to resonate with her body, reducing inflammation and helping her sleep better at night. Her experience echoes countless testimonials from others who have found a new lease on life through frequency-based therapies.

Case Study 2: Mental Health and Frequency Healing

In Toronto, Canada, a clinic specializing in mental health has integrated sound frequency therapy into its treatment protocol. Mark, a 34-year-old artist with severe anxiety, found that regular sessions with binaural beats and TMS helped to calm his mind and improve his mood. "It felt like my brain was being retuned," he said. This case highlights how mental health professionals are embracing frequency medicine as an alternative or complement to pharmaceutical interventions.

Frequency Medicine Around the World

The embrace of frequency medicine is not limited to any single region—it is becoming a global movement.

- In Europe, clinics in Germany, Switzerland, and Austria are renowned for their holistic approach to medicine, often blending conventional treatments with frequency therapies. Bio-resonance machines, which assess the body's frequencies to detect imbalances, are popular in these clinics.

- In Asia, Japan and South Korea have been at the forefront

of frequency research. Japanese hospitals use low-frequency ultrasound to aid in the healing of bone fractures and soft tissue injuries, with impressive results. South Korean spas offer frequency-based treatments that combine relaxation with subtle healing.

- In Africa, traditional healing practices involving rhythmic drumming and chanting resonate with the principles of frequency medicine. These methods, long used in communities for physical and spiritual healing, are being studied for their physiological impact, potentially bridging traditional and modern healing methods.

- In the Americas, from TMS centres in the United States to sound healing retreats in South America, frequency medicine is growing rapidly. The combination of Indigenous sound healing traditions with modern research has sparked a renewed interest in holistic wellness.

How to Explore Frequency Medicine in Your Life

The rising popularity of frequency medicine has also made it more accessible. Here are some practical ways to integrate frequency healing into your wellness routine:

1. Sound Baths and Sound Healing: Seek out sound healing sessions at wellness centres or online. These sessions typically use gongs, singing bowls, or tuning forks to create a resonant environment that helps you relax and recalibrate.

2. PEMF Devices: Available in a range of price points, these devices emit electromagnetic pulses that promote healing at the cellular level. If you suffer from chronic pain or inflammation, consider trying PEMF therapy, but consult a professional for guidance.

3. Frequency Apps and Playlists: Apps and streaming services offer playlists with binaural beats and healing frequencies. Listen to these sounds during meditation or while

unwinding at the end of the day to reduce stress and improve focus.

4. Home-based Tuning Fork Therapy: Tuning forks, which produce specific frequencies, are used to target areas of physical or emotional discomfort. Beginners can find resources and guides online to help start their own tuning fork practice.

The Future of Frequency Medicine

As frequency medicine continues to gain recognition, the possibilities for its application are boundless. Imagine wearable technology that tracks your body's frequencies and adjusts them to enhance your mood, energy, or focus in real-time. Scientists are already exploring the potential of biofeedback and AI to make frequency medicine even more personalized. Perhaps one day, hospitals will house frequency suites, where patients can be immersed in targeted sound environments for everything from bone healing to mental health care.

But even as frequency medicine advances, it remains a field grounded in ancient wisdom, reminding us that healing is not always about intervention—it's about resonance, balance, and harmony. Frequency medicine is as much a spiritual practice as it is a scientific one, urging us to attune to the rhythms of our bodies, and in doing so, to discover a new depth of connection with ourselves.

A Global Symphony of Healing

The movement toward frequency medicine is, in essence, a global symphony. It's a harmonious merging of science and spirituality, modern innovation and ancestral wisdom. As more people around the world discover the power of frequencies, they're not just finding relief from physical ailments—they're rediscovering the resonance within themselves.

In this global movement, there is a revelation to be found: that health is more than the absence of illness; it's a state of being in tune with oneself and the world. Frequency medicine offers a pathway to that state, inviting us all to become participants in a grand orchestra where every life vibrates at its unique, healing frequency.

CHAPTER 95: COMMUNITY HEALING CIRCLES AND SOUND THERAPY

In an age where the individual often faces health and healing alone, the rise of community healing circles brings us back to an ancient wisdom: healing is most powerful when shared. Community healing circles, particularly those centred around sound therapy, offer an approach to health that resonates both with modern science and ancestral practices. In these circles, sound is not just an isolated experience; it becomes a powerful collective journey, a tapestry woven from shared frequencies and intentions.

The Return to Collective Healing

Imagine stepping into a dimly lit room, a warm glow from candles lining the walls. You are surrounded by others, some familiar, some strangers, all united by an unspoken need for healing. As you sit and close your eyes, a gentle hum begins to fill the air, followed by the soft sounds of Tibetan singing bowls, gongs, and chimes. In moments, you feel the vibrations moving through your body, soothing muscles and quieting your mind. You are not alone in this journey—

every vibration is amplified, both by the instruments and the collective presence of those around you.

Healing circles have been used by Indigenous and ancient cultures across the world for thousands of years. They offer participants a safe space to be vulnerable, to release emotions, and to support each other's healing journey. Sound is a particularly powerful tool in these gatherings, as its vibrations can touch each person uniquely, dissolving boundaries between the physical, emotional, and even spiritual realms.

How Sound Therapy Works in a Group Setting

Sound therapy uses instruments like crystal bowls, gongs, drums, and tuning forks to produce vibrations that resonate with different parts of the body and mind. In a group setting, these sounds create a "harmonic field" that everyone can connect to. Think of it as an orchestra—each instrument plays its note, but together they create a powerful, immersive experience. In a community sound healing circle, every individual's unique energy and intentions contribute to this collective harmony, amplifying the healing effects.

Science shows that sound frequencies can influence the brain's electromagnetic activity, inducing states of relaxation and enhancing feelings of well-being. When this is done collectively, the vibrations interact with the group's collective energy, reinforcing each person's healing response. Studies reveal that participating in group sound therapy can reduce stress and anxiety, lower blood pressure, and even boost the immune system. The benefits multiply as each person's healing energy resonates through the group, creating a cycle of shared, amplified wellness.

Real-Life Stories of Transformation

Case Study 1: Maria's Journey with Grief

After the loss of her mother, Maria struggled with overwhelming grief that left her feeling isolated and drained. Traditional therapy provided some relief, but it was not until she joined a community sound healing circle that she experienced a profound shift. Surrounded by others who shared her openness to healing, Maria felt enveloped by the sounds of Tibetan bowls and chimes. In the months that followed, she returned to the circle, each session helping her release layers of sorrow. For Maria, the shared experience was transformative. "I felt my grief lifted by others, as if we were all carrying it together," she recalls. "It was a space where I felt seen and supported, even in silence."

Case Study 2: Healing from Burnout

Mark, a 40-year-old teacher, had reached a breaking point with chronic burnout. A friend invited him to a community sound healing circle, though he was initially sceptical . But during the session, as the vibrations from gongs and chimes moved through him, he felt the tension dissolve. Over the next few weeks, he joined several sessions, each time feeling more rejuvenated. "I didn't realize how much I was carrying until I let it go in those circles," Mark says. "It was like therapy without words—a chance to recharge with others who understood what it was like to feel completely depleted."

The Power of Intention in Collective Healing

In a healing circle, intention plays a crucial role. When people come together with a shared purpose—whether it's to heal, find peace, or seek clarity—their collective intention amplifies the effect of the sound frequencies. This isn't just a philosophical idea; research on intention and consciousness suggests that group focus can influence emotional and physiological outcomes. In a sound healing circle, participants are encouraged to set personal intentions, which are then woven into the fabric of the group's collective energy, creating a powerful ripple effect.

Consider the analogy of a pebble dropped into a still pond. Alone, each pebble creates a small ripple. But in a healing circle, each participant's intention is like a pebble dropped simultaneously into the same pond. The result is an overlapping of ripples, intensifying and spreading further than any single pebble could achieve alone.

How to Create Your Own Healing Circle

Creating a community healing circle can be a simple yet impactful way to bring people together. Here's a step-by-step guide to help you create a space for shared sound healing:

1. Set an Intention: Decide on a theme for the gathering, such as relaxation, emotional release, or gratitude. Invite participants to bring their own intentions, which will be woven into the collective experience.

2. Choose the Instruments: Sound healing can involve anything from Tibetan singing bowls and chimes to drums and vocal toning. Even simple instruments, like bells or rain sticks, can create a harmonious atmosphere.

3. Create a Safe, Comfortable Space: Arrange the seating in a circle to symbolize equality and unity. Use low lighting, soft cushions, and grounding scents like sage or lavender to create a calming environment.

4. Begin with Grounding Exercises: Start with deep breathing exercises or a brief meditation to help everyone settle into a peaceful state.

5. Guide the Sound Journey: Begin with soft tones and gradually introduce deeper, more resonant sounds. Allow the group to experience each instrument's vibration fully, encouraging them to let go and immerse in the sounds.

6. Close with Reflection: After the sound session, offer

participants a few minutes to reflect on their experiences. Invite them to share, if they feel comfortable, fostering a sense of community and mutual support.

The Future of Community Sound Healing

As sound therapy gains recognition worldwide, we can envision a future where community healing circles become more commonplace in wellness centres, hospitals, and even corporate settings. Imagine sound healing circles designed to address specific needs—helping veterans process trauma, supporting mothers in postpartum recovery, or providing emotional relief for healthcare workers.

These community sound circles are more than just a wellness trend; they are a reminder of the deep connections that sound and vibration can foster. Through shared resonance, we discover that healing is not just a personal journey but a collective one. Together, we resonate in harmony, becoming co-creators in a healing process that transcends the individual and nurtures the whole.

A Global Movement of Healing Through Sound

In a world that often feels fragmented, the rise of community sound healing circles is a powerful reminder of our interconnectedness. As more people come together to share in these vibrational journeys, we witness a revival of collective healing. The truth revealed in these gatherings is profound: we are all resonant beings, and by aligning our frequencies, we amplify our power to heal. It's a truth both ancient and modern, a revelation that healing—true, lasting healing—happens best when we come together in harmony.

CHAPTER 96: FREQUENCY MEDICINE IN TRADITIONAL AND ALTERNATIVE MEDICINE

Imagine for a moment the human body as a symphony, with each cell vibrating at its own unique frequency, creating a harmonious orchestra of life. When our body's cells are healthy, they resonate in unison. However, when illness strikes, this harmony falters, and discord arises. Frequency medicine, with its roots in both ancient wisdom and cutting-edge science, seeks to restore this balance by resonating with the body's natural frequencies. It bridges traditional healing practices and modern medical science, offering a path to wellness that is as holistic as it is effective.

A Meeting of Worlds: Traditional Healing and Frequency Medicine

From ancient Tibetan singing bowls to the rhythmic chanting in indigenous rituals, traditional medicine has long embraced sound and vibration as healing tools. In many cultures, sound was—and still is—believed to align the spirit, calm the mind, and heal the body. This practice rests on the idea that sound waves and vibrations interact with the energy fields within us, restoring balance.

The principles behind traditional healing and frequency medicine converge in this understanding of the body as an energetic system. Modern frequency medicine, through modalities like ultrasound therapy and pulsed electromagnetic field (PEMF) therapy, builds on these ancient insights. Using scientifically precise frequencies, it targets cells, tissues, and organs to encourage healing, relieve pain, and reduce inflammation.

In a traditional healthcare setting, sound healing is seen as complementary, providing non-invasive support that augments conventional treatment. Dr. Emma Fields, a physician who has integrated sound therapy into her practice, describes it this way: "While medication and surgery are powerful tools, they can overlook the subtler imbalances within us. Frequency healing helps us tune the patient's entire body, allowing for a more complete form of healing."

Case Study: Traditional Wisdom Meets Modern Science

Consider the story of Marcus, a 65-year-old man with chronic arthritis. After years of traditional treatments with limited relief, he turned to a clinic specializing in complementary therapies. There, he was introduced to PEMF therapy and Tibetan singing bowl sessions. During the sessions, the low frequencies of the bowls eased his mental stress, while PEMF therapy helped reduce the pain in his joints. Over time, Marcus found himself relying less on pain medication as the synergy of these treatments offered lasting relief.

How Frequency Medicine Aligns with Alternative Therapies

Alternative medicine has often emphasized the body's inherent ability to heal, viewing interventions as a way to support and amplify this process rather than to control or suppress symptoms. Frequency medicine aligns seamlessly with this philosophy, as it works to stimulate natural cellular processes rather than introduce foreign substances.

One modality that highlights this synergy is bio resonance therapy, which identifies energetic imbalances by analysing the frequencies emitted by the body. Bio resonance therapy, used in integrative medicine, adjusts these frequencies to restore harmony within the body, much like acupuncture or Reiki. Practitioners often find that frequency treatments enhance the benefits of other alternative therapies. Acupuncturist Sarah Langston, for instance, notes, "When I combine acupuncture with frequency treatments, patients often report quicker relief and a sense of profound calm. It's as if their bodies respond more deeply."

Bringing Frequency Medicine into Conventional Healthcare

Although frequency-based healing is gaining acceptance, it still faces barriers in mainstream healthcare. Part of the challenge lies in the limitations of clinical research in capturing the benefits of vibrational therapies. However, studies are emerging that validate the efficacy of frequency medicine in various areas, from pain management to mental health.

For example, transcranial magnetic stimulation (TMS) has been approved by the FDA to treat depression, showing the therapeutic potential of frequency modulation in mental health. TMS, which uses magnetic fields to stimulate nerve cells in the brain, echoes ancient practices in which rhythm and resonance influenced emotional states. By using electromagnetic frequencies, TMS targets specific brain regions associated with mood regulation, offering hope to those for whom traditional medications have been ineffective.

The acceptance of TMS in conventional healthcare shows a growing openness to frequency-based treatments, particularly as they are backed by rigorous research. As healthcare providers witness their patients benefit from such therapies, the scepticism surrounding frequency medicine is gradually being replaced by curiosity.

Practical Tips for Integrating Frequency Healing into Your Life

If you're interested in exploring frequency healing for yourself, there are simple and effective ways to start:

1. Begin with Sound Therapy: Explore sound baths or use binaural beats to introduce yourself to the healing power of sound. These sessions often use specific frequencies to encourage relaxation, stress relief, or focus.

2. Use PEMF Devices for Pain Relief: PEMF devices are increasingly available for personal use. If you suffer from chronic pain or inflammation, a portable PEMF device can deliver low-frequency electromagnetic waves to target areas, helping to reduce discomfort and speed recovery.

3. Experiment with Bioresonance Devices: Some bioresonance devices are designed for home use, allowing you to track and adjust your body's frequency balance. While they're not a replacement for medical treatment, these devices can provide supportive insights into your energetic health.

4. Combine with Other Therapies: If you practice yoga, acupuncture, or meditation, introduce frequency elements to enhance these experiences. Many find that using sound therapy or bioresonance devices before a meditation session deepens their focus and enhances relaxation.

A Vision for the Future: Frequency Medicine as Standard Care

Imagine a future where frequency healing is as accessible as an annual physical check-up, where hospitals incorporate sound healing rooms alongside surgical suites, and where sound baths are part of physical therapy routines. This is a vision of healthcare that embraces the body's natural energy systems and acknowledges that true healing addresses not just the physical but also the emotional and energetic aspects of health.

In this world, we could envision hospitals with dedicated vibrational therapy rooms. Patients recovering from surgery might use PEMF to accelerate tissue repair, while those experiencing anxiety or chronic pain could benefit from sound immersion therapy. This model isn't far-fetched; it represents a growing global movement that blends ancient wisdom with technological innovation.

Reflecting on the Power of Frequency Medicine

As frequency medicine continues to grow, it encourages a deeper reflection on the nature of health and healing. By seeing ourselves not just as physical bodies but as dynamic systems of energy, we open doors to treatments that resonate with us on every level of being.

While frequency medicine may not replace traditional treatments, it offers an invaluable complement. It serves as a reminder that healing is not a singular path but a symphony, a blend of instruments, harmonies, and notes, each contributing to the greater whole. Frequency medicine asks us to consider our health as an art as much as it is a science. It calls us to rediscover our own capacity to heal, guided by vibrations that have been here since time began, waiting only for us to listen.

In Healing Harmonies, we explore this profound journey,

where science and spirit meet, and where every note has the power to restore balance. We are part of an exciting moment in medicine, standing on the threshold of a revelation that connects us to a deeper understanding of health—one where frequency is a bridge between what we have known for centuries and what we are still learning to embrace.

CHAPTER 97: THE IMPORTANCE OF PRACTITIONER TRAINING AND CERTIFICATION

Imagine a painter, meticulously preparing their canvas, carefully selecting colours and brushes. With every stroke, they bring a vision to life. Now, consider a frequency healing practitioner. They, too, are artists, using sound waves, electromagnetic pulses, and subtle vibrations to "paint" on the canvas of human health. Just as an artist refines their skill through years of study and practice, so must a practitioner of frequency healing. The training and certification process, often overlooked, forms the foundation for effective, safe, and transformative healing.

The Complexity of Frequency Medicine

Frequency medicine, though rooted in ancient traditions, has evolved into a scientifically rigorous field. In the past, healers relied on intuition, chants, and natural resonance, attuned to the world around them. Today, practitioners must combine this intuitive understanding with knowledge of physics, biology, and medical science. Just as a doctor requires years of study to understand the body, a frequency healer must grasp complex concepts, like how specific frequencies interact with cellular structures or impact mental health.

Training provides the technical and theoretical foundation necessary for practitioners to wield these tools effectively. For instance, sound frequencies like binaural beats might be used to alleviate anxiety, while pulsed electromagnetic fields (PEMF) target physical pain. Each technique demands not only knowledge but precise execution to achieve the desired effects.

Why Certification Matters

Certification in frequency healing ensures practitioners have a deep understanding of both the benefits and risks. Frequency healing is not without its complexities. A trained practitioner knows that, much like administering medication, frequency treatments require careful calibration. Overuse, misuse, or inappropriate application can lead to unintended side effects or simply be ineffective, resulting in a loss of trust in frequency medicine itself.

Certified practitioners adhere to standards and ethical guidelines, ensuring clients receive the safest and most effective care. When a practitioner is certified, clients can trust that they've undergone rigorous training, tested their skills, and demonstrated a commitment to responsible practice. Certification also helps practitioners stay up-to-date with advancements in technology, evolving techniques, and new scientific discoveries, all of which contribute to enhanced healing outcomes.

Case Study: Training's Impact on Real-World Practice

Take, for instance, Rachel, a sound therapist who initially approached frequency healing with limited knowledge. She had a deep desire to help her clients but found her early sessions inconsistent in results. After undergoing formal training, which covered topics from acoustics to bioenergetics, she gained a far deeper understanding of how different frequencies affect the body.

One of her clients, Paul, suffered from chronic insomnia. With her newfound knowledge, Rachel chose frequencies specifically suited to calm Paul's nervous system, helping his brain enter a state of restful relaxation. Within a few weeks, Paul began sleeping through the night for the first time in years. Rachel attributes her success with Paul to the structured training that helped her refine her techniques.

Elements of a Comprehensive Training Program

So, what does practitioner training in frequency healing look like? A thorough program includes both theoretical knowledge and practical application, often covering areas such as:

1. The Science of Frequencies: Understanding how sound, electromagnetic, and vibrational frequencies interact with the body. Topics may include cellular biology, resonance, and the electromagnetic spectrum.

2. Hands-On Techniques: Training practitioners to use specific tools and devices, from tuning forks to PEMF machines, with precision and care. This practical knowledge ensures that each session is safe and tailored to the individual.

3. Ethics and Safety: Like any healthcare field, frequency healing requires practitioners to understand the ethical considerations, particularly regarding vulnerable clients or those with severe conditions. Training helps them recognize when frequency therapy is appropriate and when conventional medical intervention is necessary.

4. Client Interaction and Assessment: Skilled practitioners

understand that healing is not only about technical knowledge but also about building trust and empathy. Learning to assess a client's needs holistically and to communicate openly about the expected outcomes, limits, and potential risks is essential.

5. Continuous Learning: Frequency healing, like medicine, is dynamic. New discoveries regularly enhance what we know about frequencies and their effects. Comprehensive training programs instil a commitment to ongoing education, so practitioners remain effective and knowledgeable.

DIY Frequency Healing: Proceed with Caution

With the rise of consumer-friendly frequency devices, there's been a surge of interest in at-home frequency healing. While these devices offer potential benefits, the lack of guidance can lead to misuse. For example, someone might purchase a PEMF mat, unaware that certain frequencies are contraindicated for conditions like epilepsy or heart issues. Practitioners with formal training can provide clients with DIY recommendations, instructing them on safe use and helping them avoid potential pitfalls.

For those interested in exploring frequency healing at home, consider starting with simple practices, like sound healing, where the risk is minimal. Sound baths, for example, can be a gentle introduction, requiring little more than an open mind and careful listening.

The Future of Frequency Healing: Standardizing Certification

As frequency healing continues to gain mainstream acceptance, the demand for standardized certification is increasing. Imagine a future where every city has licensed frequency medicine practitioners who are recognized alongside conventional healthcare providers.

This integration would elevate frequency healing from a complementary practice to an integral component of holistic health care.

Standardizing certification also paves the way for frequency healing to be included in health insurance plans, making these treatments accessible to a broader population. The establishment of official training and certification programs will also lend credibility to frequency healing, dispelling myths and grounding it firmly in science-backed practices.

Practical Tips for Finding a Qualified Practitioner

For those considering frequency healing, choosing a certified practitioner can make all the difference. Here are a few practical tips:

1. Check Credentials: Ask about certifications and training. A qualified practitioner should be transparent about their background and willing to discuss their expertise.

2. Seek Referrals: Often, the best practitioners come recommended by satisfied clients. Personal referrals or testimonials can offer insight into a practitioner's effectiveness and approach.

3. Evaluate Their Approach: A skilled practitioner will be attentive to your needs, willing to answer questions, and clear about what you can expect. Trust your instincts—working with someone who is empathetic and knowledgeable can enhance your healing experience.

4. Understand the Process: A professional practitioner will explain their approach, the frequencies they plan to use, and how they're tailored to your unique needs.

Closing Reflection: Certification as a Path to Healing Integrity

In many ways, certification is about more than just

credentials. It's about setting a standard of care that respects the transformative power of frequency medicine. Certified practitioners not only contribute to their clients' well-being but also to the credibility and future growth of the field itself.

As frequency healing continues to evolve, it reminds us that healing requires both knowledge and empathy, science and spirit. This commitment to training ensures that those who practice frequency healing do so with integrity, respect, and a deep understanding of the powerful tools at their disposal. It's a promise to clients that they're in safe, capable hands, guided by practitioners who are devoted to the harmonious blend of art and science that defines frequency medicine.

With these standards in place, frequency healing can reach its full potential, inspiring trust, transforming health, and bringing the healing harmonies of sound and vibration into the lives of all who seek it.

CHAPTER 98: FUTURE RESEARCH DIRECTIONS IN FREQUENCY MEDICINE

In the ever-evolving landscape of medical science, frequency medicine stands at an exciting frontier. This chapter explores the future of frequency-based healing by highlighting emerging areas of research, from advanced bioresonance studies to novel applications of sound and light therapies. As scientists, doctors, and alternative practitioners continue to uncover the profound impacts of frequencies on human health, we are inching closer to a future where frequency-based treatments may be as routine as surgery or pharmaceuticals, but far less invasive.

A New Era of Bioresonance

Imagine a world where each cell in your body resonates like an instrument in an orchestra, producing a harmonious vibration when healthy. This concept, known as bioresonance, suggests that each organ and tissue has its unique frequency. When disease disrupts these natural frequencies, we experience symptoms of illness. Advanced bioresonance research aims to identify these "ideal" frequencies and create treatments that restore harmony at the cellular level.

Scientists have already started using bioresonance technology to detect anomalies in cellular vibration. Some studies show that subtle frequencies can influence cellular activity, such as promoting the repair of damaged tissue or inhibiting the growth of cancerous cells. For instance, a recent clinical study used frequency therapy to stimulate the immune response in patients with weakened immune systems, achieving promising preliminary results. Although much is still unknown, the potential for bioresonance to play a significant role in disease diagnosis and recovery is substantial.

The Promise of Sound and Light Therapies

Sound and light have long been recognized for their potential to influence mood and health. Emerging research is now diving deeper, revealing new ways these elements can enhance cellular function, relieve pain, and even promote mental health.

1. Sound Therapy Advancements: As researchers explore the impact of sound on the human body, they are discovering that certain frequencies can produce therapeutic effects beyond relaxation. Studies on binaural beats, for instance, reveal that listening to specific tones can induce a meditative brain state, reducing anxiety and improving focus. New research is also exploring how ultrasound waves can penetrate deep tissues, potentially offering non-invasive treatments for chronic pain and muscular issues. Researchers envision a future where personalized sound frequencies could be applied to treat specific physical or mental conditions, helping to alleviate everything from migraines to PTSD.

2. The Power of Light Frequencies: Light therapy, already

used for conditions like seasonal affective disorder (SAD), is expanding into new territories. Red and near-infrared light, in particular, have shown promise in wound healing and anti-inflammatory applications. Scientists are exploring how specific light frequencies can enhance mitochondrial function, boosting cellular energy and promoting recovery. A recent pilot study found that a specific wavelength of red light significantly improved tissue repair after surgery. This research could lead to new light-based treatments that accelerate recovery, reduce scars, and even counteract aging effects by revitalizing skin cells.

Quantum Biology: A Revolutionary Field in the Making

The growing field of quantum biology examines how quantum mechanics, a branch of physics that deals with the smallest particles, might influence biological processes. Quantum biology suggests that particles within our cells operate not just as independent units but as interconnected "quantum systems," communicating through vibrational signals.

In this field, researchers are investigating how the body's natural frequencies could be harnessed to repair damage at a molecular level. One promising area is the study of electron tunnelling, where particles move between molecules through vibrational energy. By aligning with the body's natural frequencies, future therapies might enable cellular repair that currently seems impossible, such as reversing degenerative diseases or regenerating damaged nerve cells. While this field is still emerging, it presents a world of possibility for frequency medicine, where vibration and resonance could support biological functions on a fundamental level.

The Role of AI and Data in Personalized Frequency Medicine

Advancements in artificial intelligence (AI) are poised

to revolutionize frequency medicine by enabling highly personalized treatment plans. AI algorithms can analyse massive datasets to identify patterns in frequency responsiveness, tailoring therapies to each individual's unique "frequency profile." For instance, a patient with chronic pain might respond best to a blend of specific sound and electromagnetic frequencies, optimized through machine learning.

Data-driven frequency medicine could provide a roadmap for personalized care that maximizes effectiveness. Imagine an app that records your body's daily frequency readings and tailors your frequency therapy to your current state, offering real-time adjustments to optimize health outcomes. The integration of AI and biofeedback technologies would make frequency medicine highly adaptive, responsive to daily changes in the body's needs, and accessible to a wider population.

Case Study: A Glimpse Into the Future

Let's envision a potential scenario. Emily, a 40-year-old with chronic joint pain, finds little relief from conventional treatments. Her physician, trained in frequency medicine, uses a device to measure the resonance frequencies of her cells, pinpointing areas where vibration is off-balance. Through AI analysis, her doctor identifies a customized regimen that combines specific sound frequencies, low-level electromagnetic pulses, and red light therapy.

After a few weeks, Emily notices her pain decreasing significantly. The frequency treatments not only relieve her pain but also improve her mobility and mood. She's now able to exercise more and regain a part of her life she'd thought was lost. Emily's story exemplifies what the future might hold when frequency medicine combines the precision of science with a holistic understanding of the body's energy.

Overcoming Barriers to Frequency Medicine

As promising as these developments are, frequency medicine faces challenges in gaining mainstream acceptance. One major barrier is the limited research funding in non-conventional medicine. Additionally, the complexity of measuring frequency-based effects on health and the potential scepticism from traditional medicine practitioners also present hurdles.

To bridge this gap, interdisciplinary collaborations are crucial. Biologists, physicists, doctors, and alternative medicine practitioners must work together to refine techniques, conduct rigorous clinical trials, and build the credibility needed for acceptance. Advances in biofeedback and bioresonance are building this bridge, providing tangible, measurable results that traditional medicine can recognize.

Practical Tips: Engaging with Frequency Medicine

For readers inspired by these potential breakthroughs, here are a few ways to engage with frequency healing:

1. Experiment with Sound Therapy: Explore sound baths or binaural beats, both accessible methods to introduce frequencies into your daily life. Experimenting with these practices can help build personal awareness of frequency medicine's effects.

2. Consider Light Therapy: Red and infrared light therapy devices are becoming more affordable. For those with inflammation or skin issues, light therapy can be a gentle way to explore frequency healing.

3. Track Your Progress: Keep a journal of how you feel before and after sessions with sound or light therapy. Reflecting on any changes over time can help you better understand your

responses to different frequencies.

Reflecting on the Potential of Frequency Medicine

The journey of frequency medicine is one of rediscovery and innovation, merging ancient wisdom with modern science. As research advances, our understanding of health and well-being continues to shift toward a holistic paradigm where the body, mind, and energy are inextricably linked. We are standing on the cusp of a revolutionary era in medicine—one that could redefine healing as we know it.

In the years to come, frequency medicine could move from an alternative option to an essential tool in health care, revealing profound truths about our bodies and our connections to the universe's vibrations. With continued dedication, curiosity, and openness, we may soon live in a world where the healing harmonies of frequencies become a mainstream solution for human health and vitality.

CHAPTER 99: PATIENT EMPOWERMENT AND FREQUENCY HEALING

In a world where the traditional model of medicine often places patients in a passive role, frequency medicine offers a revolutionary shift. By its nature, frequency healing is participatory, inviting individuals to take an active role in their health. Whether through sound therapy, light exposure, or the use of electromagnetic frequencies, patients are discovering that their health can be enhanced by their own hands. This chapter explores how frequency medicine empowers individuals to embark on a self-guided journey of healing, making informed, intentional choices in their wellness.

Understanding the Body as an Orchestra

To understand the promise of frequency healing, we need to reimagine the body. Rather than seeing ourselves as a collection of separate organs, think of the body as an orchestra, where each cell, tissue, and organ resonates at a unique frequency. In this orchestra, every cell has its own "note," creating a harmonious symphony when healthy.

When illness strikes, it's as though some instruments are playing out of tune. Frequency healing, in essence, seeks to "re-tune" the body to its natural state.

This concept is powerful because it brings healing into a space that patients can influence directly. Unlike medications or surgical interventions, which are often out of the patient's control, frequency healing encourages self-application and observation. It allows individuals to listen to their bodies, to feel the subtle shifts that frequencies can bring, and to play a role in the adjustments needed for health and balance.

Real-World Stories of Empowerment

Consider the story of Sarah, a 45-year-old who suffered from chronic back pain for years. Traditional treatments offered her only temporary relief, leaving her frustrated and resigned to a life of discomfort. When Sarah discovered frequency healing, she was sceptical but curious. Guided by a practitioner, she began using a pulsed electromagnetic field (PEMF) device designed for home use. She also incorporated sound therapy, using frequencies known to stimulate tissue repair and reduce inflammation. Within months, Sarah noticed significant improvement—not only in her back pain but also in her overall sense of well-being. For Sarah, frequency healing wasn't just a new treatment; it was a path to reclaiming control over her health.

Stories like Sarah's are not isolated. Across the world, people are discovering the benefits of personalized, frequency-based treatments. From sound baths that reduce anxiety to light therapies that boost mood and cellular health, patients are realizing that these methods don't just treat symptoms; they empower individuals to connect deeply with their bodies, fostering self-awareness and long-term wellness.

Practical Tips: Getting Started with Frequency Healing

For readers inspired to begin their journey, here are practical tips to make the first steps accessible and enjoyable:

1. Identify Your Health Goals: Start by defining your specific wellness goals. Are you looking to reduce stress, improve sleep, alleviate chronic pain, or support recovery from an injury? Understanding your goals will help you choose the most suitable frequency-based techniques.

2. Start with Sound Therapy: Sound therapy is one of the simplest ways to explore frequency healing. Consider using binaural beats, which involve listening to two slightly different frequencies in each ear. Studies show that binaural beats can help induce relaxation, enhance focus, and even aid in sleep. Experiment with different frequencies to discover which resonates with your needs.

3. Use a Frequency Journal: Track your experiences. Each time you try a frequency therapy—whether it's a sound session or a light therapy treatment—make note of how you feel before, during, and after. Over time, patterns may emerge, revealing the specific frequencies and modalities that are most effective for you.

4. Explore PEMF Therapy: If you're interested in more advanced methods, pulsed electromagnetic field (PEMF) devices can be effective for pain relief and cellular health. Some devices are available for home use, providing a convenient way to experiment with frequency healing under professional guidance.

5. Stay Open and Reflective: Healing is a journey, and frequency medicine is a field that invites self-discovery. Approach each session with an open mind, and take time to reflect on subtle changes in your physical and emotional state.

Vision: A Future Where Patients Co-Create Their Healing

Imagine a world where patients are no longer passive recipients of medical advice but empowered co-creators of their health. Frequency healing offers this vision—a world where individuals engage deeply with their bodies, using tools and technologies that allow them to actively support their well-being.

In such a future, healthcare might look more like a partnership between patients and practitioners. Doctors and frequency specialists would work together to educate patients, equipping them with personalized plans that blend scientific precision with individual intuition. Patients would have access to AI-driven apps that recommend frequency-based treatments tailored to their changing health needs, offering a level of personalization unheard of in today's healthcare.

This future is closer than we think. Already, biofeedback devices and wearable technologies allow people to monitor their bodies in real-time, responding immediately with sound, light, or electromagnetic therapies as needed. In this way, patients can prevent illnesses from taking root, practicing a form of proactive, preventive care that empowers rather than overwhelms.

The Role of Practitioners in Patient Empowerment

While frequency healing empowers patients, trained practitioners play a crucial role in guiding this journey. Experienced practitioners bring knowledge of specific frequencies, helping patients select the right ones for their needs. They also offer support in interpreting the body's responses, teaching patients to understand the sometimes-subtle language of frequency healing.

For those interested in exploring frequency healing under professional guidance, it's essential to seek certified practitioners. As this field grows, so does the need for standardized training, ensuring that practitioners are equipped to provide safe and effective guidance.

Reflecting on the Journey of Self-Healing

Frequency healing challenges us to rethink what it means to be healthy. It suggests that wellness isn't something handed to us in the form of a prescription but something we cultivate, layer by layer, frequency by frequency. By tuning into our own "symphony," we can harmonize with our bodies in ways that encourage healing from within.

For some, this journey begins with simple practices like listening to a sound frequency or sitting under a light therapy lamp. For others, it may involve more intensive treatments, guided by the insights of a frequency practitioner. Yet, in all cases, the power lies in the patient's hands, in their ability to listen to their body, respond to its needs, and choose a healing path that resonates with their own life's rhythm.

Closing Reflection: Embracing Empowerment in Frequency Medicine

As we conclude this chapter, let us take a moment to reflect on the essence of healing harmonies. Frequency medicine does not replace conventional treatments; it complements them, offering patients a new way to engage with their health. It invites us to honour the body as a vibrant, dynamic system, capable of profound self-healing when given the right tools.

Patient empowerment is about reclaiming the knowledge that healing is not something that happens to us but something we participate in, a journey where each of us becomes our own healer. Through frequency healing, we have the chance to move beyond passive health management and into a world where wellness is an active, creative process —a journey guided by the harmonies within.

CHAPTER 100:
THE VISION FOR A WORLD OF NON-INVASIVE HEALING

Imagine a world where invasive surgeries are rare, and instead, healing happens harmoniously through gentle vibrations, sounds, and light frequencies. Where once we relied on scalpels and medications to address ailments, we now embrace a holistic approach that seeks to tune the body like an instrument. This chapter envisions a future where frequency-based treatments are not an alternative but a cornerstone of mainstream medicine, transforming our relationship with health and wellness.

A Symphony of Healing

To understand the profound potential of frequency medicine, think of the body as a magnificent orchestra, where each organ, cell, and tissue resonates at a unique pitch. When healthy, these parts perform together in perfect harmony, creating a symphony of vitality. Yet, life's stressors, injuries, and illnesses cause certain "instruments" to fall out of tune. Traditional medicine often intervenes forcefully —cutting, prescribing, or altering—like trying to fix a delicate instrument with blunt tools. Frequency medicine, however, aims to "re-tune" the body, gently guiding it back to resonance through sound, light, and electromagnetic frequencies.

This vision of frequency-based healing is not new. Ancient cultures, from the Egyptians to the Greeks, used sound, vibration, and light to restore balance. Tibetan singing bowls, chanting, and even sunlight therapy were ancient methods of aligning body and spirit. Today, with modern technology, we are rediscovering these timeless practices and refining them to fit contemporary healthcare, making non-invasive healing a reality.

Realizing the Potential: Case Studies from Today's Pioneers

In this world of non-invasive healing, frequency medicine holds promise across diverse fields. Consider John, a cancer survivor whose radiation treatments left him with chronic fatigue and pain. Desperate for relief, he turned to sound therapy. Guided by a trained practitioner, he listened to specific frequencies known to stimulate cellular repair and reduce inflammation. After several months, he found his pain had diminished, his energy returned, and he felt more connected to his body's needs. For John, frequency healing became more than therapy; it was a path to reclaiming his life.

Another story is that of Maria, a 32-year-old mother with postpartum depression. Standard antidepressants made her feel numb, disconnecting her from herself and her child. Maria discovered neurofeedback, a frequency-based mental health therapy that subtly influenced her brain waves, helping her relax without side effects. Over time, Maria regained emotional stability and reported feeling "more herself." For her, frequency medicine was not just a treatment but a bridge to mental clarity and renewed hope.

Practical Steps Toward a Non-Invasive Future

For those ready to incorporate frequency healing into their lives, here are practical steps to empower readers in becoming pioneers of this movement:

1. Seek Out Certified Practitioners: As with any health modality, working with a trained professional ensures you receive the correct guidance. Seek out certified frequency medicine practitioners who are well-versed in tailoring frequencies to individual needs.

2. Educate Yourself: Frequency medicine is evolving, and knowledge is empowerment. Familiarize yourself with different modalities—Pulsed Electromagnetic Field (PEMF) therapy, Low-Level Laser Therapy (LLLT), and sound therapy. Each offers unique benefits depending on individual health goals.

3. Experiment with Sound and Light Therapy at Home: Many frequency therapies are accessible for home use. Sound baths, binaural beats, and red light therapy can be integrated into daily routines. Track your responses in a journal to see what resonates best with your body.

4. Collaborate with Healthcare Providers: Frequency healing works best when integrated with traditional medicine. Open a dialogue with your healthcare provider, expressing interest in non-invasive methods and how they might complement conventional treatments.

5. Join Community Healing Circles: Group sound therapy sessions and healing circles are a growing movement, allowing individuals to experience frequencies in a shared space, amplifying their effects through collective energy.

Envisioning Tomorrow: Frequency Medicine as a Cornerstone of Health

In a world that fully embraces frequency medicine, hospitals might look and feel very different. Instead of bustling corridors and sterile waiting rooms, imagine serene spaces filled with the calming hum of healing frequencies. Patients could walk into "frequency suites" where light and sound therapies create an environment of calm, minimizing the need for sedation before surgery. Specialized frequency machines could target cellular repair in real-time, enabling the body to heal faster and more naturally.

In this visionary future, invasive procedures would become rare. Electromagnetic therapies might replace certain surgeries altogether, using targeted frequencies to break down tumours, dissolve clots, or repair tissue without the need for a scalpel. Chronic pain, inflammation, and even some psychological conditions could be managed through frequency therapies, reducing reliance on medications and their often taxing side effects. The transformation would not only benefit individual patients but would also lessen the burden on healthcare systems, allowing for a more sustainable, holistic model of medicine.

The Ethical Landscape of Frequency Medicine

While the potential of frequency healing is inspiring, the path forward must be approached thoughtfully. Questions of access, safety, and regulation are central to this journey. As more people discover the transformative power of frequencies, it becomes vital to ensure that treatments are scientifically validated and accessible to all, not just a privileged few. Ethical standards and practitioner certifications should be established to protect patients and promote informed decision-making.

Imagine a future where every person, regardless of location or income, can access frequency-based treatments. Clinics, hospitals, and wellness centres around the world would provide these non-invasive options, supported by government policies that prioritize preventive and holistic healthcare.

Philosophical Reflection: Harmony as Health

Frequency medicine also challenges us to rethink what it means to be healthy. If health is harmony, as frequency healing suggests, then wellness becomes more than the absence of disease; it is an active state of balance, a daily practice of aligning body, mind, and spirit. This view shifts the role of medicine from simply treating symptoms to supporting a holistic harmony that encompasses our physical, mental, and emotional landscapes.

With frequency medicine, we are not just "patients" but co-creators in our healing journey. By tuning into our bodies and adjusting our internal frequencies, we can actively participate in maintaining health. This participatory model restores agency, encouraging individuals to become attuned to their own needs, responses, and rhythms.

Moving Forward: A Personal Call to Action

As we conclude this exploration of frequency healing, consider the journey you might take in exploring this transformative field. Frequency medicine is more than a series of techniques; it is a philosophy, a way of engaging with health that is gentle, empowering, and profoundly human. Each time you engage in a sound bath, meditate with binaural beats, or explore light therapy, you're not only exploring healing—you're also learning to listen to your body's unique language.

To begin, experiment with one frequency-based practice. Maybe it's as simple as dedicating ten minutes to listening to a specific healing frequency or incorporating a low-level light device into your routine. Over time, as you explore these subtle tools, take note of the shifts they create within you. You may find that frequency healing enhances not only your physical health but also your sense of peace, balance, and connection to yourself.

Final Thoughts: A World Aligned with Harmony

In a future shaped by frequency medicine, healing becomes more than intervention; it becomes a journey of reconnection. We stand on the cusp of a healthcare revolution that honours both science and spirit, offering tools that resonate with our deepest needs for harmony, balance, and wholeness. The vision for a world of non-invasive healing is not just a dream; it is a tangible future we are building today.

As you explore frequency medicine, may you find not only health but also a deeper alignment with the symphony of life that exists within you. This is the promise of healing harmonies—a future where health is a state of balance, a song we can all learn to sing.

APPENDICES AND RESOURCES

1. Glossary of Terms

Here, you'll find definitions of essential terms used throughout this book, helping you navigate the fascinating landscape of frequency medicine, sound therapy, and vibrational healing.

- Amplitude: The strength or intensity of a sound wave or vibration, often perceived as volume in sound therapy.

- Binaural Beats: An auditory illusion created when two slightly different frequencies are played in each ear, resulting in a perceived third frequency, which is thought to influence brain wave states.

- Biofield: The field of energy and information surrounding and permeating the human body, often referred to as the human energy field or aura.

- Brainwave Entrainment: A method of using auditory or visual stimuli to guide the brain into specific brainwave states, such as relaxation (alpha) or focus (beta).

- Cellular Resonance: The concept that each cell in the body has its own natural frequency and can be "re-tuned" for optimal health.

- Chakras: Energy centres in the body according to ancient

Eastern traditions; each chakra is thought to resonate with specific frequencies and influence physical and emotional health.

- Electromagnetic Field (EMF): The field created by electrically charged objects, used in therapies that influence cellular health and repair, such as PEMF.

- Frequency: The number of vibrations or sound wave cycles per second, measured in Hertz (Hz); different frequencies are believed to influence various aspects of physical and mental health.

- Hertz (Hz): The unit of frequency; one Hertz equals one cycle per second. In frequency medicine, specific Hertz levels are associated with particular healing effects.

- Isochronic Tones: Single tones that are pulsed on and off at a specific rate, used for brainwave entrainment and mood enhancement.

- Low-Level Laser Therapy (LLLT): A form of light therapy that uses specific wavelengths of light to promote healing, reduce inflammation, and stimulate tissue repair.

- Pulsed Electromagnetic Field (PEMF) Therapy: A type of therapy that uses electromagnetic fields to stimulate cellular repair, reduce inflammation, and enhance overall well-being.

- Resonance: The phenomenon by which a vibration in one object induces a similar vibration in another object, used in frequency medicine to restore cells to their natural frequency.

- Sound Bath: A therapeutic session in which participants are "bathed" in sound waves from instruments like gongs, Tibetan bowls, or tuning forks, promoting relaxation and healing.

- Theta Waves: Brainwave frequencies (4–8 Hz) associated

with deep relaxation, meditation, and enhanced creativity, often induced through sound therapy.

- Vibrational Healing: A holistic healing approach that uses sound, light, and other vibrations to restore balance to the body's energy systems.

This glossary provides a foundational understanding of the terms and concepts in frequency medicine. As you explore further, these definitions will help demystify the language of this emerging field.

2. Resource List

To further explore the world of frequency medicine and vibrational healing, here's a curated list of books, websites, apps, and organizations that offer valuable insights, tools, and community connections.

Books
- "The Healing Power of Sound" by Mitchell L. Gaynor, M.D.
   This book dives into the therapeutic benefits of sound and music, providing a scientific yet accessible approach to sound healing.

- "Tuning the Human Biofield" by Eileen Day McKusick
   A practical guide to sound balancing using tuning forks, exploring how vibrational healing can affect physical and emotional health.

- "Vibrational Medicine" by Richard Gerber, M.D.
   Often considered a foundational text in energy medicine, this book explores vibrational healing, frequency-based therapies, and the science behind them.

- "Sound Medicine: How to Use the Ancient Science of Sound

to Heal the Body and Mind" by Kulreet Chaudhary, M.D.

Dr. Chaudhary blends her expertise in integrative medicine with ancient sound practices, offering insights on sound's effects on health and consciousness.

- "Frequency: The Power of Personal Vibration" by Penney Peirce

A guide to understanding and elevating your personal frequency, providing both practical tools and philosophical insights.

Websites

- The Sound Healing Research Foundation (www.soundhealingresearchfoundation.org)

A comprehensive resource for research, articles, and events related to sound healing and frequency medicine.

- Centre for Biofield Sciences (www.biofieldsciences.com)

Focused on research into the human biofield, this site provides scientific resources and studies on vibrational and frequency-based therapies.

- Institute of Noetic Sciences (www.noetic.org)

Founded by astronaut Edgar Mitchell, IONS explores the science of consciousness and healing, offering insights into biofields and vibrational medicine.

- The British Academy of Sound Therapy (www.britishacademyofsoundtherapy.com)

Offers information, courses, and certification in sound therapy, making it an excellent resource for professionals and enthusiasts alike.

- Sonic Wellness Institute (www.sonicwellnessinstitute.com)

Dedicated to promoting the wellness benefits of sound and vibration therapy, with articles, research updates, and event information.

Apps
- Insight Timer

This popular meditation app includes a library of sound baths, binaural beats, and guided meditations that use healing frequencies.

- Brain.fm

Brain.fm uses scientifically designed music to influence focus, relaxation, and sleep, using frequencies to enhance mental states.

- MyNoise

Offers customizable soundscapes, including isochronic tones and binaural beats, which can be tailored to focus, relax, or energize the user.

- Solfeggio Frequencies Healing

This app provides soundtracks based on Solfeggio frequencies, believed to promote healing and spiritual well-being.

- Tide

An app focused on sleep and relaxation, using natural sounds and frequency-based music to help users relax and improve sleep quality.

Organizations
- International Sound Therapy Association (ISTA)

ISTA promotes awareness and research in sound therapy, offering training and resources for sound healing practitioners.

- American Association of Integrative Medicine (AAIM)

AAIM supports integrative health practices, including frequency medicine, offering certification and education in various alternative therapies.

- National Centre for Complementary and Integrative Health (NCCIH)
   Part of the U.S. National Institutes of Health, NCCIH provides research and resources on alternative therapies, including frequency-based treatments.

- World Sound Healing Organization
   Dedicated to promoting the use of sound for healing, this organization provides events, training, and resources for sound therapists worldwide.

- Association for Comprehensive Energy Psychology (ACEP)
   ACEP is a professional organization promoting energy psychology and vibrational healing, offering resources, conferences, and research on various techniques.

These resources will provide readers with tools and knowledge to continue their journey in frequency medicine, exploring the science, applications, and communities that support this transformative approach to healing.

3. DIY Sound Therapy Exercises

These simple exercises introduce beginners to the world of sound therapy, allowing you to experience frequency-based healing in the comfort of your home. Each exercise can help you connect with the healing properties of sound, offering moments of relaxation, focus, and well-being.

1. Breathing with Sound

Purpose: To relax the mind and body by combining deep breathing with soothing sound frequencies.

Instructions:

1. Find a comfortable and quiet place to sit or lie down.

2. Choose a soft sound or low-frequency tone (e.g., 432 Hz, known for its calming properties).

3. Close your eyes and take a deep breath in for four counts, hold for four, then exhale for four counts.

4. As you breathe, focus on the sound, letting it guide you into a state of relaxation.

5. Continue this process for five to ten minutes, allowing the sound and breath to harmonize.

2. Humming for Resonance

Purpose: To experience self-generated sound and its calming, grounding effects on the body.

Instructions:

1. Sit comfortably with your spine straight, shoulders relaxed.

2. Take a deep breath in, and as you exhale, hum softly. Notice how the vibration feels in your throat and chest.

3. Experiment by humming at different pitches, noticing where you feel resonance in your body.

4. Spend five to ten minutes humming, focusing on any calming effects the sound may bring.

3. Solfeggio Frequency Meditation

Purpose: To experience the potential healing properties of Solfeggio frequencies, specifically 528 Hz (often associated with transformation and DNA repair).

Instructions:

1. Find a quiet place to sit or lie down, and play a 528 Hz Solfeggio frequency tone (available on most music or meditation apps).

2. Close your eyes and focus on the sound, allowing it to fill your space and mind.

3. Visualize the sound waves gently flowing through your body, bringing a sense of calm and balance.

4. Continue for ten to fifteen minutes, breathing slowly and deeply as you immerse yourself in the tone.

4. Binaural Beats for Focus

Purpose: To enhance focus and mental clarity using binaural beats.

Instructions:

1. Use headphones for this exercise, as binaural beats require two different frequencies in each ear.

2. Find a binaural beat track at 10 Hz (beta wave range), which can improve concentration.

3. Set a focus intention, then listen to the track for 15–20 minutes while working or studying.

4. Notice any increase in focus or clarity as you listen.

5. Sound Bath Visualization

Purpose: To relieve stress and promote relaxation through sound immersion.

Instructions:

1. Lie down in a comfortable position and play a sound bath recording (try Tibetan singing bowls or chimes).

2. Close your eyes and imagine each sound wave gently washing over you like waves in the ocean.

3. Visualize the sound removing stress and tension from each part of your body, from head to toe.

4. Continue for 20–30 minutes, focusing on the sensation of sound as it calms your mind and body.

6. Chanting for Grounding

Purpose: To connect with the body and reduce anxiety through vocal expression.

Instructions:

1. Sit in a comfortable position with your eyes closed.

2. Begin by chanting "Om" (or any syllable you feel comfortable with) in a steady, comfortable tone.

3. Notice the vibration in your body as you chant, particularly in your chest and throat.

4. Repeat the chant for five to ten minutes, allowing yourself to feel more grounded with each repetition.

7. Guided Frequency Meditation

Purpose: To explore a deeper state of relaxation using a specific frequency (e.g., 432 Hz for relaxation or 639 Hz for harmony in relationships).

Instructions:

1. Find a guided meditation that incorporates a specific frequency aligned with your goals.

2. Listen to the meditation in a quiet space, focusing on the frequency's effects on your body and mind.

3. Allow yourself to relax fully, staying present with the sound and letting go of any thoughts or tensions.

4. Finish with a few deep breaths to ground yourself back in the present.

These exercises are a great starting point for experiencing the subtle yet powerful effects of frequency-based healing. Whether you seek relaxation, focus, or a sense of calm, each practice offers a different way to connect with the world of sound therapy. Remember, consistency enhances benefits, so consider adding these exercises to your routine to deepen your connection to frequency healing.

4. Practitioner Directory: Finding Certified Practitioners and Reliable Sources for Treatment

As the field of frequency-based healing continues to grow, finding qualified, certified practitioners who can guide your journey is crucial. This directory offers practical tips and resources for locating reliable professionals and understanding their credentials.

Why Certification Matters

Just as with any healthcare practice, certification ensures that frequency healing practitioners have undergone proper training, understand safety protocols, and are committed to high ethical standards. Certified practitioners are more likely to have a strong foundation in anatomy, physiology, and the technical aspects of frequency medicine, which can enhance the quality and safety of your experience.

Tips for Finding Certified Practitioners

1. Research Certification Programs: Look for practitioners certified by reputable organizations such as the International Sound Therapy Association (ISTA), the Biofield Tuning Institute, or the Association for Comprehensive Energy Psychology (ACEP). These bodies set standards for training, ethics, and practical experience.

2. Verify Credentials: When contacting a practitioner, ask about their credentials and certifications. Most certified practitioners should be transparent about their training, experience, and approach to treatment.

3. Look for Reviews and Testimonials: Search for reviews or testimonials from previous clients, which can provide insights into the practitioner's expertise and the

effectiveness of their treatments.

4. Attend Free Introductory Sessions: Many certified practitioners offer free consultations or introductory sessions. This can be an excellent way to assess their approach and determine whether you feel comfortable working with them.

5. Ask About Safety Protocols: Since frequency-based treatments involve sound and electromagnetic waves, practitioners should be well-versed in safety protocols to ensure sessions are conducted responsibly. Don't hesitate to ask about the measures they take to maintain a safe environment for clients.

Types of Frequency Healing Practitioners

The field of frequency medicine includes various specializations, each with unique techniques and approaches. Here's an overview of common types of practitioners you may encounter:

- Sound Therapists: Trained in using sound waves, such as tuning forks, singing bowls, and binaural beats, to promote relaxation, focus, and emotional balance.

- Biofield Tuning Practitioners: Certified to work with the body's biofield (energy field) using tuning forks to identify and release energetic blocks.

- PEMF (Pulsed Electromagnetic Field) Technicians: Specialize in using PEMF devices to stimulate cellular repair, improve circulation, and support healing.

- Vibrational Medicine Practitioners: Often blend sound, light, and other vibrational tools to harmonize the body's frequencies. Their training may include various vibrational modalities, offering a holistic approach.

- Energy Medicine Practitioners: May incorporate a combination of sound, light, and touch-based techniques to balance the body's energetic pathways, including acupressure points and chakras.

Questions to Ask Potential Practitioners

When considering a practitioner, asking specific questions can help you understand their qualifications and approach:

1. What certification or training have you completed in frequency medicine?
2. What specific modalities do you specialize in, and how might these benefit my condition or goals?
3. How do you assess a client's needs before starting a session?
4. What safety protocols do you follow?
5. Can you provide any case studies, testimonials, or examples of past success?
6. Are there any risks or side effects associated with the treatment you offer?

Recommended Directories and Online Resources

Several online directories and resources can help you locate certified practitioners in your area:

1. The International Sound Therapy Association (ISTA) – Provides a directory of certified sound therapists worldwide.

2. Biofield Tuning Practitioners Directory – Lists certified practitioners who have completed training in biofield tuning and are recognized by the Biofield Tuning Institute.

3. Association for Comprehensive Energy Psychology (ACEP) – Offers a directory of energy medicine practitioners who

may incorporate sound and frequency techniques in their practices.

4. Integrative Practitioners Directory – A comprehensive directory of certified alternative health practitioners, including those specializing in frequency medicine.

5. Local Wellness Centres and Holistic Health Clinics – Many wellness centres have on-staff frequency healing practitioners or can provide referrals to certified professionals.

Considerations for Remote Sessions

With advancements in technology, some practitioners offer virtual sessions for sound healing, frequency guidance, and biofield tuning. If you choose a remote session:

- Ensure Clear Communication: Discuss the session format and what to expect, including how sound frequencies will be transmitted (e.g., through headphones, speaker setups).
- Verify Tech Requirements: Ensure you have the necessary equipment, such as high-quality headphones, to experience the session fully.
- Schedule a Pre-Session Chat: Consider a brief introductory call to ensure comfort and connection with the practitioner.

Final Thoughts on Choosing a Practitioner

Finding a certified practitioner is an empowering step in your journey with frequency healing. Remember, the right practitioner will listen to your needs, explain their approach, and prioritize your well-being. Trust your intuition, and don't hesitate to explore different practitioners until you find one whose expertise and approach resonate with you. With the right support, frequency-based healing can become a transformative part of your wellness journey.

5. Frequency Chart: Therapeutic Frequencies and Their Known Effects on Health

A frequency chart can serve as a quick reference guide, helping readers identify the therapeutic frequencies that may aid in various health and wellness goals. Each frequency resonates with specific body systems and tissues, potentially influencing cellular function, emotional balance, and overall vitality. Below is a list of commonly used therapeutic frequencies, along with their associated benefits and suggested applications.

Frequency (Hz)	Known Effects	Applications
40 Hz	Promotes cognitive function and memory; used in neurostimulation research for Alzheimer's disease	Brainwave entrainment, cognitive enhancement
174 Hz	Pain relief and stress reduction	General relaxation, pain management
285 Hz	Cellular repair and rejuvenation ↓	Tissue healing, recovery from injury

396 Hz	Liberates guilt and fear, encourages emotional release	Emotional healing, stress relief
417 Hz	Facilitates change and undoing negative situations	Supporting transition, breaking negative cycles
432 Hz	Said to promote harmony and clarity; aligns with Earth's natural frequency	Meditation, relaxation, grounding
528 Hz	Known as the "Love Frequency"; associated with DNA repair and transformation	Heart-centered healing, emotional release

639 Hz	Improves communication and relationships	Emotional balance, improving social connections
741 Hz	Detoxifies the body, boosts immunity	Cellular health, detoxification support
852 Hz	Awakens intuition and mental clarity	Enhancing creativity, spiritual insight
963 Hz	Aligns with higher consciousness; often called the "Frequency of the Divine"	Meditation, spiritual practices, clarity

Binaural Beats (e.g., Alpha 8-14 Hz)	Induces relaxation, reduces anxiety, supports creativity	Meditation, focus, sleep enhancement
Gamma Waves (30-100 Hz)	High-level information processing, memory enhancement	Cognitive boost, alertness, learning

How to Use This Chart

1. Select a Frequency for Your Goal: Identify the frequency associated with your desired effect, whether it's physical healing, emotional release, or mental clarity.

2. Choose a Delivery Method: Frequencies can be applied through various methods, such as sound baths, tuning forks, binaural beats (for brainwave entrainment), or through therapeutic devices for more targeted applications.

3. Set a Comfortable Environment: Find a quiet space where you won't be disturbed. Creating a comfortable, calming environment can enhance the experience of sound and frequency therapy.

4. Use Duration as Needed: Generally, sessions of 10 to 30 minutes are effective for frequency-based practices, but

you may adjust based on how you feel and the specific application.

5. Reflect on Results: Track how you feel before, during, and after exposure to the frequency. Noticing subtle shifts over time can help guide you to the frequencies that work best for you.

This chart is intended to empower you with accessible tools for exploring the potential of frequency-based healing in everyday life. Remember, consistency and mindful engagement with each session can deepen your connection to the practice and enhance the therapeutic effects.

www.ingramcontent.com/pod-product-compliance
Lightning Source LLC
Chambersburg PA
CBHW071443220526
45472CB00003B/639